Die-Attach Materials for High Temperature Applications in Microelectronics Packaging

Kim S. Siow
Editor

Die-Attach Materials for High Temperature Applications in Microelectronics Packaging

Materials, Processes, Equipment, and Reliability

Springer

Editor
Kim S. Siow
Institute of Microengineering
and Nanoelectronics
Universiti Kebangsaan Malaysia
Bangi Selangor, Malaysia

ISBN 978-3-319-99255-6 ISBN 978-3-319-99256-3 (eBook)
https://doi.org/10.1007/978-3-319-99256-3

Library of Congress Control Number: 2018958713

This Springer imprint is published by the registered company Springer Nature Switzerland AG
The registered company address is: Gewerbestrasse 11, 6330 Cham, Switzerland

"This timely and well-written/edited book is a great reference for those researchers and practitioners engaging in high-temperature Pb-free die attach materials. It is highly recommended."

– Simon S. Ang, Professor and Director of High Density Electronics Center, University of Arkansas, USA

"Rich and detailed analysis of high temperature die attach materials especially sintered silver joint."

– Erich Kainhofer, Manager Techsupport Singapore, Besi Singapore

"For those working in high temperature die attach materials, this book is a timely arrival to address the Pb-free issues. Read this well-edited book to find out more from the different thought-leaders."

– Hu Anming, University of Tennessee, Knoxville

"It was yet time to put some order among the sintering die attach available information, and this book definitely does."

– Francesc Masana, Barcelona Semiconductors S.L.U.

"We must be thankful to Kim for bringing a group of thought-leaders in this collection of high temperature die attach materials especially sintered metals."

– Hidetomo Asami, Nihon Handa

"The book well compiles extensive studies regarding high temperature die attach materials especially using sintered Ag or Cu."

– Toshitaka Ishizaki, Toyota Central R&D Labs., Inc.

"A book on high-temperature die attach materials which iNEMI eagerly waiting for, since we have a workgroup on it in 2017. Hopefully, the industry will benefit from the solutions discussed within this edited book."

– Masahiro Tsuriya, iNEMI

"High temperature die-attach materials, in particular Ag-sintering, is at present an enabling technology for power electronics in EV automotive applications and as such this book is at the heart of current industrial developments."

– Marco Koelink, Advanced Packaging Center B.V. / Boschman Technologies BV

"It is a must read book for packaging and assembly engineer who want to get start and practice in the subject matter on Silver Sintering. It encapsulated the subject from fundamentals science to application of Silver Sintering for practice to produce silver sintered packages."

– Eric Kuah, ASM Technology

"This book really filled the gap on high temperature die attach materials with in-depth coverage. It definitely added a great value to the industry."

– NC Lee, Indium Corporation

"Kim Siow has edited a must-read guide for anyone considering high temperature die-attach sintering. . .Read this book, and learn from one of the best."

– Giulio Locatelli, GM Locatelli Meccanica, Italy

Foreword

Technology development in microelectronics continues at a historically unprecedented pace. Given the complexity of development and manufacturing—from the many different components and functionalities comprising any single system, to the range of processing methods and materials, through to competing application requirements like reliability, temperature resistance, miniaturization, and cost-efficiency—this may seem astounding. However, perhaps it is exactly this need to approach development of any one component, process or material with a view to its compatibility with the system as a whole, that accelerates development across the board of research specialties in this area.

This book focuses on a key demand in microelectronics application areas—along with miniaturization and robustness, high temperature is a key demand driving development. Industries relying on high-temperature microelectronics are economically key and include automotives (e-mobility, power electronics in hybrid, and electrical cars), sensors systems, and the lighting industry (LED packaging). Extreme demands, operating temperatures of 150°C, are placed on electronic components and the interconnections in these areas, and in order to do justice to the harsh operating conditions, new packaging materials are needed that are above all corrosion and temperature resistant, as well as suitable, cost-efficient manufacturing processes.

Die attach, the process of making the electrical connection between the semiconductor device die and its package, plays a key role in performance and reliability of high-temperature applications. Because it is the first packaging layer in contact with the die, optimizing die-attach materials and processing techniques is crucial. We are now beginning to obtain useful insight and understanding of the design, material, process, equipment, manufacturing, quality, and reliability issues of die attach for high-temperature applications. Important parameters such as the thermodynamics, kinetics, microstructure, failure mechanisms, manufacturing, and reliability of die-attach materials have been now been widely studied. However, the results are spread across individual publications in diverse journals, conference proceedings, and specialist workshops. No single source of information devoted to state of the art of high-temperature die-attach technology has been previously published. This book aims to remedy this and present, in one volume, a timely summary of progress in all aspects of this fascinating field.

This collection of essays on high-temperature die attach presents a systematic and detailed discussion of the microstructural build-up comprising the die-attach materials (particularly silver sintering), the material deformation, as well as the progress of material damage that leads, in the final instance, to component failure. As part of this, in particular, the interrelationship between these three aspects of thermomechanical reliability is explored, in order achieve a comprehensive overview of cause-effect relationships, a prerequisite for a comprehensive understanding of die-attach modules in high-temperature microelectronic technology.

Three different die-attach technologies—soldering, silver sintering, and transient liquid phase bonding (TLPB)—are discussed, as are relevant aspects of epoxy-based die-attach materials. Of the interconnection surface joining technologies (SMT) in use today, soldering is perhaps the most popular. As a rule, Sn-based soft solders are used, which at thermomechanical stress at temperatures above 150°C fail so quickly, that even improving strength by adding antimony as doping agent, they are not suitable. Thus, the main focus of the book is in fact silver sintering, which is another, extremely popular and reliable joining technology for operating temperatures over 150°C but which also has clear disadvantages due to its cost as raw material and the complicated processing technique. A further alternative to create high-temperature-resistant joining layers is the complete transformation of the joints in intermetallic phases. This approach is known as transient liquid phase bonding (TLPB).

The first seven chapters focus on silver sintering—an innovative technology that offers superior thermal and electrical conductivity. In the first chapter Chen and Zhang provide a basic understanding of the kinetics and mechanisms of sintering versus the soldering process. Subsequently, the advantages of silver sintering are outlined using LEDs as application example. The third chapter discusses the recent introduction of several materials and technologies that have led to the successful industrialization of silver sintering and discusses those compared to solder-based and epoxy-based die-attach materials. Paret, DeVoto, and Narumanchi then explore the thermal mechanical modelling of sintered silver processes necessary to obtain accurate results and offer a deeper understanding of fracture mechanics-based modelling methodology. The fifth chapter discusses reliability and failure mechanisms of silver sintering in the context of power applications. In the sixth chapter, Mannan et al. focus on the growing importance of diffusion and electromigration in silver sintering interconnect reliability, particularly the effects of sinter silver microstructure on electromigration. The seventh chapter deals with the controversial "doctrine of equivalents" (DOE) within the context of silver paste formulations in three patents. DOE states that the patents claims can extend beyond their literal meanings to include those which behave in "substantially" equivalent function-way-results, otherwise known as the triple identity test. While this chapter only serves as a guide, not a substitute for legal advices, it highlights the different tools and limitations used in such DOE analysis.

The final chapters focus on alternatives to silver sintering, and any discussion on silver sintering is incomplete without discussing the copper sintering. In chapter "Sintered Copper: Chemistry, Process, and Reliability", Yamada discusses copper sintering as an extremely promising low-cost alternative to silver, with excellent mechanical robustness due to its lower coefficient of thermal expansion (CTE) and very high electrical and thermal conductivities. In chapter "Transient Liquid Phase Bonding", transient liquid phase bonding (TLPB), which involves the complete transformation of the solder layer into intermetallic phases, is presented in detail, and its suitability as a new die-attach technology for high temperature and high reliability applications is evaluated. Chapter "Die-Attach Materials for Extreme Conditions and Harsh Environments" takes a different approach of including adhesives and solders as die-attach materials used in extreme and harsh conditions, in addition to the well-established TLPB approach. These various materials systems are discussed within the applications commonly found in energy, oil, and gas industry

This book is primarily aimed at three groups of specialists:

- Those involved or interested in research and development of sintered and soldered die-attach interconnects
- Those facing practical die-attach challenges and in need of more reliable methods and solutions
- Those seeking to identify the right high-performance and cost-effective packaging technique for their interconnect system

I hope this book will serve as a valuable source of reference to all those faced with the challenging problems created by the ever-more-expanding use of die-attach materials in high-temperature applications. I trust it will stimulate further research and development on materials themselves and the processes needed to apply them in the microelectronic packaging environment.

This special issue would not have been possible without the enthusiasm of the publishing editor Kim S. Siow to get all the experts together, the valuable input of the authors, and the advice of Anita Lekhwani and Abhishek from Springer Nature.

Fraunhofer IZM, Berlin Rolf Aschenbrenner

Preface

When I wrote the first review paper on sintered nano-silver (Ag) for die-attach applications more than seven years ago, little did I know that I would be editing a book on this topic, as well as on other high-temperature die-attach materials, let alone with a group of prominent researchers from Australia, China, Japan, Malaysia, the UK, and the USA. What was even more surprising was that I had not met most of these researchers, yet they placed their faith in me to edit this book and bring it to fruition. As such, I am deeply indebted to them for creating this valuable information in time to meet the challenges of bonding the semiconductor dies in today's demanding conditions, i.e. the need for technologically superior, cost efficient, yet environmentally friendly bonding techniques.

The keywords for the title of this book are "high temperature" and "die attach". So, given the advent of wide bandgap semiconductors into mainstream manufacturing in the last few years, what is considered "high temperature" in die-attach applications?

Coincidentally, I had the honour of hosting Dr NC Lee (Indium) during the 37th International Electronics Manufacturing Technology conference and the privilege to be part of the iNEMI consortium[1] which addressed exactly this topic of high-temperature die attach in 2017. After agonizing for many hours with various definitions, I decided to take the open-ended definition for "high temperature" as operating continuously at above 200°C, though others preferred to consider this definition to be dependent on the specific applications or market segments of high-temperature die-attach materials.

---

[1]iNEMI is a not-for-profit consortium of more than 90 electronics companies, associations, government agencies, and universities, formed with a mission to "forecast and accelerate improvements in the electronics manufacturing industry for a sustainable future." http://community.inemi.org/hightemp_pbfree_dieattach. The first phase of this project is published here: .SP Lim, B Pan, H Zhang, W Ng, B Wu, KS Siow, S Sabne, M Tsuriya, High temperature Pb-free Die Attach Material Project Phase 1: Survey Result, International Conference on Electronics Packaging 2017, Yamagata, Japan, 19–22th April 2017, pg. 51–56. https://doi.org/10.23919/ICEP.2017.7939323.

Besides high temperature, the die-attach material also needs to be Pb-free to align with the EU directives to use environmentally friendly products in the electronic products; this requirement is currently exempted under End of Life Vehicle Annex 2, Exemption 8e, and RoHS7a. However, this exemption may be cancelled if an alternative is available and proven to be technologically feasible as a Pb-free die-attach material. Hence, there is an incentive to "corner" the market for Pb-free die-attach materials with new bonding material but several past attempts failed to impress during large-scale testing.

Based on these twin requirements of being Pb-free and high temperature, Die Attach 5 (DA5) group[2] proposed four main alternatives, i.e. conductive adhesive, silver (metal) sintering, transient liquid phase sintering/bonding (TLPS/ TLPB) and solder, though conductive adhesive clearly did not fulfil the definition of continuous high-temperature operation mentioned earlier, while most available Pb-free solder did not operate reliably at this temperature due to creep failure at operation temperature above 0.5 Tm (homologous temperature). Hence, the initial winner in this race for high-temperature die-attach materials will be the sintering technologies, be it silver (Ag) (or copper (Cu)) sintering and transient liquid phase sintering because of their high melting temperatures.

Silver sintering technology forms the bulk of this book content, not because of the my personal involvement in this technology, but because of its availability and its adoption by several companies to produce power modules and light emitting diodes (LED) with sintered Ag in a mass manufacturing environment. This early manufacturing experience deserved a special chapter as manufacturers still grapple with the dilemma of whether to invest in the pressure sinter equipment or to wait for the next generation of sintered Ag paste to be more robust in producing reliable joints (chapter "Process Control of Sintered Ag Joint in Production for Die-Attach Applications"). Another chapter (chapter "Sintered Silver for LED Applications") addresses the use of sintered Ag in LED applications because of the unique sintering requirements, i.e. LED die sizes and interfaces.

However, the fundamentals of Ag sintering remain the same in spite of this application-specific requirement in power modules or LEDs. This sintering science forms the first chapter of this book (chapter "Silver Sintering and Soldering: Bonding Process and Comparison"). It is really important to establish the differences between sintering and soldering in the initial adoption of this technology. Sintering is a solid-state reaction which does not go through a liquid-to-solid transition, resulting in the absence of self-alignment. Unlike soldering, sintering also does not have any possibility of rework after the bonding steps. A common question amongst solder engineers is "what about intermetallics formation in the sintered silver joint?". Since most of the die-attach bonding interfaces in the power electronics or modules are Ag, Cu, or gold (Au), no intermetallics are formed because of the Ag forming

[2]DA5 consortium consists of Bosch, Infineon Technologies, NXP Semiconductor, ST Microelectronics and Nexpria, formed to look for alternative Pb-free technologies to comply with EU Directive ROHS by 2021.

solid solution with these three elements. The absence of intermetallics would suggest a reliable bond but the inherent "instability" of the sintered Ag microstructures demands the application of pressure during sintering, a process step which is somewhat "unthinkable" in the die-attaching process. Based on these general scientific principles and observations, soldering is not sintering in terms of manufacturability.

Besides the operating conditions, what is the difference between conductive adhesive filled with Ag flakes and sintering Ag paste per se? While some sintered Ag pastes incorporate polymeric resin in their formulation, most Ag pastes are pure Ag fillers with binders, solvents, and capping agents[3]. These Ag paste formulations do not produce any residue fluxes and hence, no cleaning is required after the bonding process. Adhesives may be added to this Ag paste to overcome the lack of adhesion and substrate-specific sintering for Ag paste, but often result in a lower thermal conductivity than a pure sintered Ag joint. These Ag adhesives also have different reflow profiles than the usual conductive adhesive to allow the incorporated Ag nanoparticles to sinter amongst the neighbouring nanoparticles and faying interfaces.

The evolving microstructure and morphology of the sintered Ag deserved a separate chapter (chapter "Morphological Changes in Sintered Silver Due to Atomic Migration") in this book to discuss the various mechanisms including the electromigration in sintered silver; which was differentiated from electrochemical migration (chapter "Reliability and Failure Mechanisms of Sintered Silver as Die Attach Joint"). Electromigration refers to the electron wind within the metal lines dislodging the atoms, resulting in whiskers and voids while electrochemical migration results from metal ion migration between adjacent metal conductors to form dendrites.[4] Chapter "Reliability and Failure Mechanisms of Sintered Silver as Die Attach Joint" also addresses other failure mechanisms in sintered Ag joints such as mechanical or thermal stresses, within the context of different mechanical properties, i.e. elastic modulus, tensile, shear, creep and fatigue (ratcheting) strength, and different stress conditions like thermal ageing, thermal cycling, and power cycling.

Common engineering practice also demands the use of modelling and simulation of any electronic package before any real manufacturing. This simulation forms the basis of chapter "Thermomechanical Modeling of High-Temperature Bonded Inter face Materials", which provides several options and strategies to understand the performance of sintered Ag before manufacturing the real Ag joints.

Besides Ag sintering, copper is the other metal being actively researched by groups around the world, to address the "electrochemical migration" problem in

[3]Siow KS, Mechanical properties of nano-silver joints as die attach materials. Journal of Alloys Compound, 2012, 514, 6–19. https://doi.org/10.1016/j.jallcom.2011.10.092.

Siow KS and Lin YT, Identifying the Development State of Sintered Silver (Ag) as Bonding Material in Microelectronic Packaging via A Patent Landscape Study. Journal of Electronic Packaging, 2016, 138, 020804-1–020804-13. https://doi.org/10.1115/1.4033069.

[4]Mackie A, Electromigration (EM) and Electrochemical Migration (ECM), http://www.indium.com/blog/electromigration-em-and-electrochemical-migration-ecm.php, accessed on 8th March 2018.

certain niche applications using sintered silver technology (chapter "Sintered Copper: Chemistry, Process and Reliability"). Several paste manufacturers and "pressure-sinter" equipment manufacturers also readied their process and equipment to include sintered copper as another possible sintered paste in their repertoire.

Chapter "Doctrine of Equivalents and Sintered Ag Paste as Bonding Materials" discusses the controversial Doctrine of Equivalents (DOE) within the context of three sintered silver paste formulations. The author created a "legal fiction" that the infringing products possessed the same formulations as stated in their US patents and proceeded to conduct the non-literal infringement, also known as DOE analysis. In a typical DOE analysis, the patent scope can be extended beyond the literal meanings stated in the claims based on the function, way, and results of the components in the process, materials formulation, or product.

Chapter "Transient Liquid Phase Bonding" is easily the most comprehensive chapter ever written on the topic of transient liquid phase bonding (TLPB) for die-attach application. TLPB, also known as diffusion soldering (by Infineon Technologies[5]), was already used as a die-attach joint for their niche products in mass manufacturing, but this chapter provides diverse information addressing the science, kinetics, as well as different TLPS schemes available in the industry and literature.

Chapter "Die-Attach Materials for Extreme Conditions and Harsh Environments" is written by two engineers with Baker Hughes, Inc. a General Electric company who bring their wealth of experiences and knowledge in oilfield and energy industry applications into a concise and easy to understand chapter on the use of solder, adhesives, and transient liquid phase bonding in high-temperature applications. While these adhesives and solders may not strictly fall into my definition mentioned earlier, they are nevertheless included in this chapter to complete the discussion on high-temperature die-attach materials. Amongst others, they spoke on various solders, such as ZnSn, ZnAl, SnSb, AuGe, as well as adhesives such as cyanate ester, and Ag-glass.

The various issues, materials, and bonding systems discussed in this book bring much optimism for the search and implementation of truly high-temperature die-attach joints in the industry. My ultimate hope is that this edited book contributes to this development; whether or not, I achieve my aim is left in your hands, our readers, as it should be. This book is dedicated to all engineers who work tirelessly, carrying out numerous evaluations to bring these technologies to fruition. This book is also not possible without the support of many colleagues, students, suppliers, collaborators, customers, and employers (including UKM research grants GGPM-2013-079 and GUP-2017-055), who continuously provided feedback, encouragement and financial support to me throughout the years.

[5]Diffusion soldering is mentioned here: Automotive power MOSFETs, https://www.infineon.com/cms/en/product/power/mosfet/20v-800v-automotive-mosfet/20v-150v-p-channel-automotive-mosfet/ipd90p04p4-05/, accessed on 14th March 2018.

Special thanks are due to my family, especially my wife Hui Min, for her understanding and patience, which made this writing possible, and for doing the heavy lifting, both figuratively and literally, with the arrival of another child since the book was first conceived in 2016. I am also thankful to Springer Nature for first approaching me to edit this book following the heavy download and citation of my review paper on this topic.[6]

Bangi Selangor, Malaysia Kim S. Siow

[6]Siow KS, Are Sintered Silver Joints Ready for Use as Interconnect Material in Microelectronic Packaging? Journal of Electronic Materials, 2014, 43(4), 947–961. https://doi.org/10.1007/s11664-013-2967-3 *(Note: Editor's Choice for 2014)*.

Contents

Silver Sintering and Soldering: Bonding Process and Comparison

S. Chen and H. Zhang

1 Introduction

Development in hybrid and electric vehicles, high-speed trains, aircraft/aviation, and deep well oil/gas extraction demands more robust semiconductor power devices that can endure harsh operating conditions such as ≥200 °C. In the meantime, recent availability of wide bandgap silicon carbide devices makes it possible to operate theoretically at temperatures above 500 °C [1]. High-temperature die-attach materials are thus in urgent need to satisfy these applications.

High-temperature electronics require operating temperatures higher than or equal to 125 °C, especially ≥200 °C [2]. High junction temperatures, >125 °C, have been observed for silicon-based diodes or thyristors used in high-voltage power distribution systems. Environmental temperature as high as 150 °C has often been observed in some automotive systems. Discrete power modules may undertake the subsequent surface mounting reflow to be bonded on the PCB or PWB board, and thus it has to endure maximum peak temperature of 245–280 °C multiple times without remelt and damage.

Different materials have been used for high-temperature die-attach applications. Figure 1 displays the relationship between operating and processing temperatures for solder, silver-filled epoxy, silver (Ag) and copper (Cu) sintering, and transient liquid phase (TLP) materials. For solder materials, as a rule of thumb, when the serviceable temperature reaches 80% of the solder's melting point, i.e., homologous temperature of 0.8 in absolute temperature scale, creep becomes severe and joint reliability is compromised. Note that the value of the operating temperature shown in Fig. 1 serves as a reference for comparison of different materials; the actual reliable operating temperature in industrial usage may be higher or lower based on the

S. Chen (✉) · H. Zhang
Indium Corporation, Clinton, NY, USA
e-mail: schen@indium.com; hzhang@indium.com

K. S. Siow (ed.), *Die-Attach Materials for High Temperature Applications in Microelectronics Packaging*, https://doi.org/10.1007/978-3-319-99256-3_1

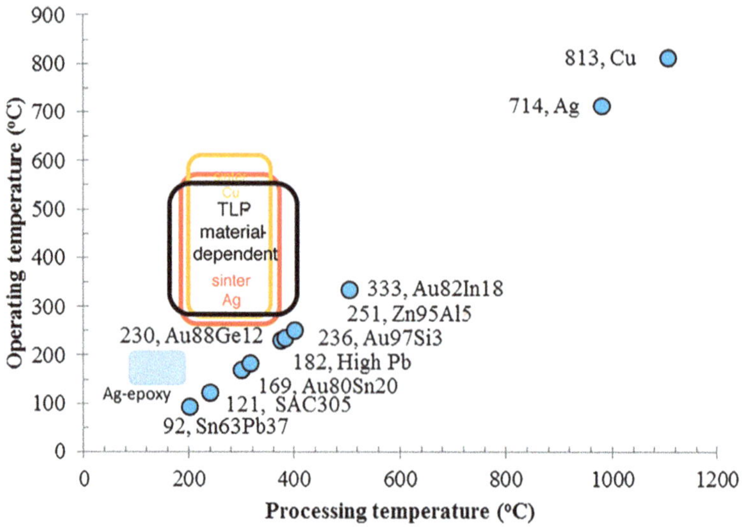

Fig. 1 Estimated operating and processing temperatures for solder, sintering, transient liquid phase, and Ag epoxy materials. Operating temperatures are calculated assuming 0.8 homologous temperatures. The processing temperatures for solders are solder melting points plus 20 °C. High Pb solder is represented by Indalloy 151 (Pb92.5/Sn5/Ag2.5)

type of alloys and their application conditions. From Fig. 1, lead-free solders that can operate under ≥200 °C condition do exist, such as AuGe, AuSi, AuIn, and ZnAl. However, these materials have drawbacks such as high price (Au-based alloys) or poor processability (ZnAl). In contrast, Ag or Cu sintering and TLP materials can withstand a much higher operating temperature due to a much higher melting point [3]. Also, these materials can be processed at a much lower temperature compared to that of solder materials, which will greatly help reduce the thermal stress during processing. Studies have also shown that pressure-assisted Ag sintering joints perform much better than that of solder under active power cycling test (16 times the statistical lifetime extension on direct bond copper (DBC) substrate), showing great promise for industrial usage [4].

The emergence of Ag sintering technology for die- or substrate-attach and bonding applications in recent years is motivated both by the requirement for replacing high Pb solder due to its harmful effects to human and environment [5] and Ag sintering joints' superior mechanical, thermal, and electrical properties compared to other lead-free alternatives in the market [6–11].

Sintered Ag has been used as interconnect material in power electronic components since 1987 [12–14]. In the early days of development, paste materials contained only micron-sized Ag particles, and a high pressure, up to 40 MPa, was used to assist bonding. This high pressure may result in die cracking, and thus low pressure [15, 16] or pressureless sintering [6, 17–20] attracted a lot of interest. Due to the pressure-assisted Ag sintering process requires large capital investment and has a very low manufacturing throughput, pressureless sintering materials through the introduction of nano-sized particles into paste formulation are under intensive research.

A lot of topics about Ag sintering material/process, such as paste formulation including dispersant, passivating layer, capping agent, binder, solvent/thinner, and different sized (nano-, micro-, or hybrid sized) particles, based on pressure/pressureless processes and application methods, have been reviewed by Siow et al. [21–23]. Properties such as shear strengths and reliability performance of sintered joints and their affecting factors, including bonding pressure, sintering temperature and time, nanoparticle size and distribution, heating rate, bonding area, and substrates, have also been reviewed [21–24].

In our day-to-day communication with industrial customers, since Ag sintering materials are relatively new, there is a lot of misunderstanding in terms of material processing and applications. For example, some people do not realize that the Ag sintering joint has porosity, which is different from that of solder joints. In the following sections, we will first discuss soldering and then Ag sintering. One of the purposes is to emphasize the difference between Ag sintering and solder formation mechanism and clarify the above misunderstanding. Finally, we will discuss the effect of porosity on the properties of Ag sintering joint since it is a unique feature compared to soldering.

2 Solder and Soldering

Soldering is commonly used to form a metallurgical interconnection by melting and flowing a solder alloy into the gap between the joining surfaces and then solidified [25–27]. In the power semiconductor application, solder was used for die attach because of its fairly good thermal, electrical, and thermomechanical performance [28–30]. Soldering to form joint involves three steps in sequence: (a) solder melting and flowing, (b) the interfacial reaction between the molten solders and the joining surfaces, and (c) solder solidification upon cooling. Soldering reactions are complicated and specific to the types of solder materials, surface metallization, and the reflow process. The nature of soldering still pertains to the basic phenomena widely recognized in materials, namely, phase transformation and atomic diffusion.

2.1 *Solder Melt*

Solder starts to melt right after reaching the solidus temperature. However, the completion of solder melting occurs only when the temperature reaches the liquidus temperature and above [26]. Normally, the reflow requires a peak temperature above the liquidus temperature and for it to be maintained for a while to allow the completion of solder melting and the escape of outgassing bubbles thoroughly from the flowing liquid. High-lead (Pb content >85 wt%) solders have been widely used for die attachment in discrete power components. As an example, Fig. 2 shows the Sn-Pb binary phase diagram, in which the high-lead solders are located at the Pb-rich end.

With the increasing Pb content (or the decreasing Sn content), both the solidus and the liquidus temperatures increase, while the pasty range temperature reduces, as shown in the phase diagram. Table 1 summarizes the solidus, the liquidus, and the pasty range temperatures of some representative high-lead solders. In practice, high-lead solders such as Pb95/Sn5, Pb92.5/Sn5/Ag2.5, and Pb95.5/Sn2/Ag2.5 have been widely used in industry because of their narrow pasty range, which has always been desired in electronics joint formation [5] (Table 1).

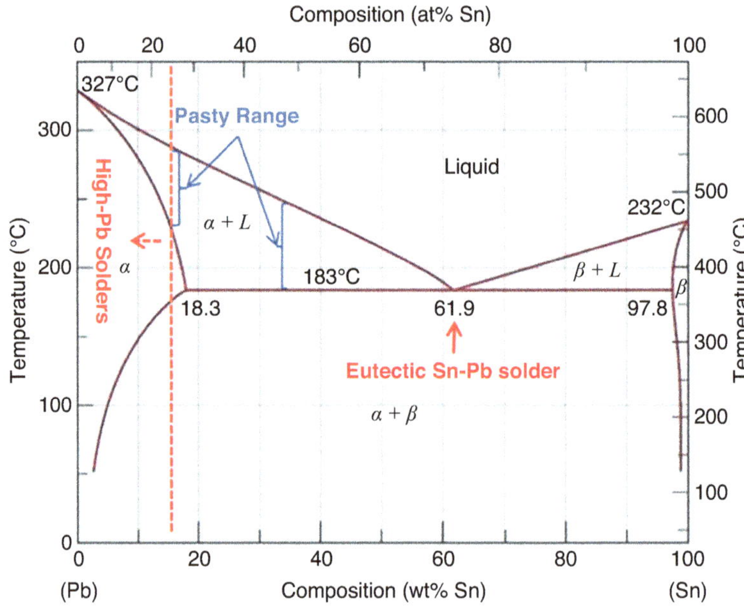

Fig. 2 Sn-Pb binary phase diagram

Table 1 Solidus, the liquidus, and the pasty range temperatures for representative high-lead solders

Solder	Solidus temperature (°C)	Liquidus temperature (°C)	ΔT (°C)
Pb85/Sn15	183	288	105
Pb88/Sn10/Ag2	268	290	22
Pb89.5/Sn10.5	275	302	27
Pb92.5/Sn5/Ag2.5	287	296	9
Pb95/Sn5	308	312	4
Pb95.5/Sn2/Ag2.5	299	304	5
Pb	327	327	0

2.2 Interfacial Reactions

After solder melt, interfacial reactions occur through two processes: (a) dissolution of the surface metallization atoms toward the molten solder and (b) formation of the intermetallic compounds (IMCs) pertaining to the joining interfaces.

The solubility of the metallization elements in the molten solder determines the maximum thickness of the metallization dissolved into the molten solder, if the solder volume is known. The solubility is dependent not only on the solder and the metallization materials but also on the temperature. Table 2 lists the solubility of Cu (at %) in three molten solders [27, 31] where the solubility of Cu increases with the increasing temperature and relies on solder composition (Table 2).

In practice, the solubility can be expressed in Eq. 1, where Q is the activation energy and C_{so} is the empirical constant, which is associated with solder alloys.

$$C_s = C_{so}\exp\left(-\frac{Q}{RT}\right) \tag{1}$$

The solubility of another common metallization, Ni in molten Sn and eutectic Sn-Pb, is extremely small at temperatures lower than 400 °C. The solubility of Ni in eutectic Sn-Pb is approximately 10^{-5} at % [32].

The solder joint has limited solder volume, and the dissolution rate of the metallization elements into the solder joint decreases as the concentration approaches its solubility limit. The dissolution rate can be modeled with the Nernst-Brunner relationship shown in Eq. 2 [33].

$$C - C_o = (C_s - C_o)\left[1 - \exp\left(-\frac{KA}{V}t\right)\right] \tag{2}$$

C_o: the metallization concentration inside the molten solder at time 0
C: the metallization concentration inside the molten solder at time t
C_s: the solubility concentration of metallization inside the molten solder
V: the solder volume
A: the contact area between the solder and the substrate
K: the dissolution rate constant with temperature and activation energy following $K = K_o\exp\left(-\frac{Q}{RT}\right)$

Table 2 Solubility of Cu (at %) in three molten solders

	Cu in Sn63/Pb37	Cu in Sn	Cu in Sn96.5/Ag3.5
232 °C	1	1.75	2.2
260 °C	1.5	2.5	3.4
Q (kJ/mol)	32.4	28.5	34.8
C_{so} (at%)	22.0	15.2	85.3

Table 3 shows the values of the parameters in Eq. 2 for Cu in eutectic Sn-Pb and eutectic Sn-Ag molten solders, respectively. Ni dissolution rate was reported to be one half that of Cu in the same solders [34] (Table 3).

Intermetallic compounds form along the interface after the metallization elements dissolve and reach the solubility along the interface. A phase diagram can be used to elaborate the interfacial reaction products. Figure 3 shows the Sn-Cu phase diagram, in which Cu_6Sn_5 starts to form when Cu content exceeds the solubility, which has been denoted from the liquidus, in molten solder locally along the bonding surface.

The interfacial IMC layer forms through nucleation and growth processes. The nucleation of the IMCs on the interface may cover the whole contacting surface immediately after solder melting, spreading, and dissolving the metallization elements [35, 36]. After the formation of a thin layer of nuclei, the growth requires the transportation of the reacting atoms toward the IMC from both the molten solder and beneath the metallization surface through the growing IMC layer. The atomic diffusion in the liquid solder is much quicker than the atomic diffusion in solid. Thus, after the nucleation, the IMC growth would be dominated by the solid atomic diffusion, which indicates the atomic transportation through IMC layers either from the molten solder toward the metallization surface or from the metallization surface toward the molten solder. The growth of the IMC layer slows down when it becomes

Table 3 Dissolution rate (K_o), activation energy (Q), and ratio of volume to area (V/A)

	Temperature (°C)	K_o (mm/min)	Q (KJ/mol)	V/A (mm)
Cu in molten Sn-Pb eutectic	255~310	2.9	13	0.33~0.56
Cu in molten Sn-Ag eutectic	254~302	14.2	14.9	0.2~0.4

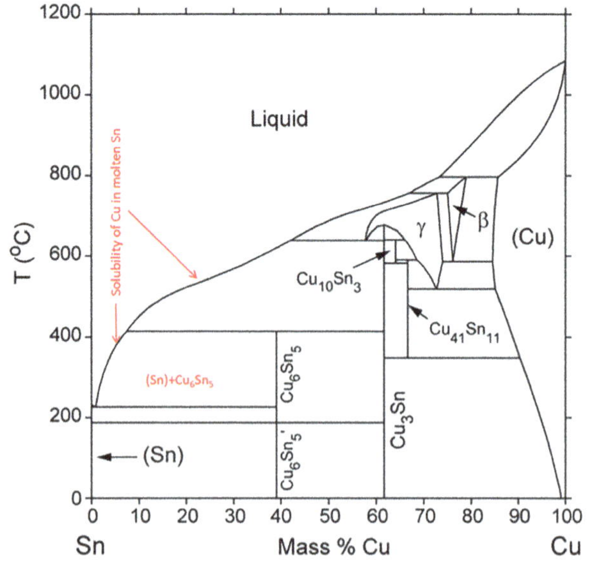

Fig. 3 Sn-Cu binary phase diagram

thick enough. When this happens, further growth of Cu_6Sn_5 cannot be supported by atomic transportation, and Cu_3Sn IMC is then formed between Cu_6Sn_5 and Cu following the reaction:

$$Cu_6Sn_5 + Cu \rightarrow Cu_3Sn \quad (3)$$

High-lead solders normally contain around 10 wt% and less of Sn (Table 1). Sn still dominates the interfacial reactions since Pb does not form IMCs with the common metallization materials of Cu, Ni, and Ag based on the phase diagrams. This limited Sn content in high Pb solder ends up with the formation of Cu_3Sn or Ni_3Sn_2 IMCs instead of Cu_6Sn_5 or Ni_3Sn_4 commonly in eutectic Pb-Sn and Sn-rich lead-free solders.

Interfacial reaction can change with solder composition [37–39]. For example, Cu of 5% and above in Sn-rich lead-free solder on Ni metallization changes the interfacial IMC formation from $(Ni,Cu)_3Sn_4$ to $(Ni,Cu)_6Sn_5$, in which the former is Ni_3Sn_4 originated while the latter is Cu_6Sn_5 associated. On Ni metallization, $(Ni, Cu)_3Sn_2$ could convert into $(Ni,Cu)_3Sn$ if enough Cu has been alloyed into the high-lead solders.

In high-lead solders with less Sn content (Sn < 2 wt%), the complete consumption of Sn may lead to IMC layer spalling if reflowed at higher peak temperature or temperature above liquidus (TAL) profile. The mechanism of spalling in high-lead solder has been rationalized: that the high interfacial energy between Cu_3Sn and Cu renders the penetration of Pb through Cu_3Sn grain or Cu_3Sn grain boundaries toward the interface between Cu metallization and Cu_3Sn layer and lifts the IMC layers [40]. Cu_3Sn IMC layer can also spall during solid-state annealing in high-lead solder joints, which is driven the interfacial energy reduction.

2.3 *Solidification*

A solder joint is completely formed only after the molten solder solidifies thoroughly. Traditional solidification theory has concluded that the undercooling drives the solid nucleation and thus the solidification always starts at a certain temperature below the liquidus temperature. After nucleation, the growth of the solid nuclei takes place immediately and quickly. Eutectic solders, having no pasty range, allow the solidification to start and end in a narrow temperature range. For off-eutectic solders, the solidification starts below the liquidus temperature and goes through the pasty range until it reaches the solidus line to finish in an equilibrium state. The wide pasty range allows the nuclei of the primary phase to grow bigger until the solder entirely solidifies. The growth of the primary phase expels the solute elements into the liquid and drives the liquid composition change along the liquidus line in an equilibrium state.

The joint interface may serve as the nucleation sites to facilitate the nucleation of solid. Sn-rich solders have been reported to have three major grain orientations because the nucleation starts from the bonding surface [41, 42]. The beach ball and the interlaced structure have been observed from Sn-rich solder as a result. No similar results have been reported for high-lead solders yet.

2.4 Microstructural Evolution

The morphology and microstructure of a solid joint include those of both the two bonding interfaces and the joint solder body. The factors influencing the microstructure of an as-reflowed joint include (a) the solder composition, (b) the metallization, and (c) the reflow profile. For high-lead solders, the limited Sn content may lead to the stable IMC layer thickness because Sn might be completely consumed to form interfacial IMC during soldering with no more IMC formation afterward.

The post-reflow microstructural evolution of a solder joint, driven by free energy reduction, is typically through solid atomic diffusion, which has been categorized as lattice diffusion, grain boundary diffusion, surface diffusion, dislocation core diffusion, etc. The microstructural evolution in a joint can be seen as grain coarsening, precipitate growth, phase growth, interfacial IMC layer thickening, and even IMC layer spalling. Solid atomic diffusion is influenced by both temperature and stress. The high service electrical current may impact the joint behavior by interfering with atomic diffusion. For most solder materials, room temperature is equal to homologous temperature of 0.5 and higher. The device service temperature can even be 0.8/0.9 homologous temperature. It is well recognized that above 0.5 homologous temperature, the atomic diffusion is favoring the dislocation movement, which is the most important and common way for permanent deformation. Also, the stress generated by the coefficient of thermal expansion (CTE) mismatch between the two joining components, together with the thermal factor, facilitates the microstructural evolution [43].

3 Ag Sintering

Unlike soldering, Ag sintering is a solid-state material transport process based on atomic diffusion driven by the reduction of total surface energy and/or interfacial energy. The starting materials are nano- and micro-sized Ag particles possessing high surface energy. The process occurs at a temperature below the material's melting point, which can be as low as 0.38 homologous temperatures (~200 °C). At the interface, the bonding mechanism is the formation of atomic interdiffusion layer between sintered Ag and substrate. In contrast, soldering is the process when solders melt into liquid state and react with substrate materials to form an intermetallic compound (IMC). The chemical reaction to form IMC is the driving force for bonding. It occurs at a temperature above the solder melting point.

3.1 *Sintering Driving Force*

The driving force of sintering is the reduction of Gibbs free energy by replacing high-energy free surfaces with low-energy grain boundaries and finally eliminating the grain boundary area via grain growth. Externally applied pressure can be used to enhance this intrinsic driving force and to eliminate residual porosity without excessive grain growth.

High-energy free surfaces refer to the atoms located at particle surface, which have higher energy than those in the bulk phase. Surface atoms possess surface energy which can be named specific surface energy γ (J/m^2) [44]. If the total surface area is A (m^2), the total surface energy is γA (J). The specific surface energy is due to the existence of an unsaturated bond of surface atoms, and it increases as the surface curvature increases, i.e., the particle size decreases. In reality, the unsaturated bonds of atoms at Ag particle surface are stabilized by organic stabilizers (aka capping agents), so that a stable Ag sintering paste can be generated. Note that we use the term "unsaturated bond" here as a tentative definition. Basically, it means that the silver surface is open for further interaction with other materials.

The stabilizers of the particles must be removed to expose the Ag atom for sintering as shown in Fig. 4; solvent normally can assist this process upon heating. Oven heating is mostly used for organic stabilizer removal. Other local heating methods such as UV [45, 46], laser irradiation [47, 48], or even non-heating ion-induced [49, 50] and solvent exchange [51, 52] processes can also be used to initiate sintering by removing the stabilizers.

There are basically two routes in sintering for reducing the total surface energy (γA). The reduction of the total surface energy can be expressed as

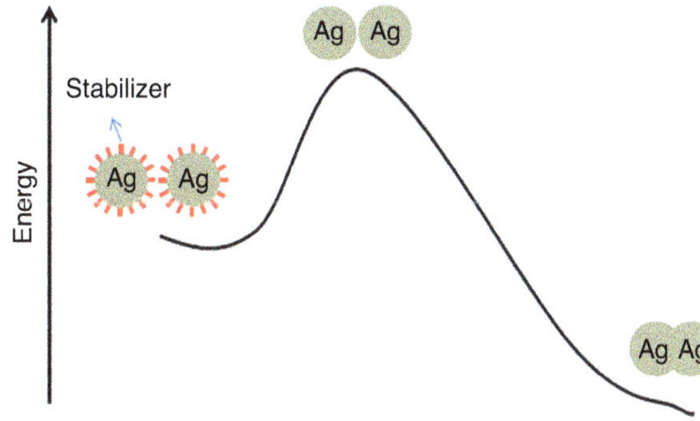

Fig. 4 Schematic diagram illustrating the initial steps in the sintering process

$$\Delta(\gamma A) = \Delta\gamma * A + \gamma * \Delta A \tag{4}$$

The first route is densification caused by $\Delta\gamma$, where interfaces between Ag and air (or solvent) or Ag grain boundary diminish to form Ag-Ag bonding. It shows the removal of pores. The second route is grain coarsening caused by ΔA, where grain growth results in decrease in total pore surface area with a small number of larger pores and bigger grains; it shows the coalescence of pores [53].

The amount of total unsaturated bonds of surface atoms, i.e., the total surface energy, determines the strength of the driving force for sintering. This shows that the total surface area (A) of the material also plays an important role for sintering. As we know, for a fixed amount of material with volume V_0, when a spherical particle size d decreases, the total surface area A increases dramatically according to Eq. 5:

$$A = \frac{6V_0}{d} \tag{5}$$

It is inversely proportional to that of the particle size; this also shows that the driving force for sintering is inversely proportional to the particle diameter. A pictorial description of Eq. 5 is shown in Fig. 5. One can see that when particle size decreases to a certain level, total surface area increases dramatically. This also explains why nanoparticles can be used for increasing the driving force and eliminating the pressure process during sintering.

On the other hand, when the particle size increases to the micrometer range, besides the reduced particle curvature, the reduced total surface area renders the driving force too weak to achieve reasonable sintering; that is why pressure is needed to assist sintering when only micron-sized silver particles are used in paste formulation [54, 55].

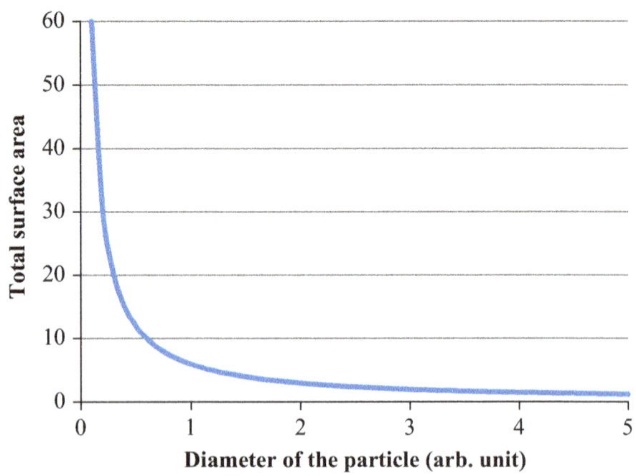

Fig. 5 Relationship between particle size and total surface area for a fixed amount of materials

3.2 Ag Sintering Path and Stage

Following the removal of the surface stabilizers, neighboring particles start to neck each other through surface and bulk paths. These mass transport paths are shown in Fig. 6 assuming a two-sphere sinter model. The details are also explained in Table 4 [53]. Surface paths are normally present at the initial stage of the sintering process. Bulk paths are responsible for densification. For crystalline materials such as Ag, it densifies through grain boundary diffusion, lattice diffusion, and viscous flow.

There are mainly three overlapping stages during solid-state sintering; the initial stage is characterized by surface atomic diffusion. Necks are formed between particles, and their contribution to compact densification is limited to a maximum of 2–3%. The intermediate stage features densification, up to 93% of relative density, and occurs before the isolation of pores. The final stage includes sintering after pore isolation [53]. Material transport paths in the latter two stages include reduction of interparticle distance, which are achieved by bulk material flow via viscous flow or by material transport from grain boundary via atom movement

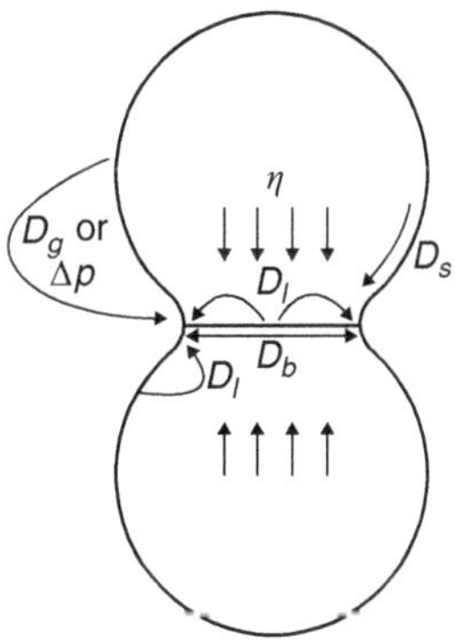

Fig. 6 Material transport paths during sintering [53]

Table 4 Material transport mechanism during sintering [53]

Material transport mechanism	Material source	Material sink	Related parameter
1. Lattice diffusion	Grain boundary	Neck	Lattice diffusivity, D_l
2. Grain boundary diffusion	Grain boundary	Neck	Grain boundary diffusivity, D_b
3. Viscous flow	Bulk grain	Neck	Viscosity, η
4. Surface diffusion	Grain surface	Neck	Surface diffusivity, D_s
5. Lattice diffusion	Grain surface	Neck	Lattice diffusivity, D_l
6. Gas phase transport			
6.1. Evaporation/ condensation	Grain surface	Neck	Vapour pressure difference, Δp
6.2. Gas diffusion	Grain surface	Neck	Gas diffusivity, D_g

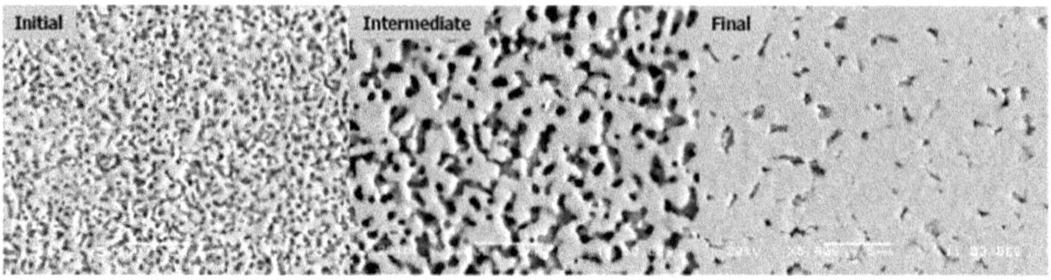

Fig. 7 Morphology of the Ag-sintered joints at different sintering stages

[56, 57]. Examples of the morphology of Ag sintered joints at different stages are shown in Fig. 7. From the microstructure, such as porosity, pore size, and grain size evolution, the balance of two basic sintering processes can be revealed: From initial to intermediate stages, pore size increases, indicating that coarsening plays a main role; from intermediate to final stages, pore size decreases, showing that the densification process dominates.

To achieve a high-density joint under a pressureless process, extending the heating time, elevating the sintering temperature, and using a smaller particle will help. Small particles possess high energy per unit volume, more surface area, and high curvature; these characteristics help speed up sintering. A classic treatment of the size effect on sintering is shown in Eq. (6), and the time (t_1 and t_2) for two particles of different sizes (d_1 and d_2) to achieve the same degree of sintering can be calculated [58]:

$$t_2 = t_1 \left(\frac{d_2}{d_1} \right)^m \tag{6}$$

Exponent m depends on sintering transport process. It can be 4 (grain boundary and surface diffusion), 3 (volume or lattice diffusion), 2 (evaporation/condensation), and 1 (plastic flow). For silver sintering materials, if a grain boundary and surface diffusion mechanism are assumed, to achieve the same degree of sintering, a 2 times difference in particle size may result in a 16 times difference in sintering time and rate.

Compared to micron-sized particles, sintering of Ag nanoparticles (NPs) occurs at much lower temperatures and might be driven by mechanisms such as mechanical rotation, plastic deformation, and evaporation/condensation, in addition to that of surface diffusion, grain boundary diffusion, and lattice diffusion [59]. In prior work, sintering of Ag NPs (4–20 nm) was numerically performed [60], showing enhanced sintering rates due to higher diffusion coefficients for nano-sized materials and a large driving force of sintering, which is caused by high curvature of the NP surfaces. At the second stage of sintering, neck to particle radius ratio increased gradually at the temperature increasing toward the surface premelting point of the sintered structure, and a twin boundary was formed. The last stage was characterized by liquid-state sintering and was driven by the surface premelting [60].

3.3 *Ag Interdiffusion Layer Formation*

The formation of a Ag sintering bond depends on the interfacial reactions between sintered Ag and surface finish of die/substrate. Figure 8 shows the heating profile and shear strength at 30-min interval for a joint of 3 × 3 mm^2 Ti/Ni/Ag surface-finished Si die on Ti/Ni/Au surface-finished Si_3N_4 active metal brazing (AMB) substrate [61]. At 60 min, when shear strength reaches 15 MPa , most Au-coated AMB surface still does not interact with sintered Ag yet, especially at the center area (Fig. 9). At 90 min, when shear strength reaches ~30 MPa and sintered Ag totally covers the Au-coated AMB surface, indicating a good bonding. This set of experiments demonstrate that sufficient interdiffusion between sintered Ag and substrate surface finish is critical for bonding.

The formation of a dense interdiffusion layer is confirmed by SEM and EDX analysis as shown in Fig. 10. Under this condition, similar Au surface finish is used but with DBC as substrate. The dense layer thickness is around 500 nm; the average composition is characterized as 65% Ag, 25% Ni, and 10% Au for the as-sintered

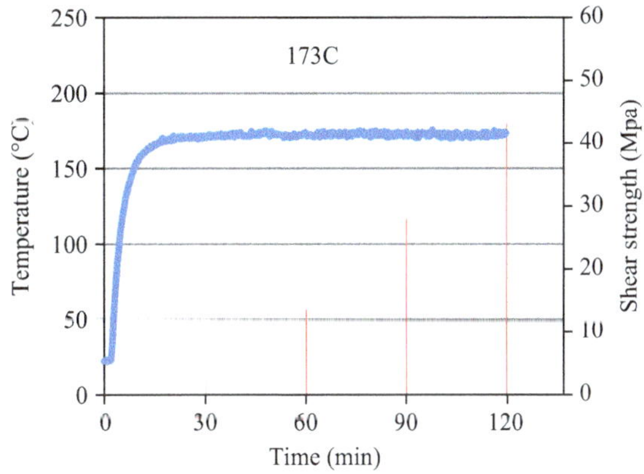

Fig. 8 Sintering profile and shear strength of the joint at times 30, 60, 90, and 120 min at 173 °C heating [61]

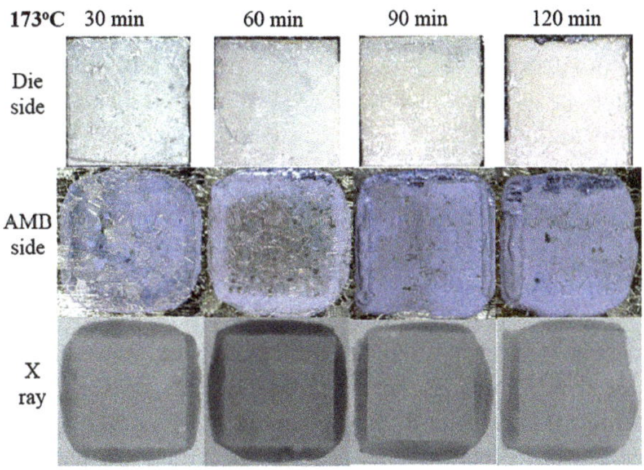

Fig. 9 Images of the sheared surfaces of die side and AMB substrate side, as well as X-ray images of the joint formed at different times at 173 °C heating [61]

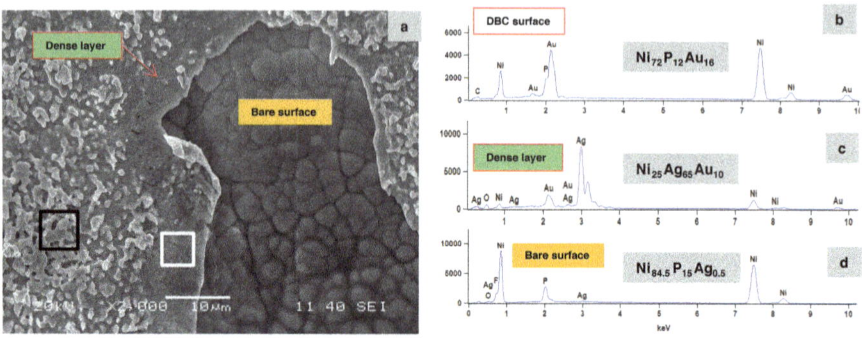

Fig. 10 (**a**) SEM image showing the dense layer structure and a bare surface spot on the sheared DBC. The point analysis results and atomic compositions were shown for (**b**) pristine Au-DBC substrate surface, (**c**) dense layer, and (**d**) bare surface spot as indicated in image (**a**) [62]

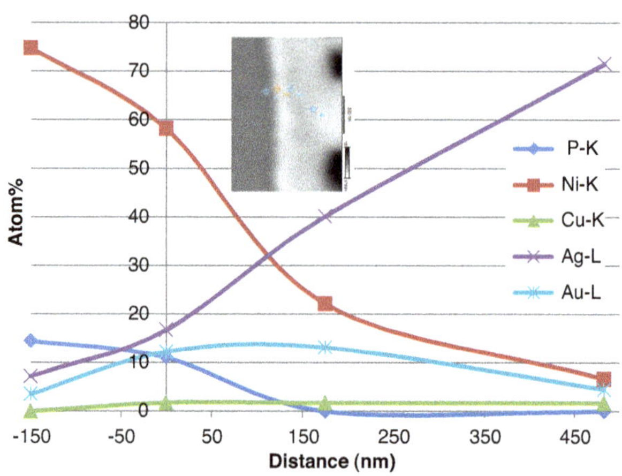

Fig. 11 Atomic percentage of P, Ni, Cu, Au, and Ag across the interface (0 at distance axis corresponding to point 2 in inset photo) from ENIG Cu (negative value on distance axis) to sintered Ag (positive in distance axis). Points 3 and 4 are in the sintered Ag phase [61]

sample [62]. It shows that this layer is generated by Ag diffusion from bulk Ag phase into surface Au layer through one direction and Ni diffusion from substrate through the opposite direction. One thing worth noting is that as the Ag diffuses from sintered bulk phase toward interface, an empty space may be created near the interface, and this space is called "depletion layer" as also observed and discussed by other authors [19] [20]. The formation of this "depletion layer" is undesired, because it becomes the weakest point of the whole joint, resulting in reliability issues. Development of a proper Ag sintering paste formulation is key to solving this issue.

The detailed elemental distribution within the dense silver layer can be further characterized by cross section and EDX analysis. Figure 11 shows an example of the result obtained at sintered Ag and electroless nickel immersion gold (ENIG)-Cu interface. Reconstruction of the point analysis data in the photo shows that, in the dense Ag layer, P, Ni, and Au contents decrease gradually, Cu content remains almost constant, and Ag content increases continuously into the silver phase. Note that this elemental variation is an indication that the dense layer is caused by interdiffusion.

It is generally recognized that Ag sintering paste can form good bonding with noble metal surface finish such as Au, Ag, Pt, and Pd [23]. However, it is difficult to generate bonding in industrially preferred DBC or direct bond aluminum surface due to the tenacious oxide layer [23]. Recent development shows that with proper choice of silver paste, it is possible to generate good bonding with shear strength 35–55 MPa between sintered Ag and Cu die even without any pretreatment of Cu die [3, 61]. However, Cu oxidation is still an issue to be solved during the aging tests [61].

3.4 Microstructure Evolution During Aging

Thermal stability of the joints is important for product reliability. One typical test is to trace microstructure evolution during aging. For Ag sintering, several parameters of the joint are critical for evaluating microstructures, including porosity, pore size, shape of pore and distribution, and grain size. Pore is the vacancy between particles during solid-state sintering; its formation depends on the necking between particles and subsequent neck growth; neck formation is strongly dependent on the sintering temperature, while prolonging the holding time and lowering the heating rate are beneficial to neck growth [63]. Pore size distribution and shape are a function of several of the above sintering conditions. It is found that smaller pores normally have a more spherical shape, while larger pores are more irregular and segregated along the interstices. The fraction of an irregular pore can be reduced by increasing the sintering temperature, while the fraction of smaller and spherical pores could be increased by decreasing the heating rate and prolonging the holding time [63]. Grain growth is the ongoing process under heat treatment. At the same time, pore size grows bigger with reduced pore number. Figure 12 displays SEM images of sintered Ag joints aged under a 250 °C oven for different times [3]. The as-sintered joints have an average pore size of around 300 nm and a grain size around 1 μm. After 336 h of aging, the pore size increased dramatically to 1.5 μm, with much less number of pores. At this time, the grain size also increased to around 5 μm. Surprisingly, further aging the samples to 3200 h did not change the morphology of the silver joints, and during this time, shear strength of the joints remained at around 70 MPa. One interesting finding is that a very dense Ag layer becomes thicker at the ENIG surface-finished DBC side during this period, suggesting that the interfacial interaction between the sintered Ag and DBC plays a critical role in stabilizing the joints. As this layer becomes thicker during aging at 250 °C, Cu atoms from DBC also diffuse into this layer; the average composition becomes 85–98% Ag, 1–5% Ni, 1–5% Cu, and 0–5% Au. Note that this composition is just an average measurement result, and does not correspond with IMC [62].

The growth kinetics of the dense silver layer can be characterized by plotting the thickness data vs $t^{1/2}$ (where t is the aging time) at 250 °C (Fig. 13). Linear regression analysis shows that $t^{1/2}$ dependency with correlation coefficient (R^2) is 0.95, which indicates that the dominant microstructure mechanism is the volume diffusion

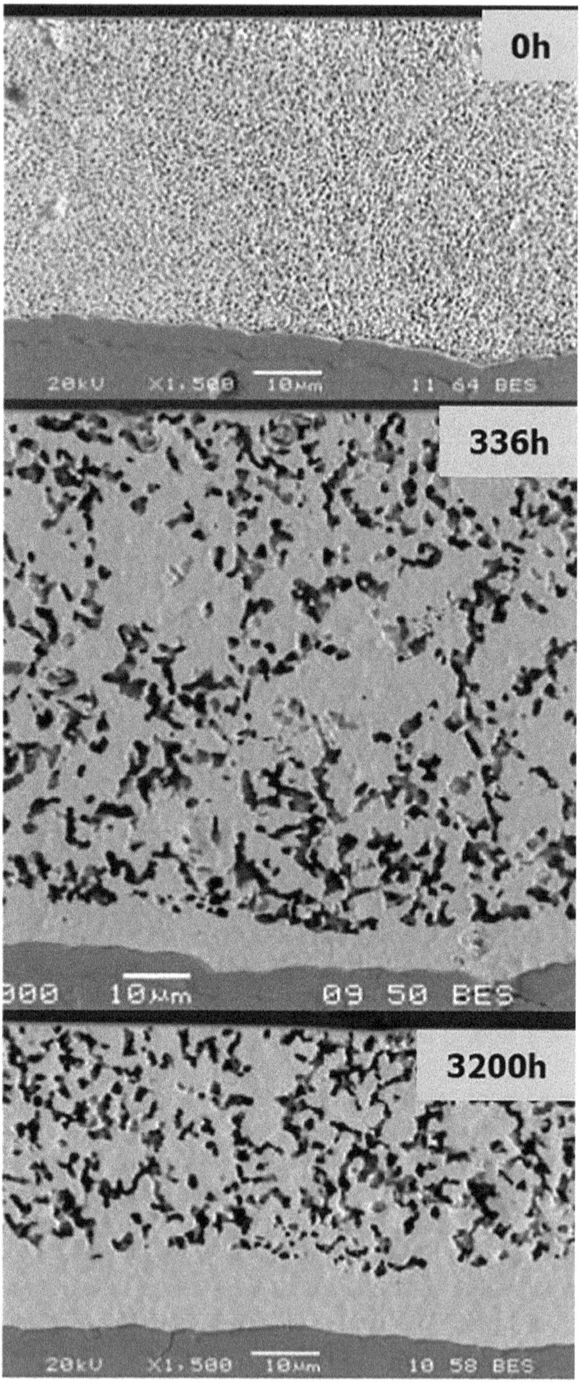

Fig. 12 SEM images of cross section of silver sintering joints aged under 250 °C for different times. Ag-Si die is attached on the ENIG-DBC surface

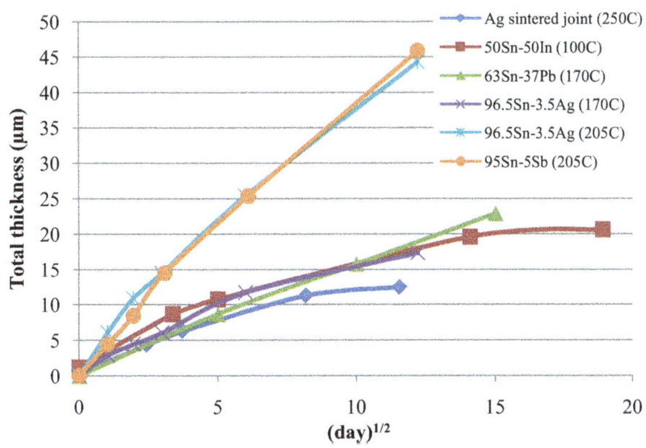

Fig. 13 Comparison of the dense layer thickness growth of silver sintering joint with that of Sn-containing solder under different temperature agings [64]

process. For comparison, intermetallic growth data of a series of Sn-containing solders were also included [64]. Note that the intermetallic growth rate is much faster than that of dense sintered Ag layer. For example, the growth rate for 50Sn-50In at 100 °C is already slightly higher than that of sintered Ag layer at 250 °C.

Intermetallic formation indicates good wetting and bonding. However, under aging conditions, excessive growth of intermetallics and their brittleness may be detrimental to the joint's reliability. It can initiate failures during deformation. In addition, this layer is very hard, making it even difficult to accommodate mechanical strain that results from the constraint of materials with differing coefficients of thermal expansion (CTE). This property is different from what we observed for Ag sintering paste. For the Ag sintering joint, assuming that the bond line thickness is thick enough (>50 μm); the growth of the dense silver layer at the interface actually helps the bond reliability because it enhances the interfacial interaction, which is normally a weak point for the sintered joints.

4 Comparison Between Silver Sintering and Soldering

From the discussion above, it can be seen that the Ag sintering processing is very different from that of soldering. A step-by-step comparison of the main events of these processes is summarized in Table 5.

Due to different formation mechanisms, the processing conditions differ between sintering and soldering. For example, for silver sintering, processing under airy conditions is preferred based on the need to burn off surface stabilizer on silver particle surface, while processing under inert gas is a better choice for soldering due to the need to prevent oxidation of the solder during joint formation. Another feature is that the solder paste normally has the self-alignment ability, i.e., when printing the solder paste with an area larger than that of the underneath Cu pad, the solder ball will collapse to wet only and form a joint on the Cu pad area after reflow. However, if silver sintering paste is used, it will not collapse since silver particles do not melt;

Table 5 Main events during joint formation for Ag sintering and soldering

Events	Ag sintering	Soldering
Initial step	Solvent evaporation, stabilizers burnt out or removed from Ag particle surface, substrate wetting	Solvent evaporation, flux flow, substrate and solder particle surface fluxing
Bulk phase reaction	Surface atom diffusion between Ag particles, necking formation, pores form connected channels, lattice and grain boundary diffusion, densification, pore isolation	Solder powder melting, solutes precipitate, and solid solution formation during solidification
Interfacial reaction	Atoms between sintered silver and substrate interdiffuse to form a dense Ag layer	Substrate atoms dissolving in liquid solder, forming supersaturated layer; intermetallic formation at interface; intermetallic growth; or new intermetallic formation
Microstructure evolution under aging	Grain growth, pore enlargement, dense layer thickening, depletion layer formation	Coarsening, precipitate growth, intermetallic growth, solid diffusion of different IMCs, Kirkendall effect

the area after sintering will be very similar or a little smaller than that previously printed. A comparison between the physical and mechanical properties and other properties of the different high-temperature bonding materials is shown in Table 6 [65].

5 Porosity of Ag Sintering Joint

As discussed above, a silver sintering joint inevitably contains pores, which affect most aspects of the joint's properties since it is located at both the bulk sintered phase and interfaces. Here we examine how the porosity affects mechanical, electrical, and thermal properties of the silver joints. These properties could further impact intended functions such as thermal transfer efficiency and mechanical stress relief, affecting reliability of the interconnect layer and its performance [66].

5.1 Clarification of the Definition of Void and Pore

In the solder material world, the terms "void" and "pore" are normally used interchangeably; they refer to the air volume present in the joint. For silver sintering materials, "pore" is generated by interparticle atomic diffusion during densification and grain growth. Porosity is percentage of "pore" by definition. Pores exist everywhere in the joint and their sizes range from nano- to micrometer with a resolution around 10–20 μm and are difficult to detect by conventional X-ray machines. "Void" refers to the big air volume formed with a mechanism similar to that observed in the

Table 6 Potential high-temperature lead-free bonding materials

	Physical and mechanical properties					Form			Cost
	Melting temp (°C)	Shear strength (MPa) @RT	Shear strength (MPa) @200 °C	Electrical resistivity (μΩ*cm)	Thermal conductivity (W/mK)	Paste	Wire	Preform	Relative to Pb5Sn2.5Ag
Pb5Sn2.5Ag	296	28	7.5	19	23	Yes	Yes	Yes	1
ZnAl	>360	120	70	7.50	100	No	Yes	Yes	0.2~1
BiAgX®	>260	45	22	86.00	14	Yes	Yes	Yes	2–5
AuSn	280	130	100	16.40	59	Yes	Yes	Yes	>2000
TLP bonding	>300	20–40	20–40	–	–	Yes	No	Yes	Varied
Ag sintering	961	20–80	20–40	5	>100	Yes	No	Yes	>40

solder materials, and the sizes of voids are in the range from micron to millimeter, which can be detected by conventional X-ray machines.

5.2 *Void Formation and Affecting Factors*

Void formation may have similar mechanism for silver sintering pastes and for solders because of their similarity as paste form. Certainly they also have differences since the solder melts during bonding process. In electronics industry that use solder as bonding material, voiding in the joint has been a severe issue for decades and also has been intensively studied. In various assemblies such as SMT, BGA, CSP, and flip chips, with the continuing progression toward miniaturization, the impact of voiding has been aggravated due to the vulnerability and small volume of the tiny solder joints [67]. This situation becomes even worse with the global move from Sn-Pb towards lead-free soldering, mainly due to the poor wetting of lead-free alloys. In addition, due to the low cost and various reliability considerations, surface mount assemblies often face complicated mixed solder alloy systems during BGA assembly. This includes a combination of eutectic Sn-Pb, low Ag, and high Ag and SnAgCu between the solder sphere and the solder paste. Such mixed solder alloy systems have experienced major issues with voiding.

Voiding could be caused by (a) solder shrinkage during solidification, (b) laminate outgassing during soldering, and (c) entrapped flux, and, most significantly, (d) they are affected by paste composition [68–71]. Detailed study has shown that the voids do not exhibit organic residue and are thus due to the flux or flux reaction outgassing. Voids are formed during vapor condensation upon cooling. Experiments have shown that void content decreases with increasing flux activity suggesting that fluxing reaction by activator is not the major source of voiding. This also suggests that the flux entrapped at the powder or substrate oxide surface takes the main responsibility for voiding. This is also supported by the fact that the good solderability of substrate reduces the void level [69].

There are a lot of factors during the soldering process that will affect the voiding. Some are listed briefly in Table 7. The readers are referred to reference [67] for further information. For the silver sintering pastes, a lot of factors obtained from solder material are also applicable. For example, Fig. 14a displays the X-ray image of a joint generated under a heating rate of 35 °C/min, which shows a lot of voids with sizes ranging from 20 μm to 230 μm. By reducing the heating rate to 3 °C/min, it is possible to avoid rapid solvent evaporation and obtain void-free sintering joints (Fig. 14b) [61].

Compared to the bulk silver metal, the biggest difference in silver sintering joint is that it contains air volume, in the form of void or pore. Sometimes, cracks are also observed, and we can include them in the void category. With abundant experience in solder materials, removal of the voids can normally be achieved with due diligence. Pores, on the other hand, cannot be removed, and the best we can do is to reduce them to the least extent.

Table 7 Affecting factors on voiding during solder joint formation [67]

Affecting factor	Comment and result
Flux activity	High flux activity results in high wetting power, thus reducing the voids
Flux volatility	Void content increases with decreasing boiling point of solvent
Sphere alloy type	SnPb more sensitive than high Pb alloy
Sphere oxide level	Voiding increases with increasing sphere oxide level
Solder powder size	Voiding increases with decreased powder size
Metal load	Voiding increases with increasing metal load due to increased oxide level
Viscosity	A high viscosity results in a lower voiding rate
Paste exposure time	Voiding increase caused by oxidation and moisture pickup
Pad dimension	Void increases with increasing pad dimension
Pad surface finishes	Voiding increases in the order of Ni/Au, HASL, OSP, Ni/Pd, and Ni/Pd with Au flash
Pad oxide level	Voiding increases with increasing pad oxide level
Board pad design	Voiding increases in via-in-pad, microvia design due to difficult void release
Quality of parts	Barrel not properly plated, leaving high porosity for volatile to blow solder
Deposition thickness	The thicker the deposition the less voids due to high flux capacity
Solder/substrate interface	Void sticks to the interface due to minimal surface energy requirement
Reflow profile	Increased profile length results in higher voiding level
Reflow atmosphere	Reflow under nitrogen yields lower voiding than that under air

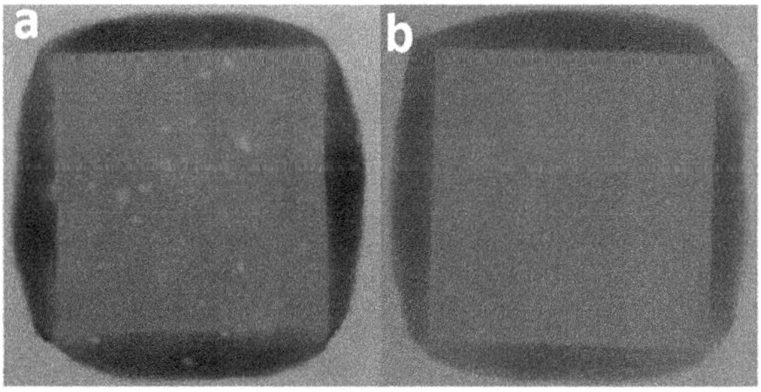

Fig. 14 X-ray images of sintered Ag joints with voids (**a**) and that without voids (**b**). Die size is 3 mm × 3 mm

5.3 *Determination of Porosity*

Understanding porosity and its effect on silver sintering material properties is key to reliability and application of this material. The first question is how to measure porosity? As we discussed above, since void (v) and porosity (p) are air volumes existing within the sintered silver phase, they are related to the density (ρ) by the following equation:

$$p + v = 1 - \frac{\rho}{\rho_0} \tag{7}$$

where ρ_0 is the bulk silver density.

If care is taken to completely remove the void, the porosity of the joints can be expressed as

$$p = 1 - \frac{\rho}{\rho_0} \tag{8}$$

If sintered silver possesses a regular shape, its volume can be accurately measured, and then its density can be easily calculated using the weight to volume ratio of the specimens. For an irregular shape, volume can also be obtained by Archimedes' methods. However, a large sample volume is normally required to reduce measurement error.

Another method for characterizing the porosity is through a cross-sectional image of the sintered silver phase. Since the SEM images a very limited area of the samples, there are several prerequisites for this method to be successful: (1) the sample porosity should be even in the whole joint, so that the small area sampled represents the whole. Otherwise, areas with different porosities should be characterized separately and calculated numerically. (2) SEM image normally captures one cross-sectional cutting plane of the sample; whether the pore area fraction in this plane is a representative of the 3D pore volume fraction is another question that needs to be answered. Experiments with focused ion beam-scanning electron microscopy (FIB-SEM) 3D imaging have shown that the difference between pore surface fraction and pore volume fraction can be as large as 6–10% [72, 73]. (3) In the process of cross-sectional sample preparation, silver is easily smeared into the pores due to the grinding and polishing, the pore is normally smaller or difficult to distinguish, and an ion milling or FIB treatment is needed to obtain the correct porosity.

Besides the above-mentioned methods, there are other ways to characterize porosity, including nanoscale X-ray computed tomography (nano-CT) [73, 74], ultrasonic wave [75, 76], or chemical replacement method [77]. However, work seldom has been systematically focused on silver sintered joints. We notice one work has focused on this topic [78], and the relation between pore surface fraction and density is shown in Fig. 15. Since the authors have carefully treated the samples to remove any void formation during sintering, the relationship between density and porosity is likely to obey Eq. (7) (red line in Fig. 15). However, the studies reported a difference between the data and the calculated equation. There are two possibilities to account for this discrepancy: (1) if surface pore fraction represents the real porosity of the materials, the density measured is smaller than the real value; (2) if density is real, then the measured surface pore fraction is systematically smaller than the real porosity. It seems the latter case is more likely if no ion milling or FIB or chemical etching process is used in the sample preparation process.

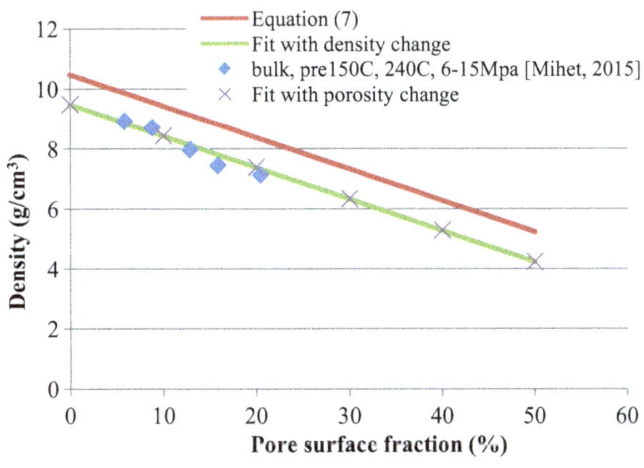

Fig. 15 Density vs pore surface fraction

5.4 *Effect of Porosity on Mechanical Property*

The reliability of the silver joints is critical for industrial application. Different reliability tests have been designed targeting different applications. For example, they can be temperature cycling (for thermal mechanical properties), power cycling (mimic real operating conditions), thermal shock (CTE mismatch tolerance), vibration test (for automation application), drop test (for mobile device application), and thermal aging test (for high-temperature operation). A joint is composed of bulk sintered silver and two interfaces, i.e., between the silver and the components (semiconductor chips such as Si or SiC) and that between the silver and the substrates (such as DBC, AMB ceramic, or Cu). These three parts are stacked and their stresses balanced to survive the reliability tests. Mechanical properties of bulk sintered silver joints are the first properties to be studied; the effect of porosity on the mechanical property is thus desired before conducting the reliability tests.

5.4.1 Elastic Modulus

Elastic modulus or Young's modulus (E) describes tensile elasticity, or the tendency of an object to deform along an axis when forces are applied along that axis. Elastic module is one of the important parameters to evaluate the life prediction of power electronic devices. As we know, die-attach devices contain stacking materials with different CTEs, which will develop thermal stress when in operation at high temperature. Sintered silver joint should have enough elasticity to accommodate this stress.

Different methods have been used to prepare samples for modulus tests, either by first milling the dried pastes into powder and then pressurizing the powders under heating [79] or by repeatedly applying paste and dry processes and sintering them together under pressure [78, 80–82]. The measurement is conducted with dynamic

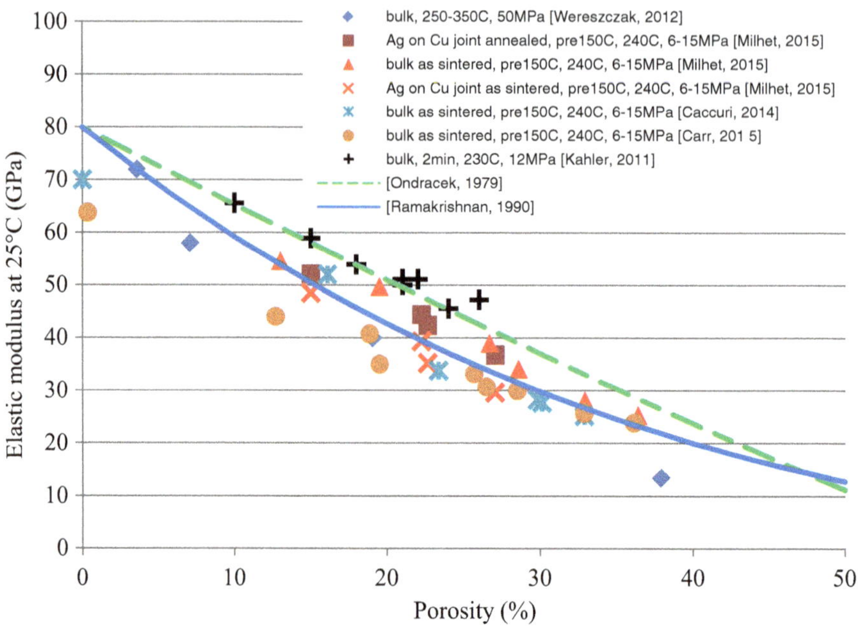

Fig. 16 Elastic modulus vs porosity for sintered Ag materials prepared from different references

resonant spectroscopy [78–81] or stress tests [82]. The elastic modulus vs porosity data of different samples are summarized in Fig. 16. The general trend is very clear, i.e., elastic modulus decreases with increased porosity. However, data scattering is also observed. For example, at an elastic modulus value of 35GPa, the porosity is observed to fluctuate within the range 20% to 30%. Two theoretical calculations from Ondracek [83] and Ramakrishnan [84] are noticed in Eqs. (9) and (10); they both calculated the elastic modulus E based on porosity p, bulk silver elastic modulus, E_{Ag}, and Poisson's ratio V_{Ag}.

$$E = E_{\mathrm{Ag}} \frac{3(3-5p)(1-p)}{9-p(9.5-5.5V_{\mathrm{Ag}})} \tag{9}$$

$$E = E_{\mathrm{Ag}} \frac{(1-p)^2}{1+p(2-3V_{\mathrm{Ag}})} \tag{10}$$

These two fitted lines are also shown in Fig. 16. The Ramakrishnan model appears to fit the data better.

One interesting finding is that when comparing Young's modulus of freestanding sintered silver and that of the Cu substrate, the latter shows an essentially lower (about 10–20%) value than the former; annealing under 200 °C for 4 h can recover it [78]. Thus, it is possible to monitor the thermal stress through Young's modulus measurement.

5.4.2 Yield Strength

Yield strength or yield stress is the material property defined as the stress at which a material begins to deform plastically. This parameter is important because it indicates when the permanent damage will occur for the joints. In the temperature cycling tests, if the thermal mechanical stress of the joint exceeds this value, stress will accumulate gradually and damage the joints. The effect of porosity on the yield strength of silver joints is shown in Fig. 17.

Overall, yield strength decreases as porosity increases. At low porosity end, yield strength can vary a lot from 60 MPa [79] to 180 MPa [85], depending on the materials and preparation processes. When porosity is larger than 20%, the yield strength decreases to around 20 MPa [80].

5.4.3 Strain to Fail

Strain to fail gives a general idea about how the joints can endure the mechanical movement under harsh conditions, such as temperature cycling or shock. This value drops to less than half when porosity changes from 0 to 3%, and then slows down upon further porosity decrease (Fig. 18) [80, 85, 86]. In this regard, a material with porosity close to that of a bulk material is preferred. An exponential fit was applied, and it fits well at a high porosity range, >15%, but deviated at a lower porosity range.

5.4.4 Ultimate tensile strength (UTS)

Ultimate tensile strength (UTS), or ultimate strength, is the capacity of a material or structure to withstand loads tending to elongate. This parameter is also important to

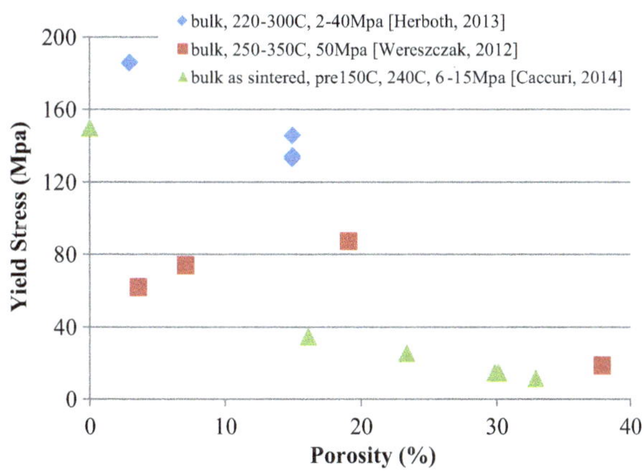

Fig. 17 Yield strength vs porosity for sintered Ag materials from different references at room temperature

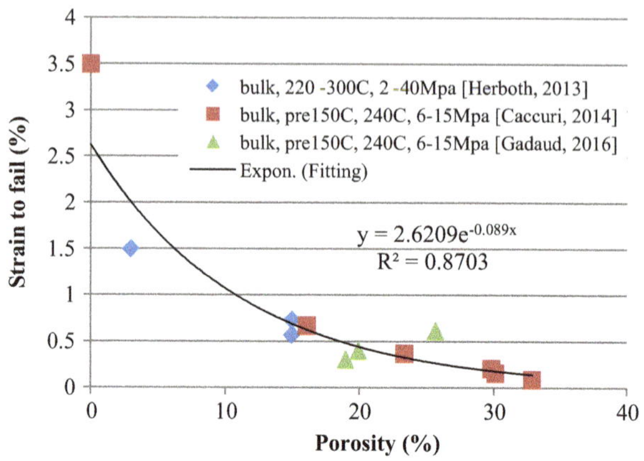

Fig. 18 Strain to fail vs porosity for Ag sintered materials from different references

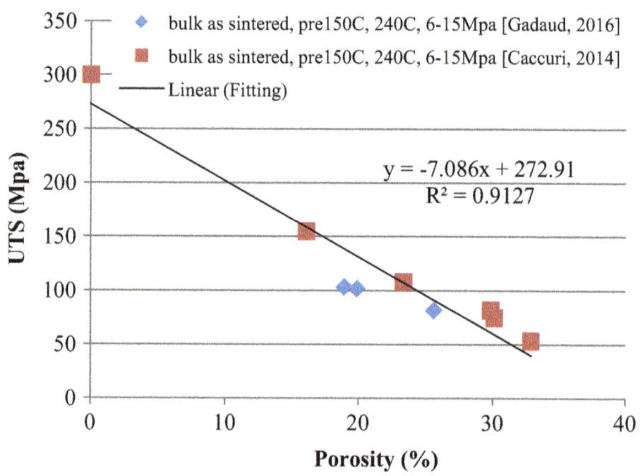

Fig. 19 UTS vs porosity for Ag sintered materials from different references

judge the joints; as can be seen in Fig. 19, 1% of increase in porosity will reduce approximately 7 MPa for the UTS [80, 86].

5.4.5 Coefficient of Thermal Expansion (CTE)

Coefficient of thermal expansion (CTE) is invariant with porosity change, according to data from reference [79]. The average value of the CTE of silver with different porosities is 20 ppm/C.

5.4.6 Poisson's Ratio

Figure 20 shows that Poisson's ratio decreases linearly with the increase in porosity [79].

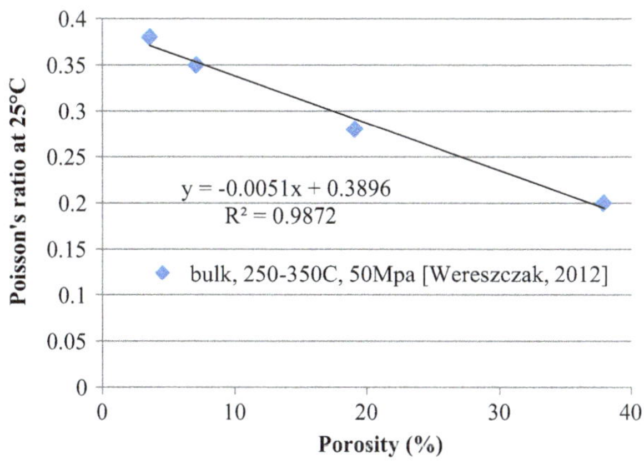

Fig. 20 Poisson's ratio vs porosity from reference [79].

5.5 *Effect of Porosity on Thermal Conductivity*

Summarized results from several studies on the relationship between porosity and thermal conductivity are shown in Fig. 21 [72, 79, 87]. Figure 21 shows that thermal conductivity of the materials decreases as porosity increases. This makes it clear that thermal energy is transferred through electrons in silver materials; the existence of pores limit thermal transfer channels.

The shape of the pore also affects thermal conductivity, and the relationship is shown in Fig. 22 [88]. Spherical pore is the most desirable pore shape for high thermal conductivity, while pancake-shaped ones degrade thermal conductivity severely.

5.6 *Effect of Porosity on Electric Conductivity*

Electrical conductivity is related to the electrons in the materials. The interior of a metal is filled up with a large number of free electrons that travel aimlessly. When the conductor is subjected to an electric force applied at its opposite ends, these free electrons rush in the direction of the force, thus forming electric current. Therefore, electrical conductivity is directly related to the amount of electron-conducting channels. Porosity limits the availability of above channels, rendering lower conductivity. Summarized results from different references are shown in Fig. 23 [72, 79, 87, 89].

Study of the sintered 40 nm nanoparticle film silver shows that measured conductivity correlates with thickness decrease and densification, i.e., the conductive channel formation [90]. Another study using 3–10 nm nanoparticle also demonstrates that when the surface stabilizer is completely removed, the conductivity increases more than 3 orders of magnitude, illustrating the importance of the conducting channel [91].

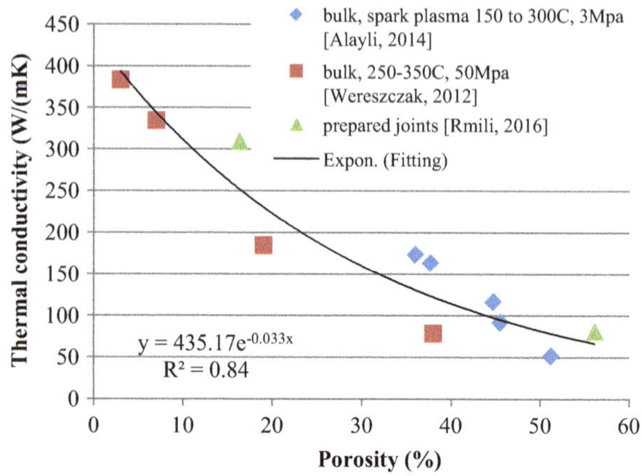

Fig. 21 Electrical conductivity vs porosity from different references [72, 79, 87]

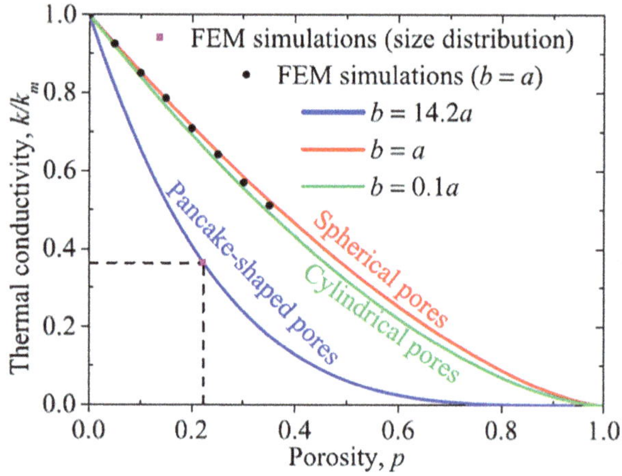

Fig. 22 Thermal conductivity vs porosity for different shaped pores [88]

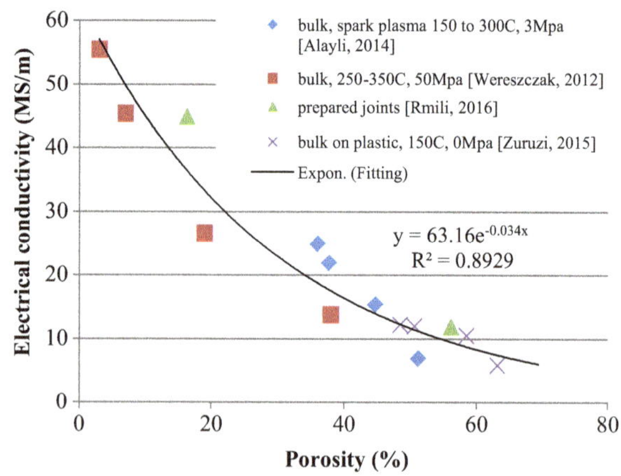

Fig. 23 Electrical conductivity vs porosity from different Ag sintering references

6 Summary and Conclusion

Silver sintering materials are one of the promising candidates for replacing high-Pb solder as joint materials in the global lead-free movement in electronics, especially when harsh conditions such as $\geq$200 °C are required for applications in hybrid and electric cars, high-speed trains, aircraft/aviation, and deep well oil/gas extraction. Mechanistic study shows that Ag sintering materials are very different materials from solders. The driving force of silver sintering is particle surface energy reduction. It occurs through solid-state atomic transportation such as various diffusions and viscous flow processes. Non-stoichiometric interdiffusion layer is generated through mutual atomic diffusion between silver and joining surface metallization. In contrast, solder joint is formed through metallurgical interaction between molten solder and joining surfaces followed by solder solidification, where stoichiometric intermetallic is normally observed at interfaces. Silver sintering joints possess intrinsic porosity, which affects their bulk mechanical properties, such as elastic modulus, yield strength, strength to fail, ultimate tensile strength, and Poisson's ratio, as well as their thermal and electrical conductivities. Ag sintering materials possess excellent thermal and electrical properties compared to that of solder joints. Their cost is higher than that of high-lead solders but much cheaper than that of Au-containing solders. Future directions for silver sintering materials in industrial application lie in several aspects: development of materials under pressureless conditions using reflow oven, which is a "drop in" replacement for high-Pb solders; materials workable on Cu surface bonding to further lower the cost of ownership; control porosity of sintered joints for realization of high reliability by passing their high temperature storage, temperature cycling, and power cycling tests.

Acknowledgment and Note We gratefully thank Dr. Ning-Cheng Lee for his continuous support during the writing of this chapter. Dr. Hongwen Zhang focused on Sect. 2; Dr. Sihai Chen was in charge of Sects. 1, 3, 4, 5, and 6.

References

1. C. Buttay, D. Planson, B. Allard, D. Bergogne, P. Bevilacqua, C. Joubert, M. Lazar, C. Martin, H. Morel, D. Tournier, C. Raynaud, State of the art of high temperature power electronics. Mater. Sci. Eng. B: Solid-State Mater. Adv. Technol. **176**(4), 283–288 (2011)
2. R. Kirschman, *High Temperature Electronics* (IEEE press, New York, 1999)
3. S. Chen, C. LaBarbera, N.C. Lee. Silver sintering paste rendering low porosity joint for high power die attach application. *IMAPS Conference & Exhibition on HiTEN* (Albuquerque, NM, 2016), pp. 237–245
4. M. Knoerr, S. Kraft, A. Schletz, Riliability assessment of sintered nano-silver die attachment for power semiconductors. *12th Electronics Packaging Technology Conference*, 2010. pp. 56–61
5. E. Bradley, C.A. Handwerker, J. Bath, R.D. Parker, R.W. Gedney, *Lead-Free Electronics* (John Wiley & Sons, Hoboken, 2007)

6. J.G. Bai, Z.Z. Zhang, J.N. Calata, G.-Q. Lu, Low-temperature sintered nanoscale silver as a novel semiconductor device-metallized substrate interconnect material. IEEE Trans. Compon. Packag. Technol. **29**(3), 589–593 (2006)
7. Y. Mei, T. Wang, X. Cao, G. Chen, G.-Q. Lu, X. Chen, Transient thermal impedance measurements on low-temperature-sintered nanoscale silver joints. J. Electron. Mater. **41**, 3152–3160 (2012)
8. G. Chen, L. Yu, Y. Mei, X. Li, X. Chen, G.-Q. Lu, Uniaxial ratcheting behavior of sintered nanosilver joint for electronic packaging. Mater. Sci. Eng. A **591**, 121–129 (2014)
9. V.R. Manikam, K.Y. Cheong, Die attach materials for high temperature applications: a review. IEEE Trans. Compon. Packag. Manuf. Technol. **1**, 457–478 (2011)
10. S. Fu, Y. Mei, G.-Q. Lu, X. Li, G. Chen, X. Chen, Pressureless sintering of nanosilver paste at low temperature to join large area ($\geq$100 mm^2) power chips for electronic packaging. Mater. Lett. **128**, 42–45 (2014)
11. J.F. Yan, G.S. Zou, A.P. Wu, J.L. Ren, J.C. Yan, A.M. Hu, Y. Zhou, Pressureless bonding process using Ag nanoparticle paste for flexible electronics packaging. Scr. Mater. **66**, 582–585 (2012)
12. H. Schwarzbauer, Method of securing electronic components to a substrate. 4810672 United States, 1987
13. H. Schwarzbauer, R. Kuhnert, Novel large area joining technique for improved power device performance, in *Conference Record of the 1989 I.E. Industry Applications Society Annual Meeting*, (IEEE, New York, 1989), pp. 1348–1351 (2016)
14. H. Schwarzbauer, R. Kuhnert, Novel large area jointing technique for improved power device performance. IEEE Ind. Appl. Soc. Annu. Meet. **27**, 93–95 (1991)
15. C. Gobl, J. Faltenbacher, Low temperature sinter technology die attachment for power electronic applications. *Proceedings of 6th International Conference on Integerated Power Electronic Systems* (Nuremburg, Germany, 2010), pp. 1–5.
16. H. Zheng, J. Calata, K. Ngo, S. Luo, and G.-Q. Lu. Low-pressure (<5 MPa) low-temperature joining of large-area chips on copper using nanosilver paste (Nuremberg, Germany, 2012), CIPS 2012. p. Paper12.3
17. J.G. Bai, G-Q Lu, Thermomechanical reliability of low-temperature sintered silver die attached SiC power device assembly. IEEE. T. Device Mat. Re. **6**, 436–441 (2006)
18. T. Wang, M. Zhao, X. Chen, G.Q. Lu, K. Ngo, S. Luo, Shrinkage and sintering behaviorof a low-temperature sinterable nanosilver die-attach paste. J. Electron. Mater. **41**(9), 2543–2552 (2012)
19. F. Yu, R.W. Johnson, M. Hamilton, Pressureless, low temperature sintering of micro-scale silver paste for die attach for 300 °C applications. *IMAPS Conference & Exhibition on HiTEN*, 2014. pp. 165–171
20. G. Lewis, G. Dumas, S.H. Mannan, Evaluation of pressure free nanoparticle sintered silver die attach on silver and gold surface. *IMAPS Conference & Exhibitionon HiTEN*, 2013. pp. 237–245
21. K.S. Siow, Are sintered silver joints ready for use as interconnect material in microelectronic packaging? *J. Electron. Mater.* **43**, 947–961 (2014)
22. K.S. Siow, Mechanical properties of nano-Ag as die attach materials. J. Alloys Compd. **514**, 6–19 (2012)
23. K.S. Siow, Y.T. Lin, Identifying the development state of sintered silver (Ag) as a bonding material in the microelectronic packaging via a patent landscape study. J. Electron. Packag. **138**, 020804-1–020804-13 (2016)
24. R. Khazaka, L. Mendizabal, D. Henry, Review on joint shear strength of nano-silver paste and its long-term high temperature reliability. J. Electron. Mater. **43**(7), 2459–2466 (2014)
25. G. Humston, D. Jacobson, Principles of soldering and brazing. Materials Park, OH, USA: ASM International, 1993
26. A. Rahn, *The Basics of Soldering* (John Wiley & Sons, New York, 1993)

27. D. Shangguan, *Lead-free Solder Interconnection Reliability* (ASM International, Materials Park, 2005)
28. M. Thomas, Die-attach materials and processes – a lead-free solution for power and high-power applications. Adv. Packag. **30**, 32–34 (2007)
29. F.P. McCluskey, M. Dash, Z. Wang, D. Huff, Reliability of high temperature solder alternatives. Microelectron. Reliab. **46**, 1910–1914 (2006)
30. X. Xie, X. Bi, G. Li, Thermal-mechanical fatigue reliability of PbSnAg solder layer of die attachement for power electronic devices. *2009 International Conference on Electronic Packaging Technology & High Density Packaging* (IEEE Xplore, 2009), pp. 1181–1186
31. I. Okamoto, T. Yasuda, Selection of optimum Cu content in Cu bearing tin-lead solder. Transaction of JWRI, 1986. pp. 245–252
32. K.N. Tu, K. Zeng, Tin–lead (SnPb) solder reaction in flip chip technology.Material Science and Engineering Report, 2001. pp. 1–58
33. E.A. Moelwyn-Hughes, *The Kinetics of Reactions in Solution* (Oxford University Press, London, 1947)
34. M. Schaefer, W. Laub, R.A. Fournelle, J. Liang, *Design and Reliability of Solders and Solder Interconnections* (The Minerals, Metals & Materials Society, Orlando, 1997), pp. 247–257
35. F. Bartels, J.W. Morris, G. Dalke Jr., W. Gust, Intermetallic phase formation in thin solid-liquid diffusion couples. J. Electron. Mater. **23**, 787–790 (1994)
36. Y. Wu, J.A. Sees, C. Pouraghabagher, L.A. Foster, J.L. Marshall, E.G. Jacobs, R.F. Pinizotto, The formation and growth of intermetallic in composite solder. J. Electron. Mater. **22**, 769–777 (1993)
37. D.F. Frear, P.T. Vianco, Intermetallic growth and mechanical behavior of low and high melting temperature solder alloys. Metall. Mater. Trans. A. **25**, 1509–1513 (1994)
38. C.E. Ho, S.C. Yang, C.R. Kao, Interfacial reaction issues for lead-free electronic solders. J. Mater Sci. Electron. **18**, 155–174 (2007)
39. T. Laurila, V. Vuorinen, J.K. Kivilahti, Interfacial reactions between lead-free solders and common base materials. Mater. Sci. Eng. R **49**, 1–60 (2005)
40. G. Zeng, S. McDonald, K. Nogita, Development of high-temperature solders: Review. Microelectron. Reliab. **52**, 1306–1322 (2012)
41. L.P. Lehman, Y. Xing, T.R. Bieler, E.J. Cotts, Cyclic twin nucleation in tin-based solder alloys. Acta Mater. **58**, 3546–3556 (2010)
42. T.K. Lee, T.R. Bieler, C.U. Kim, H.T. Ma, *Fundamentals on Lead-free Solder Interconnect Technology from Microstructures to Reliability* (Springer, London, 2015)
43. T.H. Courtney, *Mechanical Behavior of Materials* (Waveland Press, Long Grove, 2005)
44. L. Vitos, A. Ruban, H.L. Skriver, J. Kollar, The surface energy of metals. Surf. Sci. **411**(1), 186–202 (1998)
45. E.C. Garnett, W.S. Cai, J.J. Cha, F. Mahmood, S. Connor, M.G. Christoforo, Y. Cui, M.D. McGehee, M.L. Brongersma, Self-limited plasmonic welding of silver nanowire junctions. Nat. Mater. **11**(3), 241–249 (2012)
46. M. Hosel, F.C. Krebs, Large-scale roll-to-roll photonic sintering of flexo printed silver nanoparticle electrodes. J. Mater.Chem. **22**(31), 15683–15688 (2012)
47. M.K. Kim, H. Kang, K. Kang, S.H. Lee, J.Y. Hwang, Y. Moon, S.J. Moon, Laser Sintering of Inkjet-Printed SilverNanoparticles on Glass and PET Substrates. *10th IEEE Conference onNanotechnology (IEEE-NANO)* (IEEE, New York, 2010), pp. 520–524
48. H. Huang, M. Sivayoganathan, W. Duley, Y. Zhou, Efficient localized heating of silver nanoparticles by low-fluence femtosecond laser pulses. Appl. Surf. Sci. **331**, 392–398 (2015)
49. S. Magdassi, M. Grouchko, O. Berezin, A. Kamyshny, Triggering the sintering of silver nanoparticles at room temperature. ACS Nano **4**, 1943–1948 (2010)
50. M. Grouchko, A. Kamyshny, C.F. Mihailescu, D.F. Anghel, S. Magdassi, Conductive inks with a "Built-in" mechanism that enables sintering at room temperature. ACS Nano **5**(4), 3354–3359 (2011)

51. D. Wakuda, M. Hatamura, K. Suganuma, Novel method for roomtemperature sintering of Ag nanoparticle paste in air. Chem. Phys. Lett. **441**(4–6), 305–308 (2007)
52. D. Wakuda, K.S. Kim, K. Suganuma, Room temperature sinteringof Ag nanoparticles by drying solvent. Scr. Mater. **59**, 649–652 (2008)
53. S.-J.L. Kang, *Sintering: Densification, Grain Growth and Microstructure* (Elsevier, 2005)
54. P. Peng, A.M. Hu, A.P. Gerlich, G.S. Zou, L. Liu, Y.N. Zhou, Joining of silver nanomaterials at low temperatures: Processes, properties, and applications. ACS Appl. Mater. Interfaces **7**, 12597–12618 (2015)
55. J.K. Mackenzie, R. Shuttleworth, A phenomenological theory of sintering. Proc. Phys. Soc. Sect. B **62**(12), 833–852 (1949)
56. J. Frenkel, Viscous flow of crystalline bodies under the action of surface tension. J. Phys. (USSR) **9**, 385–391 (1945)
57. V. Tikare, M. Braginsky, D. Bouvard, A. Vagnon, Numerical simulation of microstructural evolution during sintering at the mesoscale in a 3D powder compact. Comput. Mater. Sci. **48**, 317325 (2010)
58. C. Herring, Effect of change of scale on sintering phenomena. J. Appl. Physiol. **21**, 301–303 (1950)
59. Q. Jiang, F.G. Shi, Size-dependent initial sintering temperature of ultrafine particles. J. Mater. Sci. Technol. **14**, 171172 (1998)
60. H.A. Alarifi, M. Atis, C. Özdoğan, A. Hu, M. Yavuz, Y. Zhou, Molecular dynamics simulation of sintering and surface premelting of silver nanoparticles. Mater. Trans. **54**(6), 884–889 (2013)
61. S. Chen, C. LaBarbera, N.C. Lee, Low temperature sinterable silver paste for high power die attach application. *Proceedings of the International Conference on Soldering & Reliability,* SMTA, Markham, 2017
62. S. Chen, C. LaBarbera, N.C. Lee, *Pressure-less* silver sintering pastes for low porosity joint and large area dies. *Proceedings of SMTA International* (Rosemont, IL, 2016), pp. 379–387
63. S. Fu, Y. Mei, X. Li, P. Ning, G.-Q. Lu, Parametric study on pressureless sintering of nanosilver paste to bond large area ($\geq$100 mm2) power chips at low temperatures for electronic packaging. J. Electron. Mater. **44**, 3973–3984 (2015)
64. D.R. Frear, P.T. Vianco, Intermetallic growth and mechanical behavior of low and high melting temperature solder alloys. Metall. Mater. Trans. A. **25A**, 1509–1603 (1994)
65. S.P. Lim, B.H. Pan, H.W. Zhang, W. Ng, B. Wu, K.S. Siow, S. Sabne, M. Tsuriya, High-temperature Pb-free die attach material project phase 1: Survey result, in *2017 International Conference on Electronics Packaging (ICEP)*, (IEEE, Yamagata, 2017), pp. 51–56
66. D.J. Green, O. Guillon, J. Rodel, Constrained sintering: A delicate balance of scales. J. Eur. Ceram. Soc. **28**(7), 1451–1466 (2008)
67. N.-C. Lee, *Reflow Soldering Processing and Troubleshooting SMT, BGA, CSP, and Flip Chip Technologies* (Newnes, 2001), pp. 127–133
68. N.-C. Lee, G.P. Evans, Solder paste – meeting the SMT challenge, 1987. SITE Magazine
69. W.B. Hance, N.C. Lee, Voiding mechanisms in SMT, in *China Lake's 17th Annual Electronics Manufacturing Seminar*, (China Lake, 1993)
70. T.A. Krinke, D.K. Pai, Factors affecting thermal fatigue life of LCCC solder joints. Weld. J. **67**, 33–40 (1988)
71. D.J. Xie, Y.C. Chan, J.K.L. Lai, An Experimental Approach to Pore-free Reflow Soldering. IEEE Trans. Compon. Packag. Manuf. Technol. Part B: Adv. Packag. **19**(1), 148–153 (1996)
72. W. Rmili, N. Vivet, S. Chupin, T. Le Bihan, G. Le Quilliec, C. Richard, Quantitative analysis of porosity and transport properties by FIB-SEM 3D imaging of a solder based sintered silver for a new microelectronic component. J. Electron. Mater. **45**(4), 2242–2251 (2016)
73. E.A. Wargo, T. Kotaka, Y. Tabuchi, E.C. Kumbur, Comparison of focused ion beam versus nano-scale X-ray computed tomography for resolving 3-D microstructures of porous fuel cell materials. J. Power Sources **241**, 608–618 (2013)
74. A. Madra, N. El Hajj, M. Benzeggagh, X-ray microtomography applications for quantitative and qualitative analysis of porosity in woven glass fiber reinforced thermoplastic. Compos. Sci. Technol. **95**, 50–58 (2014)

75. L. Vergara, R. Miralles, J. Gosálbez, F.J. Juanes, L.G. Ullate, J.J. Anaya, M.G. Hernández, M.A.G. Izquierdo, NDE ultrasonic methods to characterize the porosity of mortar. NDT&E Int. (Elsevier) **34**(8), 557–562 (2001)
76. V.S. Maalej, Z. Lafhaj, M. Bouassida, Micromechanical modelling of dry and saturated cement paste: Porosity assessment using ultrasonic waves. Mech. Res. Commun. **51**, 8–14 (2013)
77. W. Shen, L. Feng, A. Lei, Z. Liu, Y. Chen, Effects of porosity and pore size on the properties of AgO-decorated porous diatomite ceramic composites. Ceram. Int. **40**(1), 1495–1502 (2014)
78. X. Milhet, P. Gadaud, V. Caccuri, D. Bertheau, D. Mellier, M. Gerland, Influence of the porous microstructure on the elastic properties of sintered Ag paste as replacement material for die attachment. J. Electron. Mater. **44**(10), 3948–3956 (2015)
79. A.A. Wereszczak, D.J. Vuono, H. Wang, M.K. Ferber, Z.X. Liang, Properties of bulk sintered silver as a function of porosity. Oak Ridge National Laboratory, 2012, pp. ORNL/TM-2012/130
80. V. Caccuri, X. Milhet, P. Gadaud, D. Bertheau, M. Gerland, Mechanical properties of sintered Ag as a new material for die bonding: influence of the density. J. Electron. Mater. **43**, 4510–4514 (2014)
81. J. Carr, X. Milhet, P. Gadaud, S.A.E. Boyer, G.E. Thompson, P.D. Lee, Quantitative characterization of porosity and determination of elastic modulus for sintered micro-silver joints. J. Mater. Process. Technol. **225**, 19–23 (2015)
82. G. Bai, Virginia Polytechnic Institute and State University PhD thesis, Blacksburg, VA, 2005
83. G. Ondracek, On the relationship between the properties and the microstructure of multiphase materials Part III: Microstructure and Young's modulus of elasticity. Z. Werkstofft. **9**, 96–100 (1979)
84. N. Ramakrishnan, V.S. Arunachalam, Effective elastic moduli of porous solids. J. Mater. Sci. **25**, 3930 (1990)
85. T. Herboth, M. Guenther, A. Fix, J. Wilde, *Failure Mechanisms of Sintered Silver Interconnections for Power Electronic Applications* (IEEE, 2013), pp. 1621–1627
86. P. Gadaud, V. Caccuri, D. Bertheau, J. Carr, X. Milhet, Ageing sintered silver: Relationship between tensile behavior, mechanical properties and the nanoporous structure evolution. Mater. Sci. Eng. A **669**, 379–386 (2016)
87. N. Alayli, F. Schoenstein, A. Girard, K.L. Tan, P.R. Dahoo, Spark plasma sintering constrained process parameters of sintered silver paste for connection in power electronic modules: Microstructure, mechanical and thermal properties. Mater. Chem. Phys. **148**, 125–133 (2014)
88. J. Ordonez-Miranda, M. Hermens, I. Nikitin, V.G. Kouznetsova, O. van der Sluis, M.A. Ras, J.S. Reparaz, M.R. Wagner, M. Sledzinska, J. Gomis-Bresco, C.M. Sotomayor Torres, B. Wunderle, S. Bolz, Measurement and modeling of the effective thermal conductivity of sintered silver pastes. Int. J. Therm. Sci. **108**, 185–194 (2016)
89. A.S. Zuruzi, K.S. Siow, Electrical conductivity of porous silver made from sintered nanoparticles. Electron. Mater. Lett. **11**, 308–314 (2015)
90. J.R. Greer, R.A. Street, Thermal cure effects on electrical performance of nanoparticle silver inks. Acta Mater. **55**, 6345–6349 (2007)
91. J. Scola, X. Tassart, C. Cilar, F. Jomard, E. Dumas, Y. Veniaminova, P. Boullay, S. Gascoin, Microstructure and electrical resistance evolution during sintering of a Ag nanoparticle paste. J. Phys. D. Appl. Phys. **48**, 145302 (2015)

Sintered Silver for LED Applications

H. Zhang and K. Suganuma

1 Short Review of Die-Attach for LED Applications

1.1 Introduction of LED

An LED (light-emitting diode) is a two-lead opto semiconductor that converts electrical energy directly into light energy. As shown in Fig. 1, when sufficient voltage is applied to the leads, electrons and electron holes recombine and release photons, which is called electroluminescence. LEDs are superior to incandescent bulbs and fluorescent lamps due to the following characteristics: higher luminous efficiency, smaller size, longer lifetime, improved physical robustness, and faster switching. For example, most LED bulbs have an average lifetime of 25,000–100,000 h of use, which equals their being lightened continuously for 3–10 years [1]. Although the device cost of LEDs is still higher than that of their traditional rivals, LEDs have been intensively developed and widely applied because of their excellent performance, environment-friendly properties, and economical operation cost. LED bulbs are able to convert 80% of electricity into light energy and the remaining 20% into heat; typical LED bulbs are cool when touched even while they are illuminated [1].

LEDs are broadly grouped into visible LEDs and invisible LEDs according to the frequency of the released photons. Visible LEDs have been developed to cover the spectrum of visible light and are mainly used for display or illumination [2–4]. Invisible LEDs, covering the ultraviolet and infrared wavelengths, are mainly used with photosensors such as photodiodes or CMOS image sensors [5, 6].

With the rapid changes in the LED industry, assemblers who can find new ways to enhance LED performance will take preemptive opportunities in the end-user

H. Zhang (✉) · K. Suganuma
Osaka University, Suita, Japan
e-mail: zhanghao@eco.sanken.osaka-u.ac.jp; suganuma@sanken.osaka-u.ac.jp

K. S. Siow (ed.), *Die-Attach Materials for High Temperature Applications in Microelectronics Packaging*, https://doi.org/10.1007/978-3-319-99256-3_2

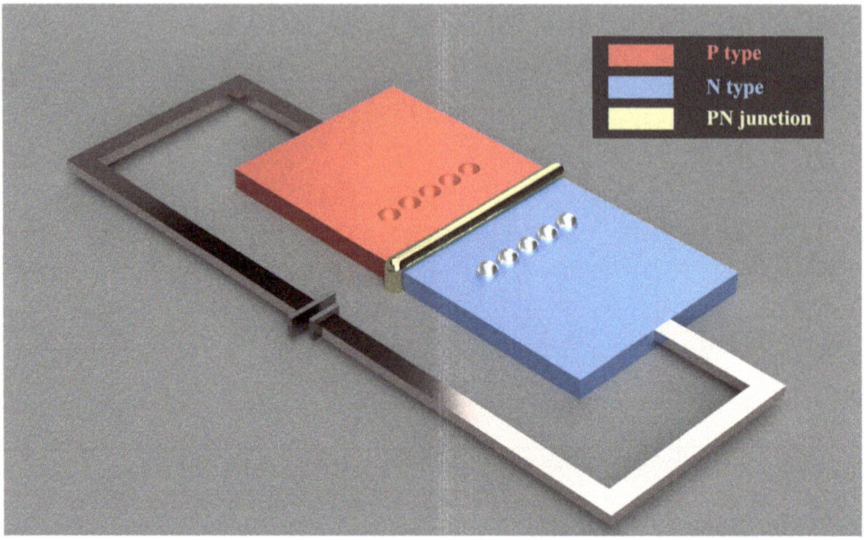

Fig. 1 Illustration of LED

market. And the real challenge is not only to bring to market a product with qualified performance but also the best cost efficiency. As one of the conclusive factors which can decide the performance and cost of LED modules, the die-attach material plays a key role in the luminous efficacy and reliability of mid-, high-, and super-high-power LEDs.

As the inevitable thermal transfer medium from the chip to the substrate or heat sink, the die-attach layer has the most significant impact on the thermal performance of the high-power LED modules. Silicones, solder, Ag-filled epoxies, eutectic gold-tin (Au-Sn), and sintered Ag materials have all been used for LED die-attach from the low-power consumption indicator to the ultra-high-power LED spotlight.

According to the particular bonding structure and specified applications, the selection of suitable die-attach material depends on several considerations. Performances such as lumen output, thermal dissipation, and reliability are commonly the primary concern; manufacturing processes, such as LED chip structures, packaging method, and throughput, also play key roles; moreover, cost performance should also be taken into account. Compromise between different attributes is commonly seen when deciding the use of particular die-attach techniques.

In this section, a brief introduction of various die-attach materials, which correspond to different LED output powers and chip structures, will be given. Among those die-attach materials, the sintered Ag technology for ultrahigh-power LEDs will be our main focus; the structure of LED chips will be elucidated at first because of their close relationship with the selection of die-attach materials.

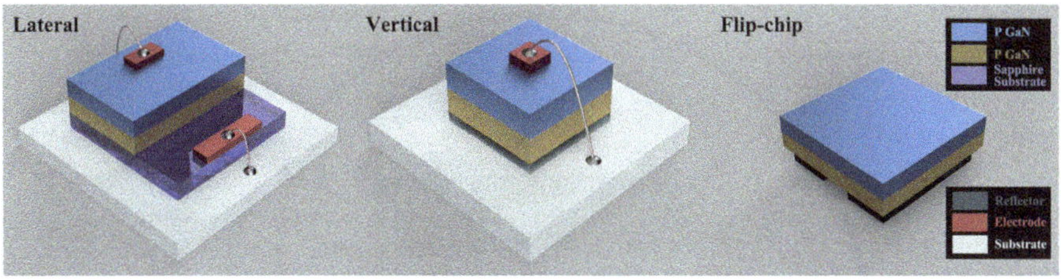

Fig. 2 Structure of commonly seen LED modules

1.2 Commonly Seen LED Chip Structures

As shown in Fig. 2, according to the arrangement of two electrodes, there are three mainstream LED chip structures: the lateral structure, the vertical structure, and the flip-chip structure, respectively. The connection between two electrodes has various orientation relationships with the P/N junction interface: intersection, vertical, and parallel, respectively.

The lateral structure consists of laterally spaced electrodes, which are bonded with one wire separately; it is commonly used in low-power applications. The vertical structure consists of a conductive bottom, which plays the role of one electrode. This structure allows a vertical current flow and is used for most high- and super-high-power applications. The flip-chip structure places both electrodes on one side and is flip settled on the substrate. It provides the highest lumen density at a cost lower than that of the vertical structure. All the three types of structures can also be mounted directly on a board to form chip-on-board (CoB) modules.

1.3 Die-Attach Technology Platforms for LED Die Bonding

1.3.1 Silicone

Silicone is being widely used in the assembly and packaging of low-to-middle power LEDs. Its usage is not limited to die-attach material; silicone encapsulants and lens are common solutions in the design of LED modules (Fig. 3) [7, 8]. The high transmittance helps to build low-loss optical systems; the yellowing and ultraviolet resistance allow long-term reliability without the decay of luminous efficacy, especially in outdoor applications; and the high resistance to moisture improves the protection of sensitive electronic components. However, when being used as die-attach material, the relatively low thermal conductivity constrains its application in high-power LEDs [9].

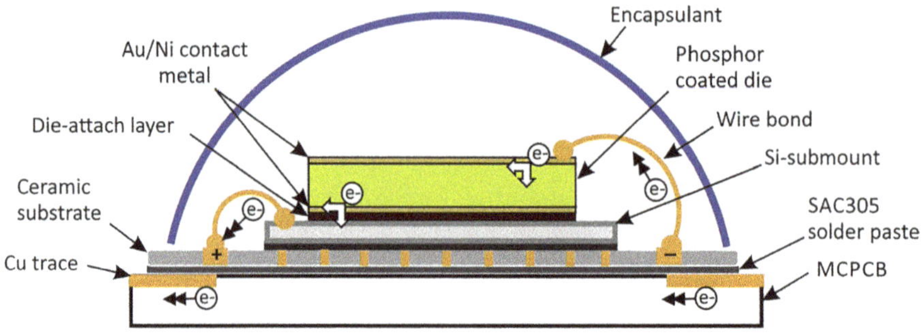

Fig. 3 Cross-section of an LED package [7]

1.3.2 Silver-Filled Conductive Adhesives

Silver-filled conductive adhesives constitute the largest class of thermal die-attach materials in semiconductor packages, not limited to LEDs. Their variable forms, including interface pads, liquid, and tape, greatly expand their applications from thermal interface materials to die-attach. They are mixtures of thermally conductive Ag fillers embedded in polymeric matrices. Various high thermal conductivity fillers, such as nanoparticles, nanowires, nanotubes, microparticles, or microflakes, are being used in their manufacturing process.

Ag-filled conductive adhesives are one type of thermally conductive adhesives (TCAs). As shown in Table 1 [1, 10, 11], changing the composition, preparation methods and morphology of conductive fillers, provide a tunable thermal conductivity up to 30 W/mK [12] for these TCAs. These options increase the flexibility of TCA selection according to the various demands of performance and cost limitations. Conductive adhesives consisting of nanofillers usually provide a lower curing temperature and higher thermal conductivity than those prepared from micron fillers. Figure 4 compares the thermal conductivity between composites constituted by Ag microparticles and nanoparticles processed under similar curing conditions [13]. An obvious difference in thermal conductivity can be observed from this comparison; the self-constructed nanostructured network that occurs during the curing process was caused by the sintering of nanofillers.

Due to the manufacturing cost and the viscosity of nanofillers, which is much higher than that of micron fillers at the same volume fractions, micron fillers are preferred to allow easy shape casting, easy surface spreading, and improved small gap-filling ability. Given that the conductivity of commercial Ag-filled conductive adhesives is less than 10 W/mK [1], a power saturation phenomenon is often encountered when the current injection level is raised because of the heat generated from LEDs unable to dissipate efficiently [14].

Moreover, the advantages of Ag-filled conductive adhesives include their compatibility with the existing back-end packaging equipment and the ability for secondary reflow compatibility. Hence, TCAs provide an attractive cost/performance balance and are commercially used in mid to high-power LED applications.

Table 1 Thermal conductivity of most recently reported TCAs [10]

Materials	Methods	Thermal conductivity ($Wm^{-1} K^{-1}$)	Fraction	Binder
Carbon nanotube	Morphology control	0.43	1 vol%	PSS/ PE/ EGMA
	Using aligned carbon nanotube	4.87	16.7 vol %	Epoxy
Graphene	Non-covalent functionalization	1.53	10 wt%	Epoxy
	Multilayer graphene composing	5.40	10 vol%	Epoxy
	Roll milling	3.15	25 wt%	Silicon
Boron nitride	POSS-boron nitride nanotube	2.77	30 wt%	Epoxy
	Cold plasma modification	2.40	55 vol%	Silicon
	High mixing speed and temperature	1.97	28 vol%	Epoxy
Aluminum nitride	Solution blending	2.02	66.7 wt%	LCP
Silver-graphene	Silver and graphene hybrid	9.90		Epoxy
Silver spheres	Employing 15 wt% nanoparticles	6.00–7.00	85 wt%	Epoxy
Silver flakes	Employing 15 wt% nanoparticle	~27.00 (in-plane) & ~4.00 (vertical)	85 wt%	Epoxy
	Employing diluents	~27.00 (in-plane) & ~5.00 (vertical)	85 wt%	Epoxy-diluents
Silver particles	Sintering the nanoparticles	27.00	45 vol%	Epoxy
	Surface modification by iodine	13.50 (isotropic)	85 wt% (~35 vol %)	Epoxy

1.3.3 Tin-Silver-Copper (SAC)-Based Solder Alloys

In general, tin-silver-copper (SAC)-based solder alloys exhibit a better thermal conduction capability than silicone epoxy and conductive Ag adhesive. SAC alloys provide exceptional value with a low-cost, fast assembly process with reasonable thermal performance (~60 W/mK) [15]. In contrast to conventional conductive adhesives, a reflow process is inevitable in LED packaging using an SAC solder. The reflowing temperature is raised above the melting point of the solder to melt and wet the backside metallization of the LED chip and the substrate. Interfacial intermetallic compounds (IMCs) are formed after the reaction between the liquefied solder and metallization, which resulted in interfacial reliability issues compared to silicone epoxy and conductive adhesives [16–18]. On the other hand, an SAC solder is unique in its broad compatibility with semiconductor packaging as well as SMT lines.

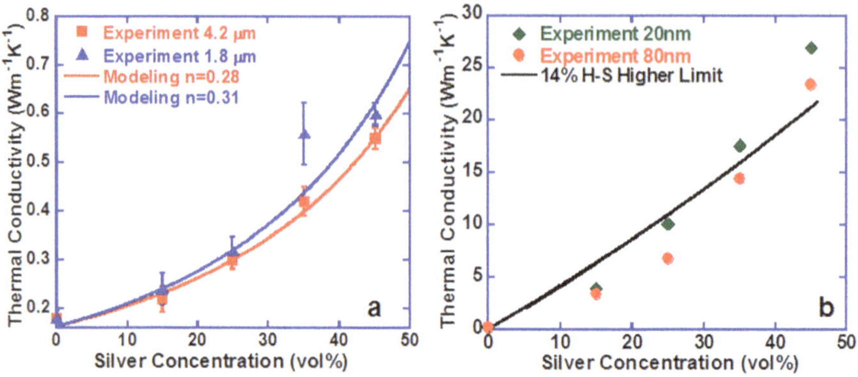

Measured and predicted thermal conductivity vs. particle volume fraction for: (a) microcomposites; (b) nanocomposites. The solid lines represent the effective κ predictions of percolation threshold model for microcomposites and ~14% of the maximum L-H theoretical limit for the ideal network structure. The dots are the experimentally measured thermal conductivities (For the nanocomposite the error bars are typically smaller than the dots). The nanocomposites have up to 50 fold larger thermal conductivities than the microcomposites.

Fig. 4 A comparison of thermal conductivity between composites constituted by Ag microparticles and nanoparticles [13]

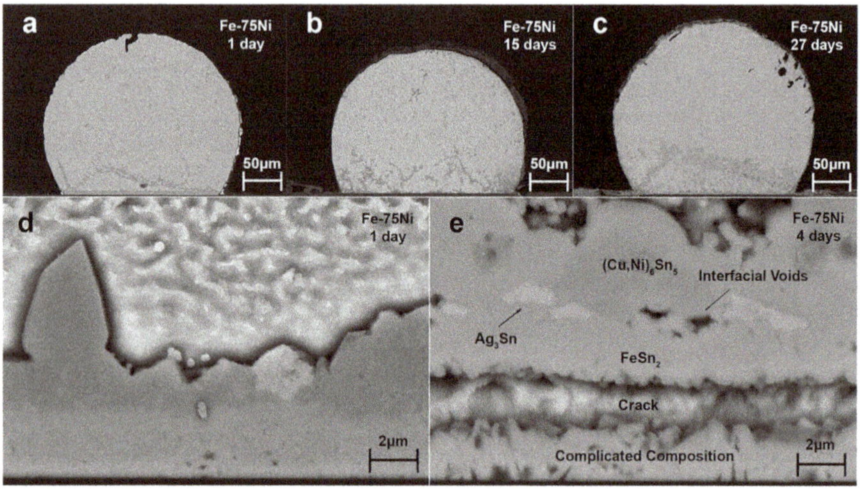

Interfacial morphology of Fe–75Ni solder joint after different days' aging in 200 °C: (a) 1 day, (b) 15 days, (c) 27 days, (d) 1 day (microscopic morphology), (e) 4 days (microscopic morphology).

Fig. 5 Aging results of an SAC/Fe-Ni structure [19]

However, an SAC solder melts in the 217–221 °C range, and its joint drastically degrades during short period aging for 24 h at 200 °C (Fig. 5) [19]. Therefore, its use should be avoided in applications where high temperature stability is required or a further heating process, such as secondary reflow, is necessary.

1.3.4 Eutectic Gold-Tin (Au80Sn20) Alloys

Eutectic gold-tin (Au80Sn20 by weight) has once been seen as the primary choice of die-attach materials for high-power LED applications where long-term high reliability is emphasized. Theoretically, the eutectic AuSn solder fulfills the challenges of acceptable electrical, thermal properties as well as the manufacturing volume needed in the high-power LED industry. Although the cost of ownership of eutectic AuSn is the highest among all the die-attach materials, it is still favorable for high-power applications. Besides suitable thermal conductivity, eutectic AuSn has high creep and fatigue resistance with secondary reflow compatibility, which matches well with the requirement of a high-power LED die attachment.

Current eutectic AuSn-bonding processes rely on the bonding equipment, and additional process complexity is inevitable due to the high melting temperature and the corresponding high void contents and high residue stress [20–22]. Alternative solutions have been proposed to modify the eutectic bonding technology. One of the solutions is to prepare an eutectic AuSn layer as a pre-coating on the LED back-metallization with a melting temperature of 300–320 °C, which is higher than 280 °C, the melting point of Au80Sn20 [21].

1.3.5 Silver Micro-/Nanofiller Sintering

Sintered Ag materials have been developed as the most promising high-temperature die-attach candidate. It is composed by micro-/nanoscale Ag fillers that fuse together at 180–300 °C through atomic diffusion to form a micro-/nano-porous, yet pure, Ag joint. This porous structure has a similar melting point (961 °C) to bulk Ag, i.e. the bonded joints have the potential of extremely high melting and thermal conductivity. Hence, the sintered Ag materials are attracting considerable research interest because of the prospect of working in harsh conditions. This is exemplified by the increasing number of research papers and patents filed over the past few years as shown in Fig. 6 [23, 24].

Although the objective operating temperature of power modules die-attached with sintered Ag seldom exceeds 250 °C, due to the limitations of metallization, wire bonding, and encapsulation, sintered Ag has no competing technologies in this temperature range. Sintered Ag materials can be applied in various formats such as paste, rolled foil, and sputtered layer [25–27]. In the case of sintered Ag paste, which is the most common product, the diversity of Ag fillers and organic solvents generates a plentiful product line that covers a wide temperature range, various joints porosity, and different manufacturing processes.

Sintered Ag materials, with costs between solders and Au80Sn20, have been proven to possess better mechanical reliability and thermal performance than Au80Sn20. For super-high-power LEDs, lumen output has been raised by up to 30% by using sintered Ag materials.

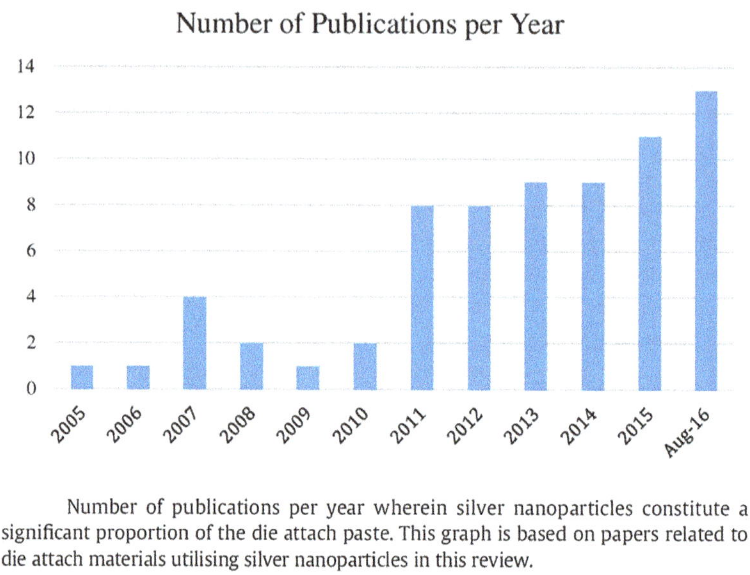

Fig. 6 The number of research papers related to Ag nanoparticles [23]

Homologous temperature is defined by the temperature of a material as a fraction of its melting point temperature using the Kelvin scale. Sintered Ag can operate even at 300 °C due to its homologous temperature of only ~0.46 (573 K/1235 K). However, the homologous temperatures that solders are required to withstand at their high-temperature applications are as high as ~0.8, which means limited mechanical strength and significant creep when they are used in harsh environments [23].

The other main advantages of Ag are its high thermal and electrical conductivities, which is critically important in the ultrahigh-power LEDs application. The developing tendency of operating LEDs at higher current densities and temperatures undoubtedly helps to gain improved light output at lower cost. As the conductive path of thermal flux and electrical current, die-attach material is increasingly becoming the bottleneck of thermal dissipation in high-power LEDs, which impedes the development of LEDs into mature technology in the general illumination space. Sintered Ag has thermal conductivity in the range of three to five times higher than Sn-based solders or two to four times higher than Au-based solders [28]. This natural advantage makes the sintered Ag displace its rivals as the die-attach materials of choice for the ultrahigh-power LEDs applications.

1.4 The Selection of LED-Bonding Materials

In the previous sections, commonly seen commercial die-attach materials were briefly introduced. In this section, criteria for selecting die attachments for LEDs will be elucidated from the aspects of the properties and performance and the manufacturing process of the technology platform. The former include the reliability requirement of the end application, while the latter includes the module design and cost control.

The end application determines the operating environment and respect lifetime. For example, outdoor applications designed to operate in harsher environments without frequent replacement prefer higher-power dies and robust packaging methods, while benign indoor applications, such as the bulb of a table lamps, may be well satisfied by low-mid-power dies and thier corresponding packaging ways.

Sintered Ag materials have the highest thermal conductivity among all the materials discussed. Their thermal conductivities are commonly higher than 100 W/mK and even as high as that of bulk Ag (~423 W/mK) based on developmental results [26]. The thermal conductivity of eutectic gold-tin has been measured around 57 W/mK, which equals those of SAC solders [15, 29]. In general, metallic solder alloys exhibit better thermal conductivity than Ag-filled conductive adhesives; the latter hardly provide a thermal conductivity higher than 30 W/mK.

The descending order of long-term reliability among the above-mentioned materials would be sintered: sintered Ag > eutectic AuSn > Ag-filled conductive adhesive > SAC solders > silicone epoxy. Sintered Ag materials, especially the pressure-aid sintered ones, provide the highest thermal reliability with a long lifetime; their excellent general properties make it possible to use them in the harshest environments, such as ultrahigh-power outdoor lighting, automobile, aviation, aerospace, and nautical signaling. Some silicon epoxy or Ag-filled adhesives may produce volatile organic compound emissions that damage the fluorescent powder and reduce the lifetime of LED devices [1]; their usage is recommended to be limited to low-mid power in benign conditions. The composition of a suitable power range for various die-attach material platforms is shown in Fig. 7.

In other aspects, design and cost control also play important roles when selecting die-attach materials. Second reflow is commonly seen in the assembly of chip-on-board (CoB) LED modules, by which the bonded dies and frame will be packaged onto the heatsink. Eutectic AuSn, conductive adhesives, as well as sintered Ag materials, can withstand the second reflow easily, while SAC solders are not able to withstand this additional reflow. However, for applications in which the CoB module is screwed to the heat sink, SAC solders can be used due to the absence of a second reflow.

The cost of the die bonders makes die-attach one of the most capital-intensive steps in LED packaging. Die-attach processes, in which special equipment and the corresponding new capital expenditure are unnecessary, are therefore preferred when determining the design and packaging method of LEDs. Solder, Ag-conductive adhesives, silicone, and pressureless Ag sintering die-attachs can be compatible

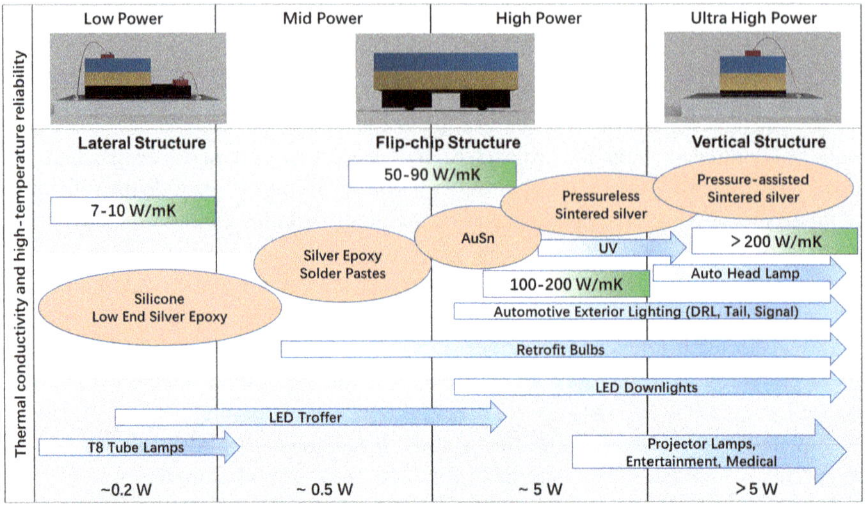

Fig. 7 Suitable power range of various die-attach material platform

with the existing die bonders, which provide more process feasibility over eutectic AuSn and pressure-assisted Ag sintering materials. The main properties of various LED die-attach materials are compared in Table 2 as a summary of this section.

1.5 Conclusion

LED-based illumination techniques are rapidly penetrating the general lighting market from the low- and medium-end market to the high-end industrial market. The long lifetime and environment-friendly properties of LED-based systems are often claimed as the most competitive advantage over other available lighting technology options. Recent advances in LED packaging have improved the thermal stability of phosphors and encapsulants, which highlights the urgency of improved die-attach performance. Thermal conductivity and long-term reliability are the core concern of properties when selecting proper die-attach materials in LED applications. Several commonly used LED die-attach materials were compared in this section. In summary, Ag sintering materials have obvious advantages over other rivals in general properties and reliabilities; they are almost the only choice in ultrahigh-power and high-reliability applications such as automotive, aircraft, space exploration, and nuclear industries. However, Ag sintered materials are less competitive in the cost-sensitive low- to medium-end market, in which silicone epoxy, SAC solder, and Ag-filled conductive adhesive competently hold their positions.

Table 2 The main properties of various LED die-attach materials

	Ag sinter paste[a]	SAC 105 solder[b, c]	Au-20Sn	Conductive adhesive[d, e]	Ag to Ag direct bonding[f]
Shear strength (MPa)	~50	~45	~65[g]	1 ~ 15	~50
Electrical resistivity (10^{-6}Omega;-cm)	~7.5	N/A	~16.4[b]	200 ~ 1000[d]	~1.59 (pure Ag)
Thermal conductivity ($W{\cdot}m^{-1}\,K^{-1}$)	120 ± 5 @25~200 °C	60	57[a] @ 85 °C	N/A	405–411 (pure Ag) @25 °C
High temperature storage	~ 40 MPa (@ 250 °C, ~200 h)	Failed within 24 h[b] (@ 250 °C)	~ 5.5 MPa[g] (@ 250 °C, ~200 h)	~ 16 MPa (@ 150 °C, ~1000 h)	N/A
Material cost	High	Lowest	Highest	Mid to high	High
Processing parameter (°C, min, MPa)	250 °C, 30 min, 0.4 MPa	250 °C, 2 min	310 °C, 1 min	140 °C, 1 h	250 °C, 60 min
Melting temperature (°C)	961	217–221 °C	280	N/A	961
CTE (ppm/K)	18–23	20 (@ 45 °C)	16	N/A	19.5 (pure Ag)
Elastic modulus (GPa)	33	95 (@ 25 °C)	69	N/A	82 (pure Ag)
Merits	Excellent properties	Most compatible with current SMT process	Compatible with solder process	Partially compatible with solder process	Perfect joints
Demerits	Ag migration & coarsening	Long-term annealing at high temperature	Cost of Au	Poor mechanical and physical properties	Not mature yet

[a]Zhang et al. [30]
[b]Zhang et al. [19]
[c]Nguyen et al. [31]
[d]Ho and Nishikawa [32]
[e]Nishikawa et al. [33]
[f]Oh et al. [34]
[g]Liu et al. [35]

2 Sintered Silver Paste for High-Power LED Applications

2.1 Introduction of Sintered Silver Paste for LED Applications

In the general illumination industry, LED is increasingly becoming a mature technology which possesses serial advantages, such as high-energy transformation ratio, high luminance, and long lifetime. Operating LEDs at high current densities and temperatures in order to gain a higher light output at a lower cost is naturally seen as a development path of LED techniques. As a promising LED die-attach material, sintered Ag allows LED modules working at a higher temperature with less energy loss and higher material robustness. Although its cost can be prohibitive—as much as four to five times more expensive than epoxy die-attach materials or solder materials—sintered Ag paste is commercially available in many high-end applications.

In the previous section, various LED die-attach materials were introduced and compared. A conclusion was drawn after the discussions, that is, as an advanced technique for LED die attachment, sintered Ag matches the requirement of ultrahigh-power LED die-attach and, therefore, attracts intensive research focus from industry and academia. In this section, a brief introduction of various sintered Ag techniques will be given first. The reliability issues and corresponding countermeasures of sintered Ag for die-attach materials will be discussed hereafter. Because sintered Ag materials are being widely used in the packaging of power devices, the cases and results included in this section are not limited to LED applications; the emerging usage of sintered Ag in WBG power semiconductors will also be covered.

2.2 *Sintered Silver: Classification, Process Condition, and Comparison*

Sintered Ag has been developed as die-attach material for power electronics packaging since the late 1980s and has been recognized as a prospective candidate for next-generation WBG semiconductors. Several types of sintered Ag techniques with excellent high temperature tolerance have been proposed such as Ag sintered paste and Ag to Ag direct bonding [34, 36–38]. Sintered Ag pastes are divided into pressure sintering paste and pressureless sintering paste according to whether pressure is required and further classified according to their filler sizes and compositions. Although the process conditions differ much from each other, they share the same merits of excellent physical properties and intrinsic high melting point compared with those of other die-attach solutions. Silver-to-silver direct bonding, which will be focused on in the next section, is a state-of-the-art technology and its mechanism and applications are still under development [27].

In the early days, micron-sized Ag fillers sintered under pressures of 9–40 MPa at a temperature of 250 °C were used to form sintered Ag joints. The invention of nano-Ag paste in the mid-2000s reduced the sintering pressure further to 1–5 MPa in the similar temperature range, which was made possible by the increase in the effective surface area of Ag nanoparticles [39, 40]. The current developing trend of sintered Ag paste, including low-cost, low-temperature, and low/none pressure sintering. Nano-Ag fillers normally provide lower sintering temperature and less sintering time; however, both the manufacturing and storage of nanofillers bring inconvenience when using them as die-attach materials [23, 41].

Unlike the grinding method of micro-Ag filler preparation, nano-Ag fillers are usually synthesized by chemical reactions, which limits their productivity and increases their cost. Moreover, to overcome the aggregation nature of nanoparticles, the surface of nanofillers is coated by an organic layer when they were synthesized. Therefore, a long preheating procedure is normally inevitable to remove the organic compounds and improve sintering, which increases the process cost and exerts negative effects on their long-term reliability [42, 43]. Given the huge specific surface area of nanomaterials, the absolute amount of this coating layer could be considerable, and its residue in sintered Ag could be significant. In order to inhibit the excessive organic compounds included in these Ag nanoparticles and maintain the low-temperature sintering property, an optimized size design of these Ag particles is necessary. For example, some polyol synthesis methods were, therefore, proposed to prepare submicron Ag particles with tailored sizes and size distributions, both of which can be adjusted freely using reaction conditions such as temperature and reaction time. These synthesized submicron particles presented rather excellent sintering ability and joints porosity; furthermore, the synthesis was accomplished by a simple polyol method, which is free of complicated instruments and expensive raw materials [44].

Similarly, the so-called hybrid paste, which is composed of microflakes and submicron particles, is developed in order to reduce the materials cost and improve paste performance. The microflakes play the role of a skeleton, and submicron particles improve the interconnection and fill the voids between flakes. The composition of Ag fillers can be adjusted to prepare pastes with a series of joint porosity and paste viscosity [45].

After the role of the organic solvent was re-evaluated, researchers had started to attach importance to the development of an advanced solvent. Some newly developed Ag sintered paste composed by microflakes and patented multi-solvent has already achieved low temperature (~200 °C), pressureless sintering with high joint quality, and thermal reliability [46].

In summary, hybrid filler sizes and mixed solvent are becoming the development trend of sintered Ag paste suppliers.

Figure 8 summarizes the sintering conditions of various Ag sintering technology platforms. Conditions selected from high-temperature solders, copper (Cu) sintering, and transient liquid-phase (TLP) sintering are also provided as references. In addition, only the optimized conditions selected by their authors have been extracted to improve clarity and understanding.

As can be seen from this figure, TLP and Cu sintering preferred higher process temperatures, which is also the main demerit of these two die-attach solutions. High-temperature solder provides a shorter process time; however, its process temperature is still obviously higher than that of Ag sintering. Micro-Ag paste sintering possesses a relatively mild sintering condition, and its pressureless branch is able to share the same instruments with the traditional solder process.

Fig. 8 Comparison of various die-attach processes
References in the graph:
1. H. Zhang et al., J. Elecron. Mater. **44**, 3896 (2015) [J]
2. M. Kuramoto et al., IEEE Trans. Compon. Pack. Technol. **33**(4), 801–808 (2010) [J]
3. T. Kunimune et al., IEEE Trans. Compon. Pack. Manuf. Technol. **2**(6), 909–915 (2012) [J]
4. S. Park et al., IEEE Trans. Compon. Pack. Manuf. Technol. **5**(7), 902–909 (2015) [J]
5. S.W. Park et al., 2014 I.E. 64th Electronic Components and Technology Conference (ECTC). IEEE, 2014: 1179–1182 [C]
6. M. Kuramoto et al., IEEE Trans. Compon. Pack. Manuf. Technol. **2**(4), 548–552 (2012) [J]
7. S. Sakamoto et al., J. Mater. Sci. Mater. Electron. **24**(4), 1332–1340 (2013) [J]
8. M. Kuramoto et al., IEEE Trans. Compon. Pack. Manuf. Technol. **1**(5), 653–659 (2011) [J]
9. J. Jiu et al., J. Mater. Sci. Mater. Electron. **26**(9), 7183–7191 (2015) [J]
10. H. Zhang et al., J. Mater. Sci. Mater. Electron. **27**(2), 1337–1344 (2016) [J]
11. J. Jiu et al., J. Mater. Sci. **51**(7), 3422–3430 (2016) [J]
12. S. Zhao, Y, et al. Microelectron. Reliab. **55**(12), 2524–2531 (2015) [J]
13. E. Möller et al., 2014 I.E. 64th Electronic Components and Technology Conference (ECTC). IEEE, 2014: 1707–1713 [C]
14. T. Hu et al., Mater. Des. **108**, 383–390 (2016) [J]
15. W. Liu et al., Mater. Sci. Eng. A **651**, 626–635 (2016) [J]
16. K. Suganuma et al., Microelectron. Reliab. **52**(2), 375–380 (2012) [J]
17. S.W. Park et al., J. Alloys Compd. **637**, 143–148 (2015) [J]
18. X. Liu et al., Scr. Mater. **110**, 101–104 (2016) [J]
19. C.X. Yang et al., Mater. Sci. Eng. A **660**, 71–76 (2016) [J]
20. J. Bai, G, et al. IEEE Trans. Device Mater. Reliab. **6**(3), 436–441 (2006) [J]
21. T. Yagishita et al., Mater. Trans. **54**(6), 866–871 (2013) [J]
22. Y. Akada et al., Mater. Trans. **49**(7), 1537–1545 (2008) [J]
23. Liu X, et al. [J]. Journal of Materials Science: Materials in Electronics, 2016, 27(5): 4265–4274.
24. Liu X, et al. [J]. Journal of Materials Science: Materials in Electronics, 2016, 27(10): 10065–10069
25. H. Shao et al., J. Mater. Sci. Mater. Electron. **27**(5), 4839–4848 (2016) [J]
26. Satoh T, et al. [J]. Journal of Electronic Materials, 2017, 46(2): 1279–1286
27. S. Zhao, Y, et al. J. Electron. Mater. **45**(11), 5789–5799 (2016) [J]
28. T. Morita et al., Jpn. J. Appl. Phys. **47**(8R), 6615 (2008) [J]

2.3 *The Reliability Issues of Silver Sintered Paste*

As a newly developed technology, reliability issues are being discovered and addressed simultaneously with process establishment and paste optimization. Industrial circles are more inclined to apply as much as their conventional technology when importing new materials and processes. However, what we must admit is that the Ag sintering technology is not highly compatible with the current lead-free solder process, not only in the process aspect but also in the related bill of materials. The higher process temperature, longer sintering time duration, and higher service temperature require a re-evaluation and redesign of materials used in the packaging schemes.

2.3.1 Metallization-Related Reliability Issues

One of the most obvious changes in the Ag sintering process is the requirement of specially prepared metallization on chips and substrates. Because atom diffusion is the driving force for bonding between sintered Ag and surface metallization, elements that cannot diffuse with Ag efficiently during a short sintering procedure are not able to play the role of the outer metallization layer. Technically, Cu, Ag, and Au are the only reported available candidates for the outer metallization layer, which can qualify as a joint connection.

A Cu substrate is the preferred substrate for the packaging of power electronics; Cu lead frames and DBC substrates are the most commonly used substrates in the power electronics industry. However, the oxidation tendency of Cu usually leads to severe growth of Cu oxide at the interface between Cu and porous Ag [43]. This kind of thick Cu oxide has a unique growth mode which causes the Ag atoms to be replaced by Cu oxide to squeeze the porous Ag to a denser state [30]. During this procedure, the interfacial stress caused by a coefficient of thermal expansion (CTE) mismatch between various materials leads to cracks or delamination and the interfacial resistance also rises.

Au is the least used choice due to its high cost; however, in some multi-reflow processes, Au is the preferred metallization because Ag could be oxidized during the

Fig. 8 (continued) 29. Roh M H, et al. [J]. Journal of Materials Science: Materials in Electronics, 2017, 28(10): 7292–7301
30. H. Zhang et al., J. Alloys Compd. **696**, 123–129 (2017) [J]
31. Muhammad K F, et al. [J]. Journal of Materials Science: Materials in Electronics, 2017, 28(13): 9351–9362
32. H. Zhang et al., J. Mater. Sci. **52**(6), 3375–3387 (2017) [J]
33. T. Wang et al., J. Electron. Mater. **36**(10), 1333–1340 (2007) [J]
34. Li J, et al. [J]. Journal of Alloys and Compounds, 2017, 709(30): 700–707
35. Maeda T, et al. [J]. Materials Science and Technology, 2017, 33(14): 1618–1625
36. L.A. Navarro, et al. Ingeniería Mecánica Tecnología Desarrollo **4**(3), 97–102 (2012) [J]

multi-reflow, which lowers the wettability of normal lead-free solder at the metallization layer [47].

Ag metallization possesses natural compatibility with sintered Ag and, therefore, has been used in most of the outer metallization. The preparation methods of Ag metallization include electroplating, electroless plating, and sputtering, all of which are able to assure the qualification of the processes between the Ag layer and porous Ag [30, 44, 48]. The substrate metallization is usually prepared by the plating method, while chip metallization is commonly manufactured by sputtering.

However, outer surface metallization is not enough to ensure the stability in a high-temperature test; a specified metallization scheme is critical, especially at the substrate side. A normal metallization scheme for the Ag sintering technology is usually composed of a stress-relieving layer, diffusion barrier layer, and surface adhesion layer.

As a suitable material for the stress-relieving and diffusion barrier layers, Ni has been widely used in electronic packaging. It has a series of advantages, such as cost-effectiveness, lower interfacial intermetallic compounds (IMC) growth rate in the Ni/solder interface, and higher oxidation resistivity, compared to Cu. A tri-layer Au/Ni/Cu metallization scheme has been used as a popular under-bump metallization structure in the soldering process. The outermost Au layer plays the role of the oxidation and corrosion resistant layer and thereby increases the solderability [16, 18, 49]. The Ni layer serves the purpose of a diffusion barrier to prevent the rapid reaction between the solder and the Cu layer [50–53]. However, in the packaging of high-temperature electronics, such as high-power LEDs, the Ni barrier layer loses the diffusion-resistant ability, and a thin layer of oxide is formed at the Ag/Ni interface because of the surrounding storage temperature [54].

Similar to Ni, Ti also has a stress-relieving effect because of its relatively low CTE coefficient (8.6×10^{-6}/K) compared to Ag (18×10^{-6}/K) and Cu (17×10^{-6}/K). An inserted Ti layer between Cu and Ag is able to adjust the CTE mismatch among various materials and increase the joint performance during a thermal shock test. Ti is also commonly used as a diffusion barrier layer, especially in chip metallization. A 500-nm thick Ti layer has the ability to inhibit Cu oxidation when stored at 250 °C for 1000 h [30, 48]. However, the only available method to prepare Ti coating in the electronic industry is the use of the high-cost sputtering process, which limits the wide application range of Ti metallization.

Systematic evaluations of various surface metallization schemes have been conducted, and all of the evaluated schemes are listed in Table 3, such as Ag/Cu, Ag/Ni/Cu, Ag/Ti/Cu, Ag/Ni/Pd/Cu, and Ag/Pt/Pd/Cu.

Table 3 Some evaluated metallization schemes

Metallization	Preparation method	Thickness (μm)	Bonding structure	Reference
Ag/Cu	Electroplating	0.2/Cu	Si-DBC	[30, 54]
Ag/Ni/Pd/Cu	Electroless plating	2/5/0.13/Cu	Si-Cu	[48]
Ag/Ni/Cu	Sputtering	2/2/Cu	Si-DBC	[54]
Ag/Ti/Cu	Sputtering	2/0.5/Cu	Si-DBC	[30, 54]
Ag/Pt/Pd/Cu	Electroless plating	2/0.3/0.05/Cu	SiC-Cu	[55, 56]

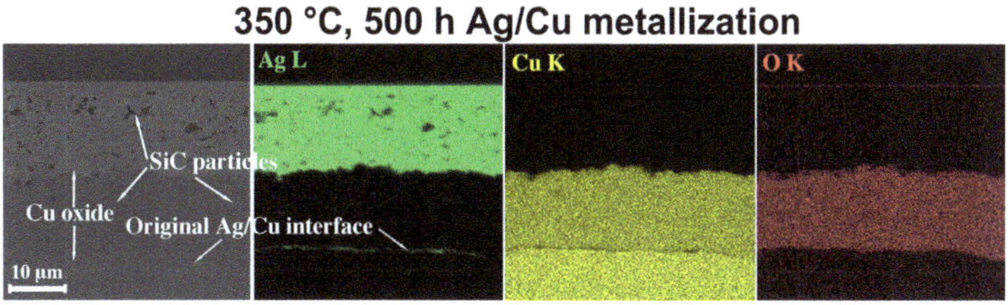

Fig. 9 EDX elemental mapping of sintered Ag joints on Ag-plated Cu substrates aged at 350 °C for 500 h [30]

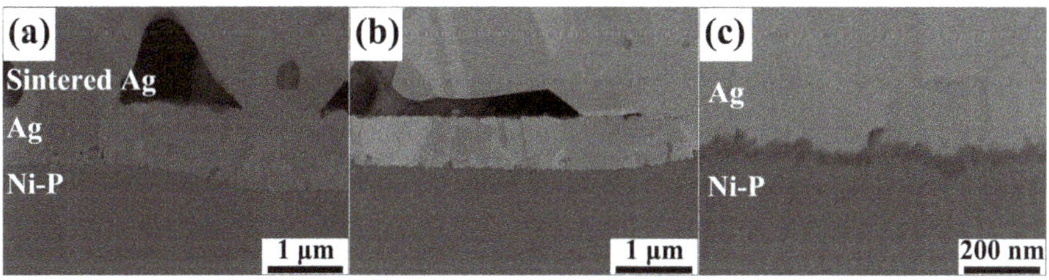

Fig. 10 Cross-sectional morphology of sintered Ag joints on the Ag/Ni substrate: (**a**) as sintered; (**b**) annealed for 200 h; (**c**) high magnification image of (**b**)

Some of the conclusions are summarized here:

1. The Ag outer layer helps to ensure the qualified bonding in the as-sintered state. Although Au and Cu are also seen as candidates of the outermost layer, Ag is still the primary choice because of its higher reliability as compared to Cu and lower cost as compared to Au. Moreover, the existence of Au increased the complexity of the interface, which may lead to unanticipated reliability issues [43, 47].
2. Oxide forms at the Ag/Cu and Ag/Ni interface during high-temperature storage. At the Ag/Cu interface, with the consumption of the Ag metallization layer during coarsening, Cu oxide forms above the naked Cu surface. This type of oxide layer grows at the expense of occupying the position of Ag atoms, and the source of oxygen was conjectured to be the residual oxygen in the porous Ag (Fig. 9). At the Ag/Ni interface, differentiated to the several micrometer thick oxidation layer, only a thin oxide can be observed at around 250 °C, while the source of oxygen is still not clear (Fig. 10). The oxide formed at the Ag/Cu and Ni/Cu interface indicates the necessity of a barrier layer of suitable thickness [30, 43, 54].
3. Ti and Pt effectively inhibit the interdiffusion at the interface and, therefore, could be used as barrier layer in high-temperature applications (Figs. 11 and 12). Sputtered Ti has good process compatibility in the manufacture of chip metallization. As a typical noble metal, Pt has obvious demerits in material cost but Pt can be coated by a relatively simple electroless method which lowers the process cost [55].

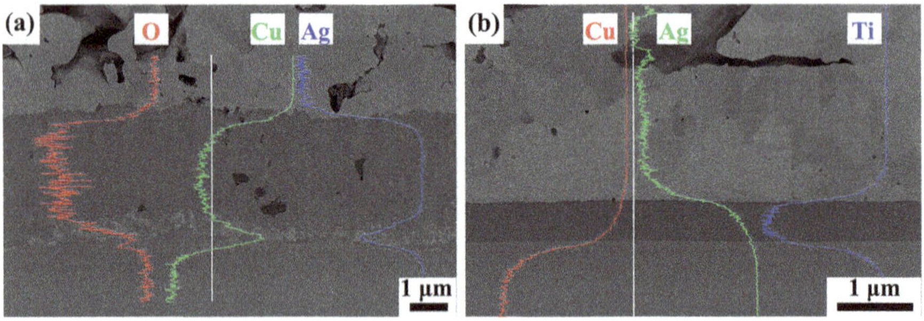

Fig. 11 EDS elemental line scans of sintered Ag joints of (**a**) Ag/Cu and (**b**) Ti/Cu metallization after 500 cycles [54]

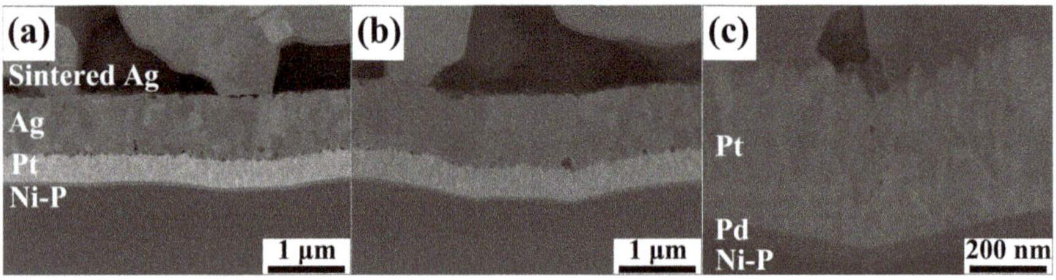

Fig. 12 Cross-sectional image of sintered Ag joints on the Ag/Pt/Pd/Ni substrate: (**a**) as sintered; (**b**) annealed for 200 h; (**c**) high magnification image of (**b**)

2.3.2 Silver Migration

Ag is a relatively active noble metal and is thus highly susceptible to Ag migration when voltage potential is applied between two electrodes. Ag migration is embodied as a directional flow of Ag ions in a conductor due to the moving of electric charges. The electric field plays the role of the driving force, and the migrated Ag ions only undergo random thermal diffusion in the absence of this electric field. Ag migration consists of two types, depending on the environmental conditions. At a relatively high temperature ($\geq$150 °C) in the dry environment, Ag electromigration is the major type of silver migration form that is involved in solid-state electron momentum transfer. On the other hand, Ag electrochemical migration (ECM) occurs at ambient temperatures (<100 °C) in moisture conditions, which means that the interplay of H_2O and Ag dominates the major mechanism.

Ag electromigration is one of the reliability concerns associated with the porous Ag itself, which creates electrical shorts in the packaged product. In a circuit that involves a metal conductor, the momentum transfer between conducting electrons and diffusing metal atoms drives ions in the conductor to transport from the cathode to the anode, which is named electromigration. As the structure size in modern electronics is becoming increasingly compact, the effect of electromigration

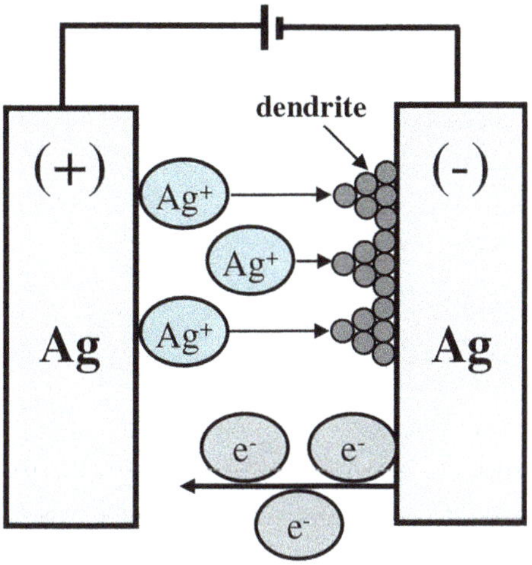

Fig. 13 Schematic of the ECM phenomenon (Ag electrode) [59]

increases significantly. Oxygen was found to play a key role in the high-temperature electromigration process; lower oxygen partial pressure leads to a longer electromigration lifetime. At 400 °C, an obviously lower growth speed of Ag dendrites was able to be observed at an oxygen partial pressure of 0.03 atm, at which the electromigration lifetime is hundred times that of 0.4 atm [57]. This result suggests the importance of packaging, which isolates joints from oxygen when the sintered Ag die-attach is being used in a high-temperature application.

Ag ECM was initially reported as early as the 1950s by Kohman who gave the definition of Ag ECM as follows: The ECM of Ag may be defined as a process by which Ag, when in contact with insulating materials under electrical potential, is removed ionically from its initial location and is redeposited as metal at another location [58]. The ECM can be divided into three processes, including electro dissolution, ion transport, and electrodeposition. The electro dissolution process requires the adsorption of water on the insulation surface such as a silicon wafer [59]. Adsorbed water on the insulating surface transforms into an electrolyte after the Ag ions are generated by electrolysis reactions [60]. In the transportation process, under the driving of the voltage potential, Ag ions move from the anode to the cathode through the electrolyte layer on the insulating surface. In the electrodeposition process, the metallic Ag is generated by accepting electrons at the cathode where the deposited metallic Ag accumulates into Ag bridges between the two electrodes. Figure 13 shows the mechanism of the ECM phenomenon [59]. This Ag bridge, which can lead to a short circuit in the electronic device, is named as dendrites because of its dendritic morphology (Fig. 14).

The potential risk of ECM failure increases because of the miniaturization trend in the pitch and spacing of conductors in electronic packages. With the shrinking dimension, Ag is able to be more easily oxidized and reduced in the absorbed electrolyte; moreover, it shows no passivation inclination, which ensures a

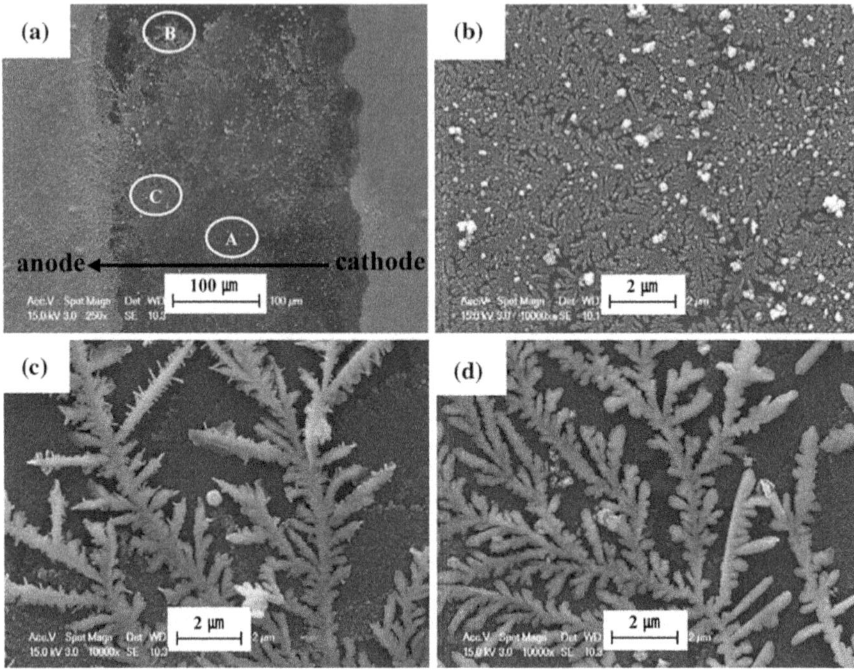

Fig. 14 SEM images of the dendrite structure on a Si substrate with an Ag electrode: (**a**) wide image, (**b**) enlarged image of region A, (**c**) enlarged image of region B, and (**d**) enlarged image of region C [59]

continuous reaction. Other metals do not present a practical migration hazard; however, the abovementioned uniqueness of Ag brings complexity when considering its applications. Hence, it is concluded that Ag should be used with great caution, especially when used in a humid environment.

2.3.3 Silver Coarsening

Silver coarsening describes the grain growth of porous Ag during high-temperature tests such as high-temperature storage and thermal cycles. Atom diffusion is the driving force of Ag coarsening. As a thermal-sensitive procedure, the coarsening of Ag can almost be neglected at 100 °C or lower; however, it can be observed in a high-temperature (200 °C) environment, which is the anticipated working condition of next-generation power devices and ultrahigh-power LEDs.

Coarsening itself does not detriment joint strength; conversely, coarsening helps to strengthen the joints. However, the excessive coarsening of porous Ag changes the joint morphology and potentially weakens the interfacial reliability. For instance, the de-wetting of interfacial Ag metallization is commonly observed in the high-temperature test. This Ag de-wetting phenomenon exposes the lower barrier layer and leads to its fracture due to the change in stress state.

Two types of strategy, resin molding and ceramic addition, have proven to be effective in inhibiting Ag coarsening.

Resin molding is a common process of module fabrication by which the nano-/micropores in sintered Ag are able to be filled and the joints morphology fixed. Nagao et al. reported a thermostable packaging method realized by a well-tuned combination of thermostable molding and Ag sintering die-attach. Imide-based polymers of silsesquioxane nanocomposite are the major ingredient of the molding, and thermal injection is conducted at 180 °C for 5 min, followed by 5 h of post-mold curing at 270 °C. Sintered Ag and a nanocomposite polymer forms an ideal microscale network structure, which is able to relax the thermal stress and inhibit the coarsening of porous Ag [56].

Another strategy for mitigating Ag coarsening is the addition of ceramic particles. Zhang et al. systematically evaluated the inhibition effect of SiC particles with various average sizes and size distribution, and SiC submicron angular particles with an average diameter of 600 nm proved to have the optimized inhibition effect and distribution state [48].

The inhibition effect of adding a submicron SiC particle can be attributed to its chemical stability in relatively low temperatures (<300 °C = and proper dispersion in porous Ag). The existence of considerable SiC particles separates Ag particles and, therefore, hinders their sintering process and grain growth. According to the published results, the mean Ag grain size of the SiC-containing joints remains unchanged whereas that for pure sintered Ag increases from 1.1 to 2.5 μm. As a result, the coarsening of porous Ag is inhibited at the expense of a slight decrease of shear strength as well as electrical and thermal conductivities (Fig. 15) [48].

The possibility of using Ag paste containing SiC submicron particles as a novel, high-temperature die attachment solution had been explored in the packaging of power devices. Actual bonding structure composed of silicon dies and a DBC substrate which were used in the high-temperature test. The SiC-porous Ag-DBC

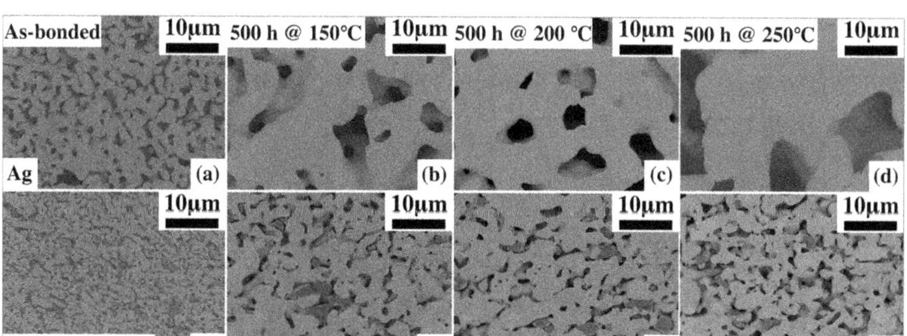

Fig. 15 Evolution of cross-sectional morphology after aging at 150, 200, and 250 °C for 500 h: (**a**)–(**d**) joints with no additive; (**e**), (**f**) joints with added SiC; (**a**) and (**e**) are the as-bonded state [48]

joint was prepared by sintering an Ag microflake paste containing 2 wt% submicron SiC particles under mild conditions (250 °C and 0.4 MPa for 30 min) [30, 54].

The results of the high-temperature storage test indicated that the SiC particle added Ag sinter paste showed better stability than the paste without SiC addition at temperatures such as 150oC, 250 °C, and 350 °C [30].

A similar bonding structure was also being used in the thermal shock test, which was from −40 °C to 250 °C. The total thermal stress generated is significant due to the high ramp rate of more than 150 °C/min, and the tested samples had experienced 1000 thermal cycles. SiC particles were distributed homogeneously in the porous Ag network and they inhibited morphological evolution during thermal shock tests. After 1000 thermal cycles, the shear strength of pure Ag- and SiC-added joints decreased by 58% and 42%, respectively, as compared with their maximum value. The coarsening of porous Ag occurred in pure Ag joints, while in SiC-added joints, morphological evolution had been mitigated during the thermal cycling [54].

In both of the high-temperature storage and thermal cycle tests, the interface in SiC-added joints exhibited better reliability than those without SiC. De-wetting of Ag metallization is seldom seen in the SiC particle added joints because of the alleviated coarsening of porous Ag.

2.4 Conclusion

In this section, reliability issues and the inhibition method of sintered Ag have been introduced. As the most promising die-attach solutions for high-temperature and high-power applications, the research focus of sintered Ag has changed from optimizing the sintering process to the long-term reliability in actual devices and modules. We hope the addressed reliability issues in this section may give rise to advance further research from both academia and industry. In the next section, a newly developed Ag to Ag direct bonding technique will be introduced.

3 Silver-to-Silver Direct Bonding and Its Application in Die-Attach for LED

3.1 Introduction of Basics of Silver-to-Silver Direct Bonding

A relatively low process temperature is required in the device fabrication of LEDs due to the commonly used encapsulating resin degrades when exposed beyond 300 °C. Low-temperature die-attach solutions, which can supply high-performance and long-term reliability, are thus pursued in power LED manufacturing. In this section, a unique Ag-Ag direct bonding method developed for die-attaching high-power LEDs

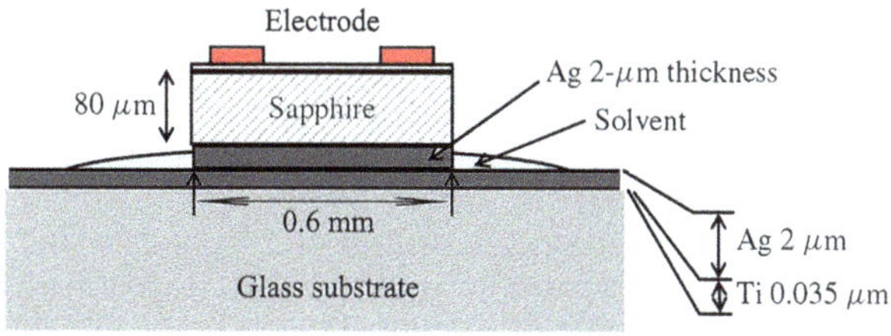

Fig. 16 Schematic illustration of die bonding using Ag direct bonding [61]

and laser diodes is presented in detail, i.e. process steps and the related mechanism. Its low-temperature and pressureless process condition is mild, while excellent thermal resistance and electric resistivity can be provided because of the bulk-like pure Ag bonding area at the interface. A delicate interplay between Ag and oxygen with the participation of residue stress plays the key role in the abnormal growth of Ag hillocks to realize the Ag-Ag direct bonding.

A typical Ag-Ag direct bonding process is shown in Fig. 16 [61]. The external dimensions of the bonded LED die were 600 × 600 μm with a thickness of 80 μm. The bottom face of its sapphire substrate was coated with an Ag layer having a thickness of 2.0 μm. The die mounting substrate was prepared with the following structures: A Ti seed layer of 0.035-μm thickness was formed on a glass slide using a sputter method, and a subsequent Ag-sputtered layer of 2.0-μm thickness was formed on the glass slide. The Ti layer plays the role of an adhesion layer to form a stable interface between the Ag layer and the glass slide [61].

After the contact surface of a light-emitting diode die and a glass substrate were coated with an Ag-sputtered layer, the Ag-Ag direct bonding proceeded by sintering the contact pair above 200 °C in air without the aid of weight. The heated Ag-sputtering layers exhibit abnormal grain growth along the bonding line, and voidless pure Ag joints are formed.

3.2 *The Role of Oxygen in Ag–Ag Direct Bonding*

Kuramoto et al. reported that the bonding strength of the Ag-Ag joints exceeded that of the low-temperature sintering of micrometer-sized Ag particles, and the bonding strength depended on the oxygen concentration in the sintering atmosphere, as shown in Figs. 17 and 18 [61]. The influence of oxygen concentration on the bonding strength is clearly observed; this suggests that the Ag-O interaction plays the key role in the abnormal hillock growth mechanism at microscopic scale. They concluded that the low-temperature sintering of Ag materials is accelerated by the abnormal grain growth of Ag with oxygen adsorption near the Ag surface. A redox reaction accelerates the surface diffusion of Ag, resulting in the abnormal grain growth of Ag along the faying faces to form the joint.

Fig. 17 Relationship between temperature and bonding strength of the Ag-sputtered layer for direct bonding of LEDs [61]

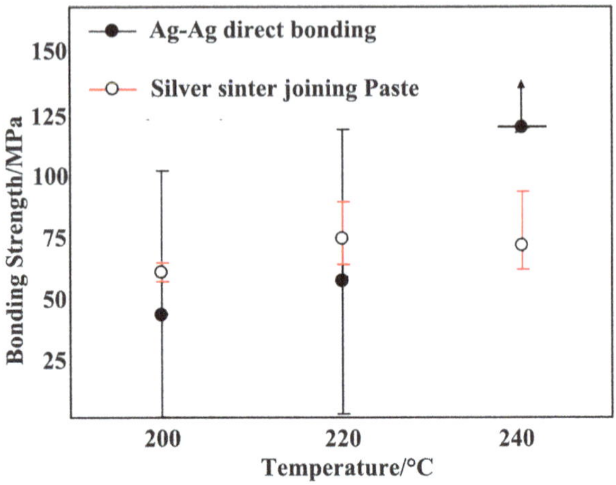

Fig. 18 Influence of oxygen concentration on the bonding strength [61]

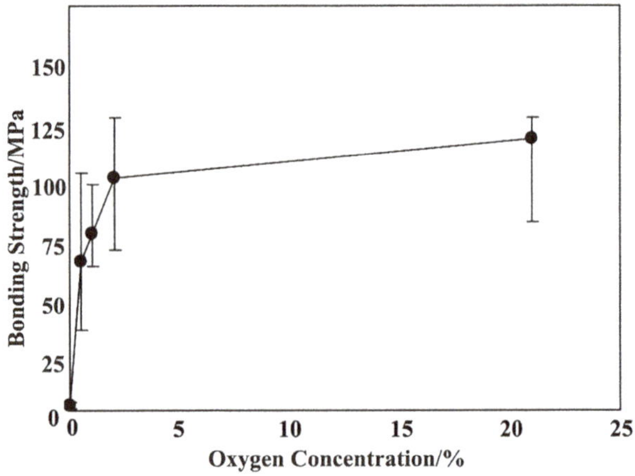

3.3 The Role of Residue Stress in Ag-Ag Direct Bonding

Oh et al. found that the abnormal growth of hillocks on sputtered Ag films is related to stress migration in the metal thin films [34, 37, 62]. The stress migration involving hillock/whisker formation often causes device failure and, hence, minimizing hillock/whisker formation is inevitable when considering the design of electronic packaging. However, by controlling the grain texture and residue stress in the deposited Ag films, hillock growth for Ag-Ag bonding is tunable in an atmospheric oxygen environment and hillock formation on Ag thin films to realize that sound wafer bonding can be maximized [34, 37, 62].

Fig. 19 Compressive residual stress in the Ag films before and after heating (**a**). The correlation between bonding strength and compressive stress changes because of heating (**b**). A higher substrate temperature at sputtering hinders the compressive residual stress of the as-deposited films, and the internal energy necessary for solid-state direct bonding is lost [38]

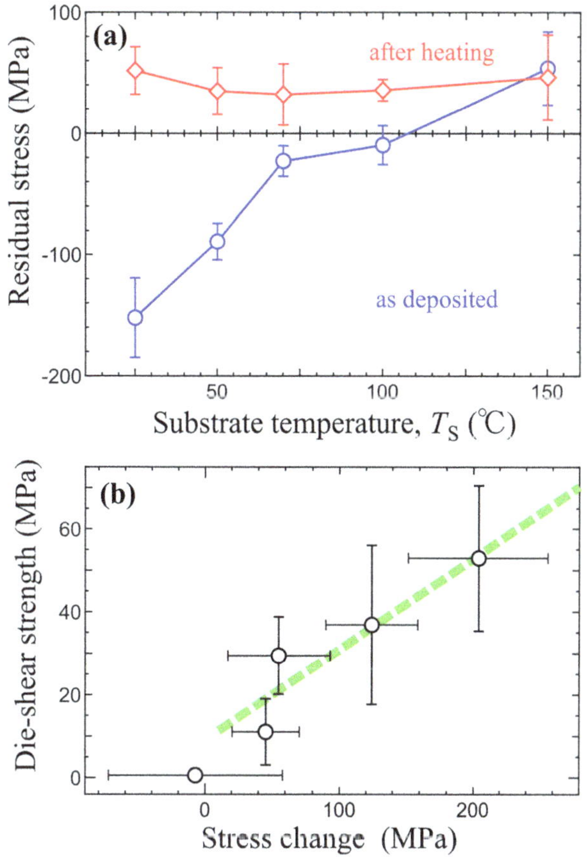

Figure 19 is a plot of the die shear strength as a function of the stress change $\Delta\sigma$ before and after the bonding process, and larger stress relaxation results in higher die shear strength [38]. This result suggests that the stored compressive stress in the films is related to the abnormal grain growth. The bonding strength was thus determined by the initial residual stress in as-deposited Ag films and, hence, by the total stress relaxation $\Delta\sigma$ under compression. Stress migration was proposed as the bonding mechanism in their study and high strength bonding could be obtained by maximizing abnormal hillock growth. They showed that this mechanism also allows a lower bonding temperature after optimizing the compressive stress relaxation. Oh et al. results suggested the influence of residue stress on the growth of abnormal Ag hillocks but they excluded oxygen in the discussion; hence, their mechanism was not able to explain the varying bonding strengths at different oxygen concentrations [34, 37, 62].

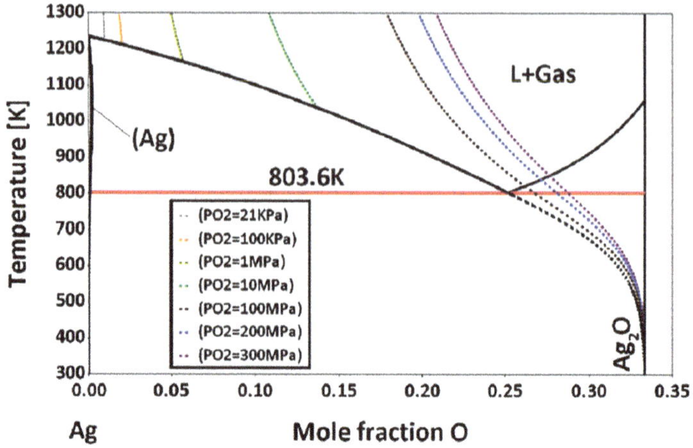

Fig. 20 Calculated Ag-O diagram [27]

3.4 Ag Nano-volcano Mechanism

To unify the two factors, i.e. oxygen concentration and residual compressive stress, for abnormal Ag hillock growth, Lin et al. used the phase equilibria of the Ag-O binary system as their unifying model [27]. The calculated phase diagram in Fig. 20 shows that the melting temperature (liquidus temperature) of the fcc-(Ag) phase declines as the partial pressure of oxygen increases.

After penetrating Ag films, O atoms are fully absorbed at the Ag grain boundaries to be in equilibrium with the ambient oxygen gas prior to sintering. Constrained space at the grain boundaries inhibit these absorbed O and Ag atoms to nucleate into the Ag_2O phase. These buckled O atoms beneath the superficial Ag atoms of each grain are able to present oxide-like structures [63]. At high partial pressures of oxygen, the melting temperature of the buckling-structured Ag-O mixtures at the grain boundaries decreases significantly. During heat treatment, the large mismatch of CTE between Ag and the Si substrate generates a very large thermal stress in the order of hundreds of MPa [38]. This great stress during heat treatment of the Ag-O mixture induces grain boundary liquation. The critical temperature of Ag-O grain boundary liquation is determined to be at approximately 150 °C, which is close to the decomposition temperature of Ag_2O. Meanwhile, the Ag-O liquids at the grain boundaries are squeezed out towards the free surface owing to the large compressive stresses. Following the sudden reduction of pressure and temperature by ejection, the Ag-O liquids are no longer stable and instantly solidify. Therefore, the formation of abnormal Ag hillocks is realized by the extremely fast mass transfer of Ag and the following crystallization.

The detailed solidification mechanism of the erupted Ag-O liquids is also elucidated. The ejected Ag-O liquids transform into O_2 gas and free Ag and Ag_2O clusters, exactly as in a volcanic eruption. As ash falls after a real volcanic eruption,

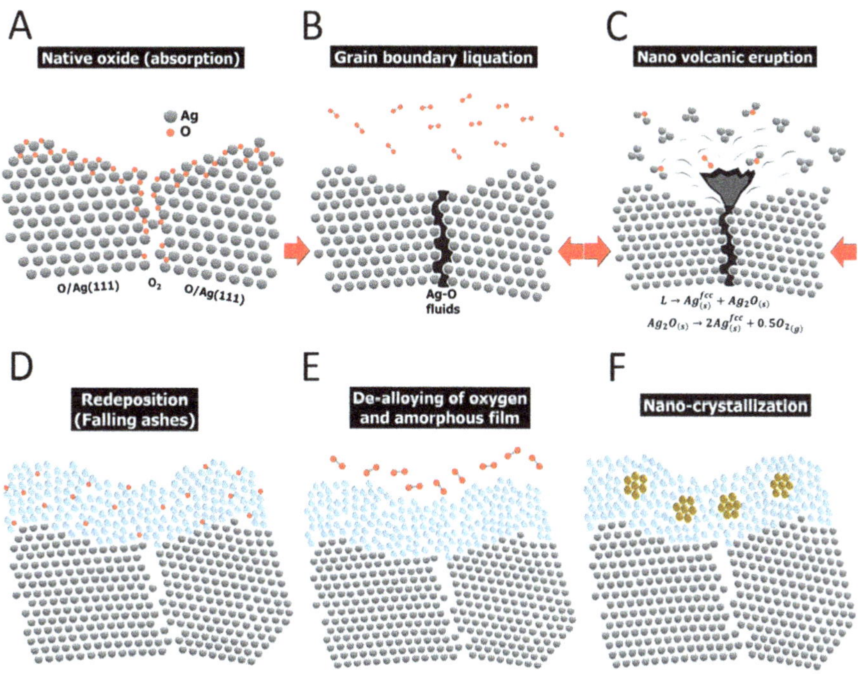

Fig. 21 Schematic diagram of Ag nano-volcanic eruption: (**a**) native oxide, (**b**) grain boundary liquation, (**c**) nano-volcanic eruption of Ag and Ag_2O clusters and O_2 gas, (**d**) redeposition of Ag-O coating (falling ashes), (**e**) de-alloying of O and formation of amorphous Ag film, (**f**) homogeneous nucleation of nano-crystalline Ag grains [27]

the suspended Ag and Ag_2O clusters are then redeposited on the Ag film. The redeposited Ag and Ag_2O transform to a conformal Ag-O coating, which decomposes immediately since the temperature of the Ag film is 250 °C. The deposition reaction releases O_2 gas, which escapes from the Ag-O coating to deposit a conformal amorphous Ag coating.

The homogeneous crystallization of an amorphous Ag coating and the subsequent Ostwald ripening of nano-crystalline grains lead to the formation of abundant Ag hillocks. The universal "Ag nano-volcanic eruption" mechanism sheds light on new applications that involve amorphous Ag hillock formation, such as Ag-Ag direct metal bonding (Fig. 21).

3.5 *Conclusion*

Ag is an important engineering material in the electronic packaging industries because of its unique physical and chemical properties. Ag thin films are extensively used in modern electronics primarily because of their oxidation resistance. By elucidating the mechanism of Ag nano-volcanic eruption, the two factors, residue stress and absorbed oxygen, played key roles in the formation of abnormal Ag hillocks and the realization of Ag-Ag direct bonding. The Ag "nano-volcanic eruption" mechanism is elaborated, shedding light on new applications of Ag-Ag direct bonding, which is able to form perfect pure Ag joints and develop as novel high-quality die-attach for power LEDs.

References

1. T.H. Chiang, Y.-C. Lin, Y.-F. Chen, E.-Y. Chen, Effect of anhydride curing agents, imidazoles, and silver particle sizes on the electrical resistivity and thermal conductivity in the silver adhesives of LED devices. J. Appl. Polym. Sci. **133**(26), 181–189 (2016)
2. G. Zhang, S. Feng, Z. Zhou, J. Liu, J. Li, H. Zhu, Thermal fatigue characteristics of die attach materials for packaged high-brightness LEDs. IEEE Trans. Compon. Packag. Manufact. Technol. **2**(8), 1346–1350 (2012)
3. J.A. Edmond, H.-S. Kong, C.H. Carter Jr., Blue LEDs, UV photodiodes and high-temperature rectifiers in 6H-SiC. Phys. B Condens. Matter **185**(1), 453–460 (1993)
4. J. Edmond, J. Lagaly, Developing nitride-based blue LEDs on SiC substrates. JOM **49**(9), 24–26 (1997)
5. M. Schneider, B. Leyrer, C. Herbold, S. Maikowske, High power density LED modules with silver sintering die attach on aluminum nitride substrates. Presented at the 2014 I.E. 64th electronic components and technology conference (ECTC), 2014, pp. 203–208
6. Z. Gong, M. Gaevski, V. Adivarahan, W. Sun, M. Shatalov, M. Asif Khan, Optical power degradation mechanisms in AlGaN-based 280nm deep ultraviolet light-emitting diodes on sapphire. Appl. Phys. Lett. **88**(12), 121106 (2006)
7. A. Jayawardena, N. Narendran, Analysis of electrical parameters of InGaN-based LED packages with aging. Microelectron. Reliab. **66**, 22–31 (2016)
8. Y.-P. Kim, Y.-S. Kim, S.-C. Ko, Thermal characteristics and fabrication of silicon sub-mount based LED package. Microelectron. Reliab. **56**, 53–60 (2016)
9. Y.S. Lee, C. Yun, K.H. Kim, W.H. Kim, S.-W. Jeon, J.K. Lee, J.P. Kim, Laser-sintered silver nanoparticles as a die adhesive layer for high-power light-emitting diodes. IEEE Trans. Compon., Packag. Manufact. Technol. **4**(7), 1119–1124 (2014)
10. H. Wu, S. Chiang, W. Han, Y. Tang, F. Kang, C. Yang, Surface iodination: A simple and efficient protocol to improve the isotropically thermal conductivity of silver-epoxy pastes. Compos. Sci. Technol. **99**(C), 109–116 (2014)
11. H. Zhang, Y. Lin, D. Zhang, W. Wang, Y. Xing, J. Lin, H. Hong, C. Li, Graphene nanosheet/silicone composite with enhanced thermal conductivity and its application in heat dissipation of high-power light-emitting diodes. Curr. Appl. Phys. **16**(12), 1695–1702 (2016)
12. Y. Tang, D. Liu, H. Yang, P. Yang, Thermal effects on LED lamp with different thermal interface materials. IEEE Trans. Electron Devices **63**(12), 4819–4824 (2016)
13. K. Pashayi, H.R. Fard, F. Lai, S. Iruvanti, J. Plawsky, T. Borca-Tasciuc, High thermal conductivity epoxy-silver composites based on self-constructed nanostructured metallic networks. J. Appl. Phys. **111**(10), 104310–104317 (2012)
14. B.-H. Liou, C.-M. Chen, R.-H. Horng, Y.-C. Chiang, D.-S. Wuu, Improvement of thermal management of high-power GaN-based light-emitting diodes. Microelectron. Reliab. **52**(5), 861–865 (2012)
15. C.-J. Chen, C.-M. Chen, R.-H. Horng, D.-S. Wuu, J.-S. Hong, Thermal management and interfacial properties in high-power GaN-based light-emitting diodes employing diamond-

added Sn-3 wt.%Ag-0.5 wt.%Cu solder as a die-attach material. J. Electron. Mater. **39**(12), 2618–2626 (2010)
16. K.N. Tu, A.M. Gusak, M. Li, Physics and materials challenges for lead-free solders. J. Appl. Phys. **93**(3), 1335–1353 (2003)
17. J. Shen, Y.C. Chan, S.Y. Liu, Growth mechanism of bulk Ag3Sn intermetallic compounds in Sn–Ag solder during solidification. Intermetallics **16**(9), 1142–1148 (2008)
18. K. Zeng, K.N. Tu, Six cases of reliability study of Pb-free solder joints in electronic packaging technology. Mater. Sci. Eng. R-Rep. **38**(2), 55–105 (2002)
19. H. Zhang, Q.-S. Zhu, Z.-Q. Liu, L. Zhang, H. Guo, C.-M. Lai, Effect of Fe content on the interfacial reliability of SnAgCu/Fe–Ni solder joints. J. Mater. Sci. Technol. **30**(9), 928–933 (2014)
20. N.-S. Lam, C.-Y. Lee, M.-Y. Wan, D. Tian, M. Li, High quality & low thermal resistance eutectic flip chip LED bonding. Presented at the 2013 14th international conference on electronic packaging technology (ICEPT), 2013, pp. 1197–1201
21. Y. Liu, J. Zhao, C.C.-A. Yuan, G.Q. Zhang, F. Sun, Chip-on-flexible packaging for high-power flip-chip light-emitting diode by AuSn and SAC soldering. IEEE Trans. Compon. Packag. Manufact. Technol. **4**(11), 1754–1759 (2014)
22. A.A. Mani, M. Arch, Direct attach led soldering by new printable AuSn paste. Presented at the 20th European microelectronics and packaging conference and exhibition: enabling technologies for a better life and future, EMPC 2015, 2016
23. S.A. Paknejad, S.H. Mannan, Review of silver nanoparticle based die attach materials for high power/temperature applications. Microelectron. Reliab. **70**, 1–11 (2017)
24. K.S. Siow, Y.T. Lin, Identifying the Development State of Sintered Silver (Ag) as a Bonding Material in the Microelectronic Packaging Via a Patent Landscape Study. Journal of Electronic Packaging **138**(2), 020804 (2016)
25. A.A. Bajwa, Y. Qin, R. Reiner, R. Quay, J. Wilde, Assembly and packaging technologies for high-temperature and high-power GaN devices. IEEE Trans. Compon. Packag. Manufact. Technol. **5**(10), 1402–1416 (2015)
26. T. Kunimune, M. Kuramoto, S. Ogawa, T. Sugahara, S. Nagao, K. Suganuma, Ultra thermal stability of LED die-attach achieved by pressureless Ag stress-migration bonding at low temperature. Acta Mater. **89**, 133–140 (2015)
27. S.-K. Lin, S. Nagao, E. Yokoi, C. Oh, H. Zhang, Y.-C. Liu, S.-G. Lin, K. Suganuma, Nano-volcanic eruption of silver. Sci. Rep. **6**, 34769 (2016)
28. F. Le Henaff, S. Azzopardi, J.Y. Deletage, E. Woirgard, S. Bontemps, J. Joguet, A preliminary study on the thermal and mechanical performances of sintered nano scale silver die attach technology depending on the substrate metallization. Microelectron. Reliab. **52**(9), 2321–2325 (2012)
29. K. Suganuma, S.-J. Kim, K.-S. Kim, High-temperature lead-free solders: properties and possibilities. JOM **61**(1), 64–71 (2009)
30. H. Zhang, S. Nagao, K. Suganuma, H.-J. Albrecht, K. Wilke, Thermostable Ag die-attach structure for high-temperature power devices. J. Mater. Sci. Mater. Electron. **27**(2), 1337–1344 (2015)
31. T. Nguyen et al., Characterizing the mechanical properties of actual SAC105, SAC305, and SAC405 solder joints by digital image correlation. J. Electron. Mater. **40**(6), 1409–1415 (2011)
32. L. Ho, H. Nishikawa, J. Mater. Eng. Perform. **23**(9), 3371–3378 (2014)
33. H. Nishikawa et al., Mater. Trans. **51**(10), 1785–1789 (2010)
34. C. Oh, S. Nagao, K. Suganuma, Pressureless bonding using sputtered Ag thin films. J. Electron. Mater. **43**(12), 4406–4412 (2014)
35. W. Liu et al., Mater. Sci. Eng. A **651**, 626–635 (2016)
36. R. Khazaka, L. Mendizabal, D. Henry, Review on joint shear strength of nano-silver paste and its long-term high temperature reliability. J. Electron. Mater. **43**(7), 2459–2466 (2014)
37. C. Oh, S. Nagao, T. Sugahara, K. Suganuma, Hillock growth dynamics for Ag stress migration bonding. Mater. Lett. **137**, 170–173 (2014)

38. C. Oh, S. Nagao, T. Kunimune, K. Suganuma, Pressureless wafer bonding by turning hillocks into abnormal grain growths in Ag films. Appl. Phys. Lett. **104**(1), 161603 (2014)
39. K. Park, D. Seo, J. Lee, Conductivity of silver paste prepared from nanoparticles. Colloids Surf. A Physicochem. Eng. Asp. **313**, 351–354 (2008)
40. K.S. Siow, Are sintered silver joints ready for use as interconnect material in microelectronic packaging? J. Electron. Mater. **43**(4), 947–961 (2014)
41. H. Zheng, K.D.T. Ngo, G.-Q. Lu, Temperature cycling reliability assessment of die attachment on bare copper by pressureless nanosilver sintering. IEEE Trans. Device Mater. Relib. **15**(2), 214–219 (2015)
42. J. Li, C.M. Johnson, C. Buttay, W. Sabbah, S. Azzopardi, Bonding strength of multiple SiC die attachment prepared by sintering of Ag nanoparticles. J. Mater. Process. Technol. **215**(1), 299–308 (2015)
43. S.T. Chua, K.S. Siow, Microstructural studies and bonding strength of pressureless sintered nano-silver joints on silver, direct bond copper (DBC) and copper substrates aged at 300 °C. J. Alloys Compd. **687**, 486–498 (2016)
44. J. Jiu, H. Zhang, S. Koga, S. Nagao, Y. Izumi, K. Suganuma, Simultaneous synthesis of nano and micro-Ag particles and their application as a die-attachment material. J. Mater. Sci. Mater. Electron. **26**(9), 7183–7191 (2015)
45. K. Suganuma, S. Sakamoto, N. Kagami, D. Wakuda, K.-S. Kim, M. Nogi, Low-temperature low-pressure die attach with hybrid silver particle paste. Microelectron. Reliab. **52**(2), 375–380 (2012)
46. J. Jiu, H. Zhang, S. Nagao, T. Sugahara, N. Kagami, Y. Suzuki, Y. Akai, K. Suganuma, Die-attaching silver paste based on a novel solvent for high-power semiconductor devices. J. Mater. Sci. **51**(7), 3422–3430 (2015)
47. S.A. Paknejad, G. Dumas, G. West, G. Lewis, S.H. Mannan, Microstructure evolution during 300°C storage of sintered Ag nanoparticles on Ag and Au substrates. J. Alloys Compd. **617**, 994–1001 (2014)
48. H. Zhang, S. Nagao, K. Suganuma, Addition of SiC particles to Ag die-attach paste to improve high-temperature stability; grain growth kinetics of sintered porous Ag. J. Electron. Mater. **44** (10), 3896–3903 (2015)
49. M.O. Alam, Y.C. Chan, K.N. Tu, Elimination of Au-embrittlement in solder joints on Au/Ni metallization. J. Mater. Res. **19**(5), 1303–1306 (2011)
50. H.G. Tompkins, M.R. Pinnel, Relative rates of nickel diffusion and copper diffusion through gold. J. Appl. Phys. **48**(7), 3144–3146 (1977)
51. K.M. Chow, W.Y. Ng, L.K. Yeung, Barrier properties of Ni, Pd and Pd-Fe for Cu diffusion. Surf. Coat. Technol. **105**(1), 56–64 (1998)
52. M.O. Alam, Y.C. Chan, K.C. Hung, Interfacial reaction of Pb-Sn solder and Sn-Ag solder with electroless Ni deposit during reflow. J. Electron. Mater. **31**(10), 1117–1121 (2002)
53. M.O. Alam, Y.C. Chan, K.C. Hung, Reliability study of the electroless Ni-P layer against solder alloy. Microelectron. Reliab. **42**(7), 1065–1073 (2002)
54. H. Zhang, C. Chen, S. Nagao, K. Suganuma, Thermal fatigue behavior of silicon-carbide-doped silver microflake sinter joints for die attachment in silicon/silicon carbide power devices. J. Electron. Mater. **46**(2), 1055–1060 (2017)
55. H. Zhang, S. Nagao, S. Kurosaka, H. Fujita, K. Yamamura, A. Shimoyama, S. Seki, T. Sugahara, K. Suganuma, Thermostable electroless plating optimized for Ag sinter die-attach realizing high T J device packaging, 2016, pp. 1–4
56. S. Nagao, T. Sugioka, S. Ogawa, T. Fujibayashi, Z. Hao, K. Suganuma, High thermal stability of SiC packaging with sintered Ag paste die-attach combined with imide-based molding. Int. Symp. Microelectron. **2015**(1), 000349–000352 (2015)
57. Y. Mei, G.-Q. Lu, X. Chen, S. Luo, D. Ibitayo, Effect of oxygen partial pressure on silver migration of low-temperature sintered nanosilver die-attach material. IEEE Trans. Device Mater. Reliab. **11**(2), 312–315 (2011)

58. G.T. Kohman, H.W. Hermance, G.H. Downes, Silver migration in electrical insulation. Bell Syst. Tech. J. **34**(6), 1115–1147 (1955)
59. B.-I. Noh, J.-W. Yoon, K.-S. Kim, Y.-C. Lee, S.-B. Jung, Microstructure, electrical properties, and electrochemical migration of a directly printed Ag pattern. J. Electron. Mater. **40**(1), 35–41 (2010)
60. S. Yang, A. Christou, Failure model for silver electrochemical migration. IEEE Trans. Dev. Mater. Reliab. **7**(1), 188–196 (2007)
61. M. Kuramoto, T. Kunimune, S. Ogawa, M. Niwa, K.-S. Kim, K. Suganuma, Low-temperature and Pressureless Ag-Ag direct bonding for light emitting diode die-attachment. IEEE Trans. Compon. Packag. Manufact. Technol. **2**(4), 548–552 (2012)
62. C. Oh, S. Nagao, K. Suganuma, Silver stress migration bonding driven by thermomechanical stress with various substrates. J. Mater. Sci. Mater. Electron. **26**(4), 2525–2530 (2015)
63. A. Michaelides, M.L. Bocquet, P. Sautet, A. Alavi, D.A. King, Structures and thermodynamic phase transitions for oxygen and silver oxide phases on Ag{1 1 1}. Chem. Phys. Lett. **367**(3), 344–350 (2003)

Process Control of Sintered Ag Joint in Production for Die Attach Applications

K. S. Siow, V. R. Manikam, and S. T. Chua

1 Introduction: Sintered Ag as a Die Attach Material

Silver sintering provides an alternative die attach (DA) process for the low-temperature processing of power semiconductors, which requires a relatively high operational temperature for the packages. As shown in Fig. 1, this requirement for high operational temperature has few alternatives which are cost-effective, reliable, and scalable. In addition, high-temperature DA material must be able to demonstrate high thermal and electrical conductivity; low coefficient of thermal expansion (CTE) mismatch between the die and the substrate; good wettability and adhesion on surfaces (primarily copper (Cu)); suitable mechanical properties which provide stress relaxation post-sintering; good lifetime behavior, i.e., creep fatigue and corrosion resistance; as well as ease of being processed and reworked (if necessary) in mass manufacturing environment [1].

In a typical power device package, for example, on a power module or discrete power device, a DA interface provides the thermal, electrical, and structural or mechanical support for the silicon or wide bandgap dies to function at their optimum range [3]. Characteristically, the DA5 consortium[1] has specified five requirements for the power DA material, such as [4]:

---

[1]DA5 consortium consists of Bosch, Infineon Technologies, NXP Semiconductors, STMicroelectronics, and Nexperia, formed to look for alternative Pb-free technologies to comply with EU Directives on Restriction of Hazardous Substances (ROHS) 2011/65/EU by July 2021.

K. S. Siow (✉)
Institute of Microengineering and Nanoelectronics, Universiti Kebangsaan Malaysia, Bangi, Selangor, Malaysia
e-mail: kimsiow@ukm.edu.my

V. R. Manikam
Robert Bosch (Australia) Pte Ltd., Clayton, VIC, Australia

S. T. Chua
ON Semiconductor (Malaysia) Sdn Bhd, Seremban, Negeri Sembilan, Malaysia

K. S. Siow (ed.), *Die-Attach Materials for High Temperature Applications in Microelectronics Packaging*, https://doi.org/10.1007/978-3-319-99256-3_3

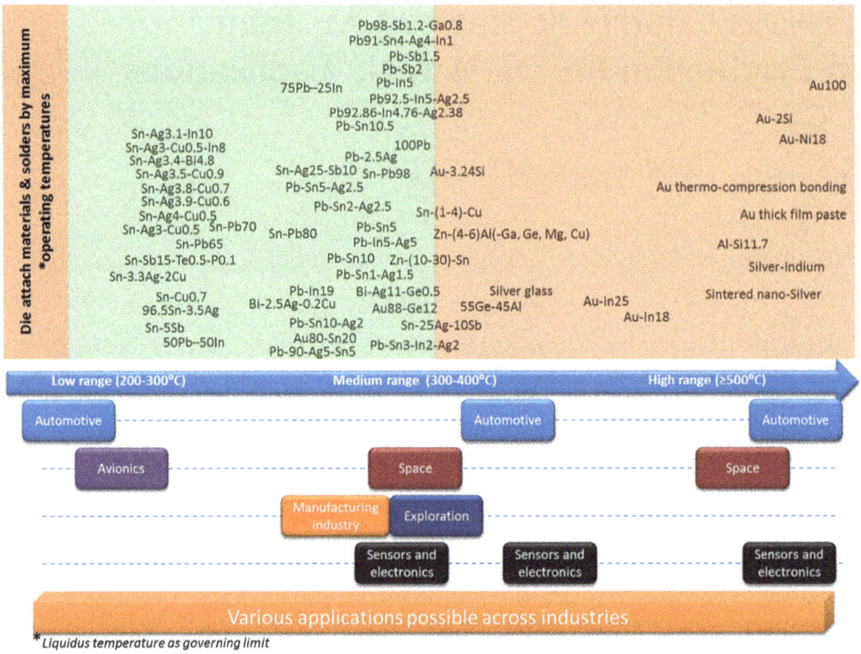

Fig. 1 Die attach materials categorized by operation temperatures and applications [2]

1. Compliant to AEC-Q100/ Q101 Grade 0
2. Typical junction temperature of 175 °C but maximum up to 200 °C
3. Reflow at 260 °C for the surface mount device
4. Moisture sensitive level 3 or better
5. Wire-bonding temperature up to 260 °C

Based on these requirements, Ag sintering is one of the leading alternatives for high-performance DA materials. This interest in Ag sintering is reflected by the number of patents owned and filed by major manufacturers in the entire value chain related to Ag sintering (Fig. 2) [5]. In this chapter, we reviewed the process of selecting and implementing Ag sintering as DA joint in a manufacturing environment.

2 Factors for Selecting Sintered Ag

Figure 3 presents our approach of selecting sintered Ag as a DA joint. The most crucial question addresses the necessity of applying pressure during sintering because applied pressure requires a relook at the material specifications, material characteristics, equipment, and process requirements in the power package. We address the first three issues before devoting Sect. 3 to compare the pressureless and pressure-assisted sintered Ag joint.

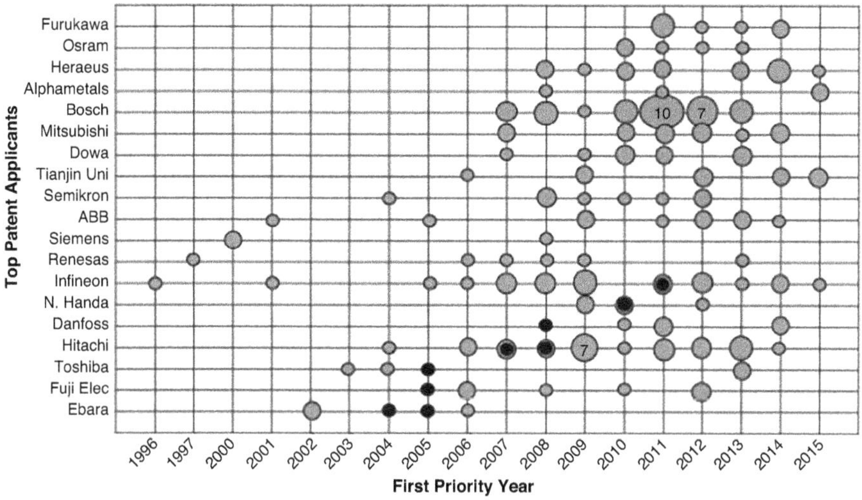

Fig. 2 Patent applications between 1996 and 2015 which depict breakdown of applications across major semiconductor and electronics manufacturers [6]

In the case of material specifications, we need to address the device requirements concerning the material specifications of DA materials. Not all power devices would need the advantages offered by sintered Ag materials. If the device performance demands the high thermal performance of the sintered Ag, the next two phases examine the material characteristics in mass manufacturing as well as investments in the development of the process and equipment. Material characteristics of the Ag sinter material are related to its reliability via the choice of processing route and equipment, shown in Fig. 4. The equipment handling capability, described in Fig. 5, would determine the impact of the chosen process, i.e., stencil printing or dispensing, and how well it would be able to handle the material characteristics. Figures 4 and 5 show that these factors are interrelated and influenced the overall performance of the packaged device. At this stage, the associated cost related to the retrofitting of the sintering oven, inert or reducing gasses (or vacuum) during sintering, and changes to the wafer backside metallization needs to be factored together with the total product volume.

3 Comparing Pressure and Pressureless Sintered Ag Joints

Pressureless sintering has the cost advantage ease of implementation and desired reliability for some niche applications, but this technique produces Ag joints of unstable and porous microstructure joints which have continuously deteriorate during the reliability testing for certain harsh applications [8, 9]. On the other hand, pressure-assisted sintered Ag joint with a uniform microstructure and density of approximately 80% is able to "absorb" the thermal-mechanical stress in a fatigue

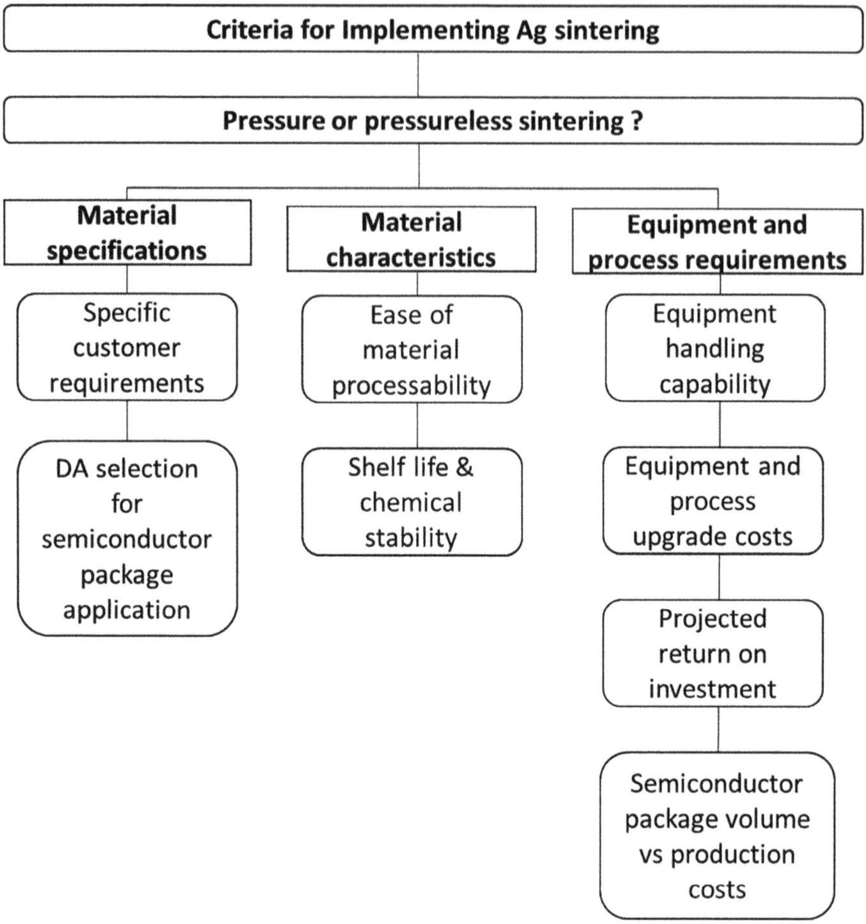

Fig. 3 Criteria for implementing Ag sintering

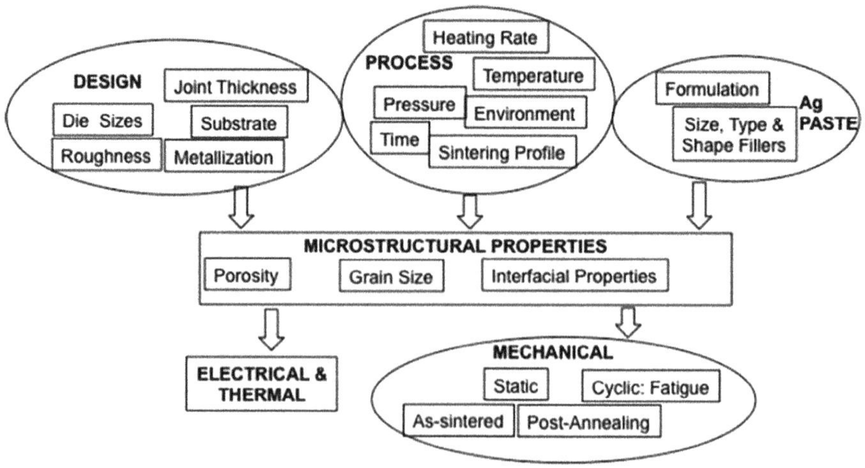

Fig. 4 Factors affecting the reliability of sintered Ag joint [5]

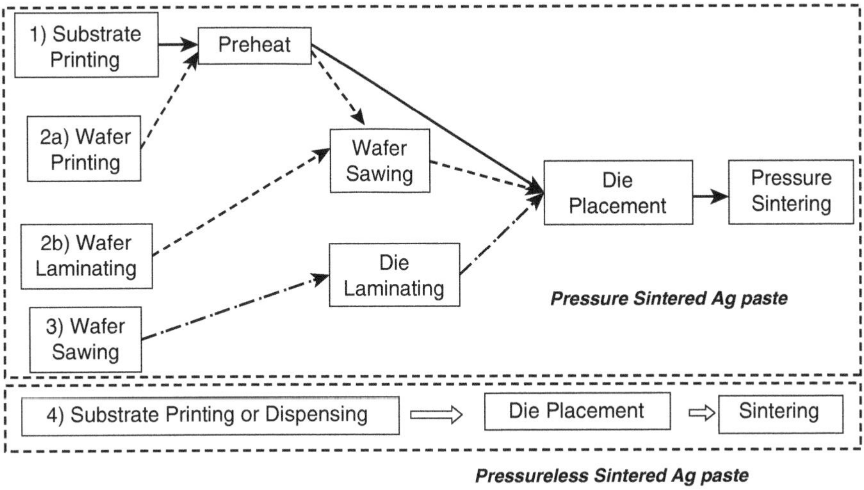

Fig. 5 Four main processes of forming sintered Ag joints [5, 7]

loading [10]. The microstructural stability of pressure-assisted sintered Ag joint is attributed to the similar ratio of final to the initial porosity of a sintered metallic joint during a sintering process [11]. Therefore, an initial structure of low porosity, induced by the pressure-assisted sintering, is likely to have a final sintered Ag joint of high density. Regarding microstructure, this applied pressure added to the inherent driving force of the Ag nanoparticles to coalescence and sinter to form the Ag joints. The equation for this sintering step is based on the Mackenzie-Shuttleworth model shown here [12]:

$$\frac{d\rho}{dt} = \frac{3}{2}\left(\frac{\gamma}{r} + P_{\text{applied}}\right)(1-\rho)\left(1-\alpha\left(\frac{1}{\rho}-1\right)^{1/3}\ln\frac{1}{1-\rho}\right)\frac{1}{\eta} \tag{1}$$

$\frac{d\rho}{dt}$ is the densification rate, while $\left(\frac{\gamma}{r} + P_{\text{applied}}\right)$ is the driving force for densification. P_{applied} is the applied sintering pressure; γ = surface energy; r = particle radius; α = geometrical constant; ρ = density; and η = densification viscosity.

The applied sintering pressure (P_{applied}) provides the additional driving force to sinter the Ag nanoparticles when this surface reactivity of nanoparticles cannot densify to the required mechanical, electrical and thermal properties. Sintering pressure, time, and temperature form the iron triangle of producing a reliable Ag joint; a reduction in one of three parameters requires an increase in other two parameters. In terms of production capacity, this iron triangle also indirectly determines the cycle time of this die-attaching process. An example is shown in Fig. 6 for package TO263; the application of sintering pressure reduces the process time from conventional pressureless sintering of 1–2 h to less than 2 min in cycle time to achieve the desired shear strength.

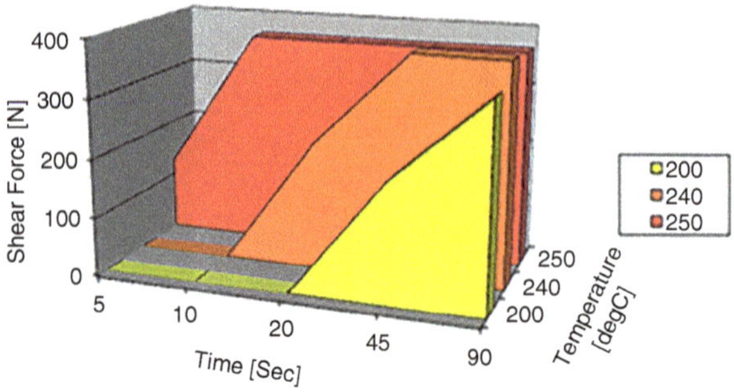

Fig. 6 Influence of sintering temperature and time on the shear strength of dies used in TO263 package when pressure sintered at 10 MPa [13]

Besides a reduction in cycle time, pressure sintering is also needed to produce reliable Ag joints for die size larger than 100 mm^2. Hitherto, reliable pressureless sintered Ag joints are restricted to die size smaller than 100 mm^2. Such small dies are typically found in LED or smaller diodes used in power modules. Larger dies hinder the diffusion of oxygen to pyrolyze the dispersants and binders of the Ag paste located near centre of a large die. Residual binders could prevent the necking of the sintered Ag nanoparticles and posed long-term reliability to the sintered Ag joint [14].

4 Key Process Steps in Ag Sintering

In a manufacturing environment, the process step of sintering Ag joints is summarized in Fig. 5. As mentioned earlier, the sintered Ag joint can be formed with assistance or absence of pressure during the sintering step depending on the Ag paste, device, and technological requirements. While the process for pressureless sintered Ag joint is similar to the current die attaching with electrically conductive adhesives, the pressure-assisted sintered Ag joints require additional steps of die placement and pressure sintering to form the Ag joint.

4.1 Substrate or Wafer Printing

While the printing process of Ag pastes is similar to solder printing, the formulation for sintered Ag paste used as DA materials is different and less well understood compared to the Sn-Pb or Sn-Ag-Cu solder used in electronic packaging. Therefore, some process developmental work is necessary for the substrate printing or wafer printing as this work is usually kept as company confidential, and this confidentiality is further exacerbated by the high material cost of the Ag paste needed for such developmental work.

4.2 Preheating

Preheating is carried out to prevent any squeeze out of the Ag paste during the pressure-sintering step while maintaining enough adhesiveness to the bonding interface at the semiconductor dies and substrate. Excessive preheating causes hardening that prevents the dried Ag paste to conform to the semiconductor dies during pressure sintering [15]. Others investigated the feasibility of preheating the Ag paste after placing the semiconductor dies and find this approach to have lower die shear strength and smaller process window [15]. Preheating after die placement entraps the solvent resulting in multiple voids and pores during pressure sintering. As shown in Fig. 7, preheating does not result in sintering but merely necking of the Ag fillers. Subsequent pressure sintering resulted in a consolidation of the Ag particles and bonding to the interface (Fig. 7b).

Regarding the appropriate parameter to preheat, some researchers specified a flow rate of atmospheric air between 0.5 and 3 l/min at the pressure of 4.0×10^{-4} Pa to 5.5×10^{-3} Pa within the temperature range of 100–150 °C for a duration of 30–60 min [17]. Others suggested a more practical range of 110–150 °C for 30–90 s to carry out this preheating. In the case of over-drying, a tacking agent was applied to increase positional accuracy and adhesion of the die before pressure sintering [18]. This tacking agent may pose residual problems, since this process and the nano-Ag tape are designed to eliminate any cleaning process after sintering. The duration of die placement can be as short as 50 ms [19], and it is typically carried out by a die bonder or flip chip bonder.

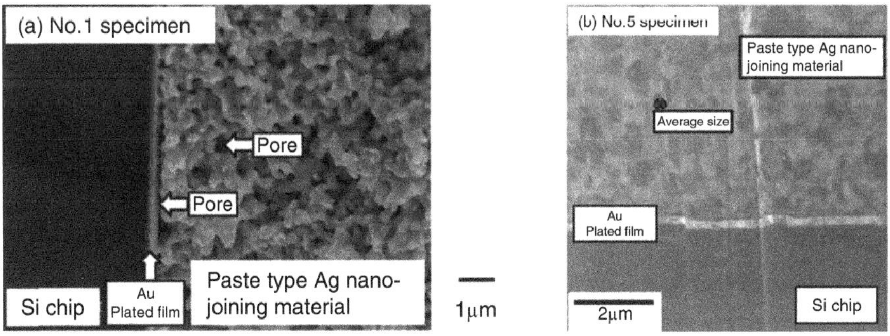

Fig. 7 SEM micrographs of (**a**) cleaved surfaces of sintered Ag joint at Au-plated silicon after preheated at 130 °C, 30 min, and (**b**) focused ion beam of the interfacial microstructure between sintered Ag joints at Au-plated silicon after preheated at 130 °C, 30 min, and pressure sintered at 250 °C for 90 s at 10 MPa [16]

4.3 Pressure-Sintering Equipment

In principle, all servo-hydraulic presses could be used for pressure sintering with an additional silicone rubber and Teflon tape to cushion the direct contact and adjust slight height differences between the semiconductor dies placed on the respective substrates. Commercially, sinter presses used in semiconductor packaging are produced by mold, oven or press suppliers like ASM Technology, Boschman Technologies, Locatelli Meccanica and PINK GmbH Thermosystem. While Locatelli pursued the hydraulic approach, Boschman used gas or better known as the dynamic insert to actuate this pressure during sintering. Interestingly, semiconductor device makers such as Infineon Technologies [20], ABB Research [21], Freescale Semiconductor [22] (now part of NXP Semiconductors), and International Rectifier Corporation [23] (now part of Infineon) also hold patents in this area though we believe that they are primarily filing the patents to obtain freedom to operate and possible patent exchange if the sintered Ag became a critical bonding technology in the future.

Regarding the process, typical pressure and temperature applied during sintering range from 3–30 MPa and 220–270 °C, respectively. Clamping force as large as 10 MN is required to hold the substrate with semiconductor dies during pressure sintering. Special tooling is required to transmit these forces to the sintering area without damaging the semiconductor dies that can be as thin as 500 μm. The parallelism between the two plates of the press is essential in ensuring equal pressure on the face of the semiconductor dies as well as the substrate. As mentioned earlier, silicone rubber (few mm) and Teflon films (50–200 μm) are used to cushion the impact between the plate and the top face of the semiconductor dies. In this case, a semiconductor die with thicker Ag paste will receive higher pressure than a die with thinner Ag paste.

Hence, inserts are used to compensate the different height of the semiconductor dies tacked on the dried Ag-printed substrates. Figure 8 shows the different inserts used in the sinter press, namely, (a) static insert, (b) spring-loaded insert, and (c) dynamic insert. Spring-loaded inserts were used to compensate these varying thicknesses, but they suffered from embrittlement during pressure sintering at temperature above 250 °C. Hence, dynamic insert technology used gasses to

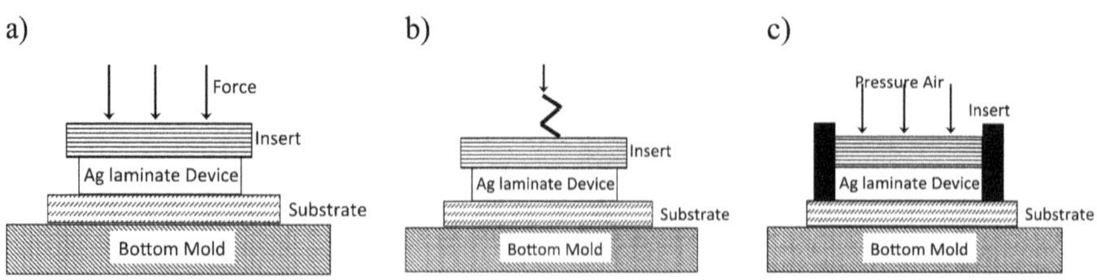

Fig. 8 Evolution of inserts used in the sinter press: (**a**) static insert, (**b**) spring-loaded insert, and (**c**) dynamic insert technology

Fig. 9 Pressure-sintering equipment showing the bottom tool (left) and sinter tool in press fixture (right) (courtesy of Boschman Technologies) [13]

compensate these height differences between adjacent dies or substrates during each sinter press cycle [24]. Photographs of the dynamic insert are shown in Fig. 9.

The units per hour (UPH) of the pressure sintering depend on the cycle time and products per cycle tool change. Current modular design allows tool change between different package designs to take less than 5 min [19]. The typical cycle time for pressure sintering is between 2 and 3 min depending on the temperature, time, and pressure mentioned earlier. The number of products to be sintered depends on the module or package area size. Effective sinter area of the sinter press is 350 mm × 270 mm for one sinter press design [25]. Therefore, an approximate UPH can be calculated based on this sinter area, package size, and "dead area." Dead area refers to the area around the package design which cannot be fully flush inside the tooling of the sintering press. One published study puts the daily output of this sinter press to be 800 cards/day [26].

4.4 Die Laminating-Placement-Pressure Sintering

The use of pressure-assisted sintering which involves laminating sawn dies with Ag tape before die placement and pressure sintering in a single step was discussed earlier (Route 3, Fig. 5). This approach is pioneered by Alpha Metals [104] with die bonder or flip chip bonders like EVO2200 (Datacon) or SET FC-150 (Smart Equipment Technology). Figure 10 compares the pores from the sublimation and decomposition of binders in Ag paste during preheating stage (10a) with the necking and bonding of the Ag nanoparticles of the nano-Ag tape during pressure sintering (10b). Pressure sintering is needed to initiate sintering and formation of bonding with the Au-plated silicon shown in Fig. 10.

As shown in Fig. 11, the sintered Ag joints from Ag paste produced Ag joints with smaller grain size than those produced from the nano-Ag tape approach because of the residual binders in the Ag paste likely to restrict the grain growth [16]. This carbon residual has been observed on sintered Ag joints formed from Ag paste on Au-Ni-plated substrate [27].

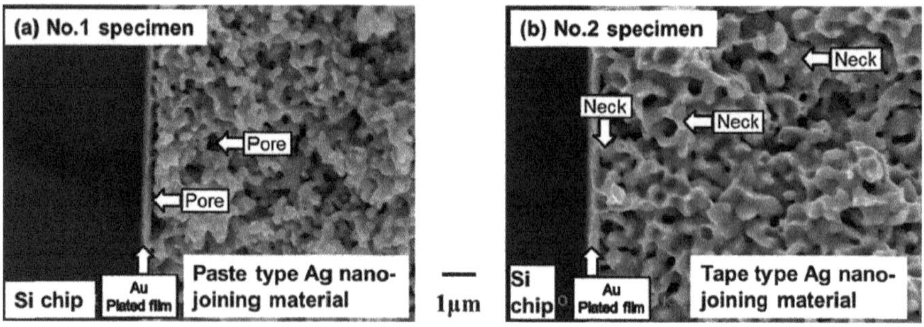

Fig. 10 SEM micrographs of cleaved surfaces of (**a**) preheated Ag paste at 130 °C, 30 min, and (**b**) nano-Ag tape pressure sintered at 250 °C for 90 s at 10 MPa [16]

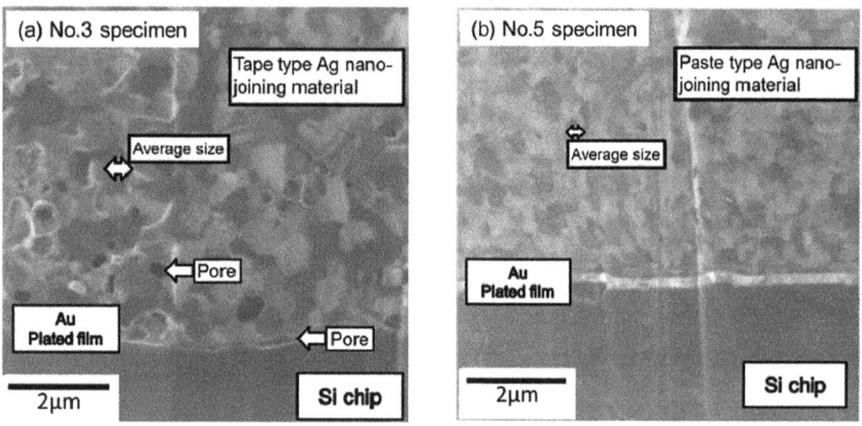

Fig. 11 Focused ion beam of the interfacial microstructure after pressure sintered at 250 °C for 180 s at 10 MPa between Au-plated silicon and (**a**) nano-Ag tape. (**b**) nano-Ag paste preheated at 130 °C, 30 min [16]

During process development of nano-Ag tape as bonding materials, the wafer is sawn as per normal and placed in waffle pack, feeder, or Gel-Pak® and remained in the saw ring. The bond head of the flip chip bonder like EVO2200 or FC150 picked up the sawn dies to laminate on the nano-Ag tape. The temperature of the bond head is elevated to more than 120 °C to promote adhesiveness in nano-Ag tapes [19]. Then the bond head will place and pressure sinter the Ag-laminated dies sequentially on designated locations of the substrates. As a result of this sequential pressure sintering, the throughput is lower than other batch process shown in Fig. 5, but this approach provides flexibility in bonding semiconductor dies of different sizes and accurate placement as small as 30 μm. Precise placement is necessary for the LED module which places the emitter to be as narrow as 100 μm. Such placement accuracy is well within the accuracy of the die bonder which has placement accuracy as small as 1 μm for SET FC-150 [28] and 7 μm [29] for EVO-2200 (Datacon). Since this approach adopts serial sintering of one device followed by another, the UPH is lower than the pressure sintering mentioned earlier (i.e., substrate or wafer printing), but this die-laminating approach provides the selectivity and accurate placement of the semiconductor dies on the substrate.

5 In-Process Control Tools for Ag Sintering in Mass Manufacturing

In a manufacturing environment, in-process control of Ag-sintering DA is needed to ensure sound quality of the sintered Ag joints produced in different package designs. Unless mentioned otherwise, similar in-process tools, discussed in Fig. 12, are used to inspect sintered Ag paste and laminate/tape/film, though minor adjustments are sometimes needed to suit the physical and chemical properties of these two forms of Ag sinter materials.

5.1 Bond Line Thickness (BLT), Porosity, and Fillet Height Control for Sintered Ag Joint

Excellent control of BLT for Ag sintering is equally as crucial as on solder (paste and preform) or epoxy-based technologies. In the case of Ag sintering, BLT will influence the formation of pores beneath the die and the Ag migration failures in the package. Pressureless sintering of Ag paste consistently results in Ag joints of higher porosity than pressure sintering. While the pores provide the compliances and reduce the CTE of the DA layer, pores, especially of irregular shape, also provide the path for crack growth through the interfaces. Excessive pores, being thermally and

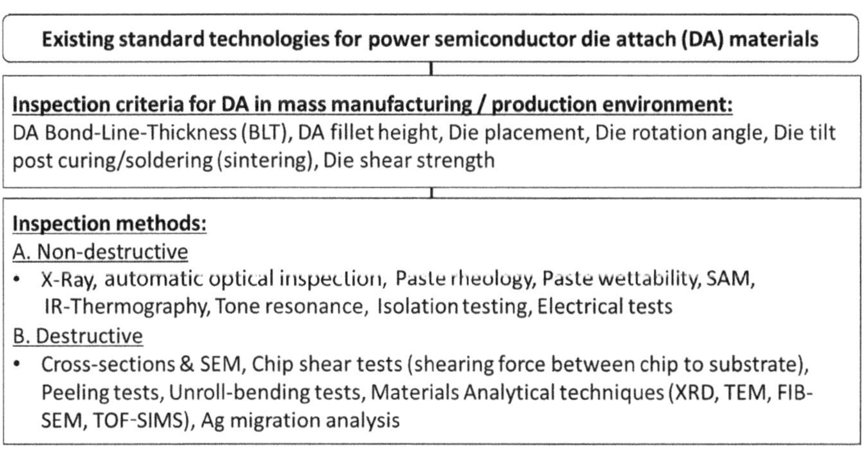

Fig. 12 Inspection tools used in DA process for Ag sintering

electrically insulative, also resulted in overheating failure in the dies or reduction in electrical performances. Therefore, effective quantification of the porosity is the first and necessary step to produce reliable sintered Ag joints.

Although the issues related to porosity have been discussed in earlier chapter of this book, it should be emphasized here within the context of process control of sintered Ag joint in production. In general, similar applied pressure during sintering may result in differently measured porosity within the sintered Ag joints because of the different quantification methodologies, location measured (center vs. side), optimized reflowed profile, and different Ag paste formulations among published results. Most researchers used image processing software (e.g., Image J) to filter unwanted noises and "binarize" the image to the desired threshold in the gray scale to obtain the percentage of porosity (e.g. in Fig. 23). This methodology can be subjective but seldom discussed between different researchers and publications. Others etched the sintered Ag joint with etchant (e.g., 25% NH_4/distilled water/ 100% H_2O_2 = 11:10:16) before the imaging step to reveal the microstructure and remove the smearing of the Ag occurring during mechanical polishing [30]. This etching step introduces another uncertainty to the percentage of porosity. Others used focused ion beam polishing to reduce this uncertainty but incurred a higher cost than mechanical polishing and limited measurement area.

In terms of location, pores tend to build up at the die edges because of the outgassing of organic volatiles from binders during the sintering step [31]. An optimized sintering profile can reduce this nonuniform density and irregular pore shape within the sintered Ag joints. Before sintering, the die placement step also squeezes the "liquid" from the Ag paste in a phenomena known as squeeze film flow, resulting in lower porosity in die center during sintering step [32]. Fortuitously, the low porosity at the die center also provides the route for heat dissipation generated from the die center.

In terms of dimension, the as-printed Ag paste typically shrinks by half of its original thickness because of the solvent removal during the preheating stage of the sintering [33]. There are some disagreements about further shrinkage during thermal aging. Chua et al. reported another 25% reduction in shrinkage caused by pore consolidation [9], while others reported negligible changes during thermal aging because the sintering shrinkage is balanced by the interfacial tensile stress with either the substrate or semiconductor dies [31]. This difference can be attributed to the different initial densification that provides the thermodynamic drive to densification. In addition, this porosity has an inverse relationship with the BLTs of the sintered Ag joint because of the higher tensile stress associated with thinner BLT suppresses the shrinkage to induce porosities within the sintered Ag joints [31]. Therefore, the initial process conditions (i.e., sintering profile and BLT) need to be well defined to obtain a balanced BLT thickness and porosity percentage [31].

BLT also affected Ag migration behavior because Ag tends to migrate toward the substrate surface that creates pores on the die side of the BLT [31]. These pores result in fractures at the die metallization surface instead of the substrate side of the BLT.

In addition, BLT and the pores within the BLT need to be monitored before the die placement stage. The process selection and characteristics of Ag sinter materials,

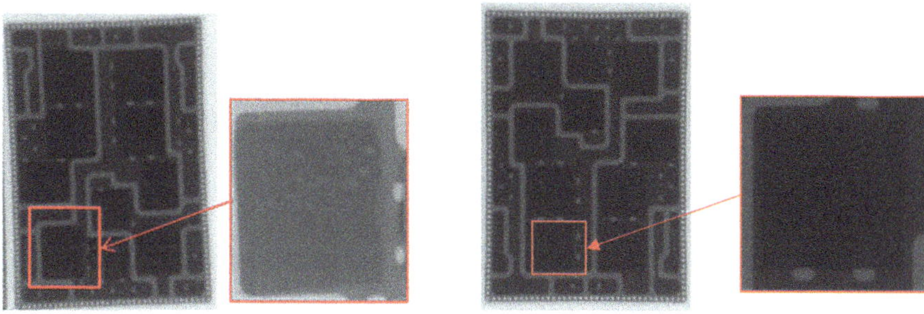

Fig. 13 Scanning acoustic microscopy images: (left) preheated Ag paste pressure sintered on bare Cu DBCs with visible outgassed pores being clearly visible; (right) Ag tape pressure sintered on bare Cu DBCs without any visible pores being observed

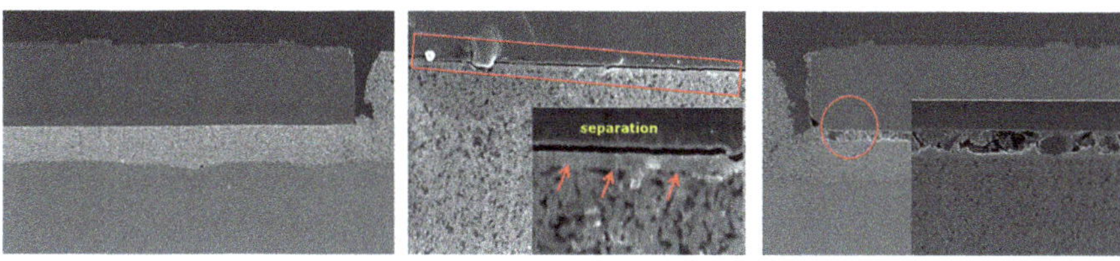

Fig. 14 SEM images of preheated Ag paste pressure sintered on bare Cu DBCs; incomplete DA fillets (left) and separation of DA from die are observed in this image (centre and right)

whether paste or tape, played a major role here [5, 7]. The use of pressure sintering to create desired and optimum BLT needs to be balanced with material utilization, material properties, and material selection, i.e., film or paste. Pressure sintering using Ag tapes provides better control of pore sizes and distribution and resistance to Ag migration beneath the die, compared to Ag pastes (Fig. 13).

Besides pore properties, fillet heights of the sintered Ag joint need to be controlled to avoid mechanical stresses on the die edges, even though the die thickness determines the mechanical stress more than the fillet height. The optimum DA fillet height is fixed at a certain percentage of the die thickness to reduce this stress [34, 35]. In the case of pressureless sintering, the fillet height would be determined by Ag paste formulation, dispensing/printing and preheating of the Ag paste to form naturally along the die edges. On the other hand, pressure sintering uses the applied pressure to control the squeeze out and density of the sintered Ag joints. Figures 14 and 15 show that improper use of this pressure and other process parameters can result in various defects, such as DA-die separation (delamination), die crack, and lifted DA fillet, for the pressure-sintered Ag paste and tape, respectively.

Here, a combination of SEM and SAM techniques can monitor the pore growth within the sintered Ag joint during pressure sintering. SAM can be implemented as an in-process control tool using a waterless immersion system for imaging the pores

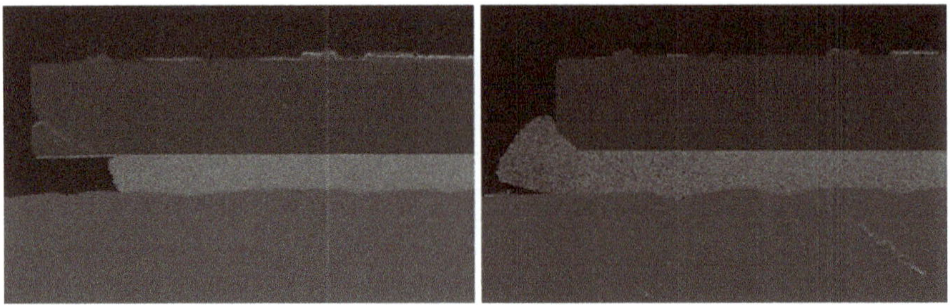

Fig. 15 SEM images of Ag tape sintered with pressure on bare Cu DBCs (left). Cracked die caused by placement errors and lifted DA fillet is visible after pressure sintering (right)

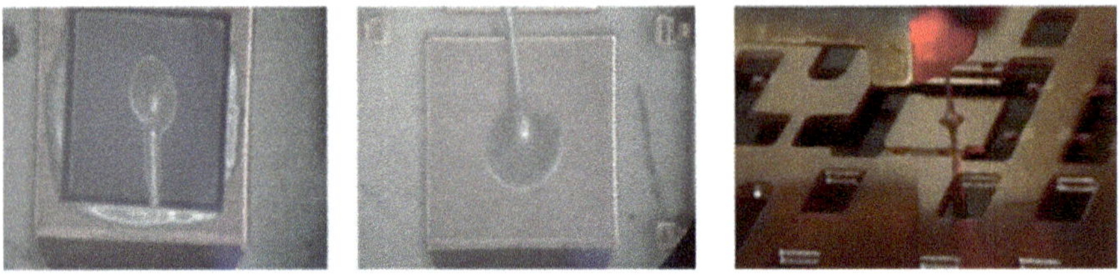

Fig. 16 Different views of Ag paste tailing caused by unsuitable viscosity and dispensing parameters

within the sintered Ag joint of the IGBT power modules [36]. Although this waterless immersion setup does not wet the die top, it may still entrap some moisture inside the pores of the sintered Ag joint, posing a reliability risk to the device. SEM complements this in-process control by providing the failure analysis of the failed sites.

5.2 *Die Equipment Control for Ag Sintering: Die Placement, Rotation, and Tilt*

Unlike solder, sintered Ag joint does not undergo a solidus to liquidus transition, i.e., no molten material forms during heating [37]. Ag sinter materials shrink during sintering when the organic matter and volatiles burn off. This gradual and slow loss of weight does not cause a shift or movement of the bonded die after placement. This lack of movement allows good control of die placement, rotation, and tilt that affect the DA material squeeze out, coverage beneath the die, and DA creep to the die top. Nevertheless, the rheology of the Ag sinter paste material needs to be adjusted to the selected processes, i.e., stencil printing or dispensing. The coverage of the Ag sinter paste also depends on the dispensing or printing parameters, i.e., speed, pattern (dispensing), etc. (Fig. 16). On the other hand, the die placement of sintered Ag tapes would have more stable process control than the printing or dispensing routes.

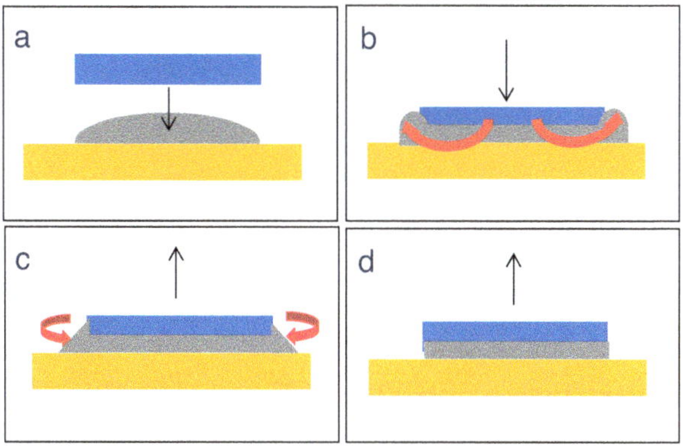

Fig. 17 Typical die-attach material responses during the die-bonding steps for adhesive epoxies, solders, and Ag sinter paste [38]

Since there is no solid-liquid-solid transition for Ag sinter materials, the die coverage needs to be adjusted before sintering step. As shown in Fig. 17, the bonding pressure creates a "cushion effect" which forms the fillets [38]. As the Ag paste does not transform to liquid during joint formation, there is no flow-back into the crevice between the die and substrate unlike solders, to form the full die coverage. This fillet height will remain constant and only reduces in height as the Ag paste suffers weight loss and shrinks during sintering. Therefore, Ag sinter paste coverage beneath the die needs to be finalized in the die bond step.

5.3 *Ensuring Die Bond Strength*

Die bond strength serves as a useful yet simple proxy for the thermal and electrical conductivity between the semiconductor die and the substrate; high die shear strength suggests high thermal conductivity because of the strong bonding at the faying surfaces and within the sintered Ag joint results in continuity of the joint at substrate-sintered Ag die. Sintered Ag material needs oxygen (O_2) to burn off the organic/binder content and allow Ag particles to coalescence and densify during the sintering process [37, 39]. Figure 18 shows the positive effects of increasing the O_2 percentage in the N_2-O_2 gasses on the bond strength in spite of the oxidation on the Cu leadframe substrate [40]. However, others showed that oxidizing the Cu substrates creates poor bonding which can only be improved by using reducing atmosphere H_2-N_2 gasses [41]. In addition, oxygen has another deleterious effects of causing the Ag migration failures at temperature above 250 °C [42].

Some researcher suggested that the increase in die bond shear strength for the pressureless sintered Ag joint was not caused by the densification effects but the continuous growth in the Ag phase, i.e., a dense Ag alloy zone just above the DBC surface, and macroscopic homogeneous sintered Ag joint that increased the bond strength [31]. The creation of this continuous Ag layer can be obtained through

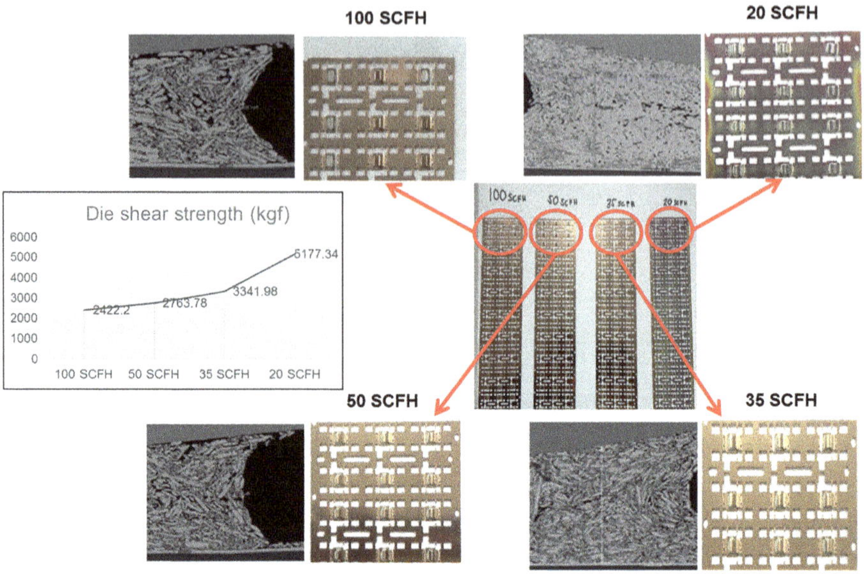

Fig. 18 Shear strength of sintered Ag joint on bare copper surfaces in an inert gas-filled oven [40]

proper process parameter selection, which involves the right sintering environment, temperature, time, and applied pressure during sintering. These parameters would be applicable to sintered Ag paste and tape.

5.4 *Electrical and Reliability Tests*

Power devices, in particular high-voltage devices such as power modules, require good isolation to create a safe and reliable environment for the device. Traditional electrical isolation techniques focus on separating the DA areas by using nonconducting epoxies, polyimides, and the silicon backside laminates which are attached on the die as the point of isolation. Power modules use silicone dielectric gels that are not electrically conductive to create this isolation environment. The creation of this isolation environment helps avoid dendrite growth and electrochemical migration, both of which lead to early life failures [43]. Therefore, standardized tests are needed to ensure these electrical isolation measures are reliable. Details of these tests methods (e.g., thermal shock, thermal cycle, vibration, high-temperature storage, etc.) are provided inside various sections of JEITA EIAJ ED4701 and their corresponding IEC68 standards [44].

6 Failure Analysis Techniques for Sintered Ag Joints

The following analytical tools are used to evaluate the different sintered Ag joints as part of process development in production as well as package reliability studies: differential scanning calorimetry-thermal gravimetric analyzer (DSC-TGA), thermomechanical analyzer (TMA), scanning electron microscopy-energy dispersive spectroscopy (SEM-EDS), focused ion beam (FIB), transmission electron microscopy (TEM), time-of-flight secondary ion mass spectroscopy (TOF-SIMS), thermography, X-ray radiography, and C-scanning acoustic microscopy (C-SAM). Besides analytical tools, finite element analysis and thermal impedance analysis also provide additional understanding to the formation and reliability of sintered Ag joint.

6.1 DSC-TGA

As mentioned in the earlier chapter, Ag paste consists of binders and passivation/capping agent that prevent the Ag fillers from agglomerating and aggregating. In the absence of the capping agent, the Ag fillers, notably Ag nanoparticles, will self-sinter. The decomposition of these organics initiates the Ag sintering process and, hence, the importance of determining this temperature accurately. DSC and TGA showed the temperatures of organics decomposition, albeit on different operating principles.

In the case of TGA, the decomposition or removal of the organics is accompanied by a distinct weight loss at the temperature ranging from 150 °C to 300 °C, depending on the type of organics used [45]. Similar to other DSC thermogram, the onset temperature sets the minimum peak sintering temperature in a sintering profile. Figure 19 further illustrates the difference in onset and offset temperature

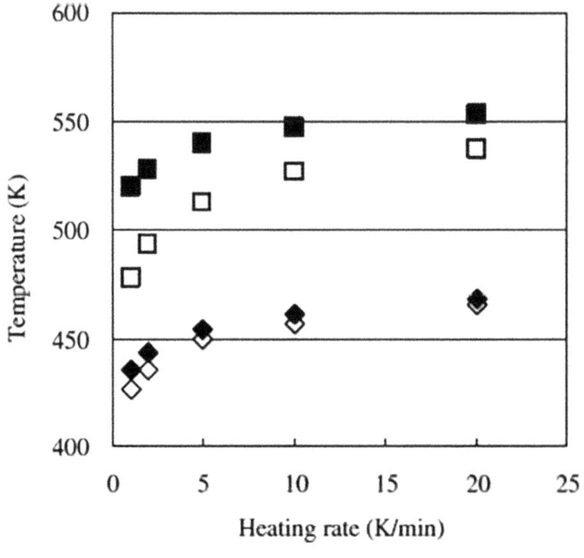

Fig. 19 The onset and offset temperature of mass loss in TGA thermograms when heating the nano-Ag paste under air (white) and nitrogen (black). Note: diamond is onset temperature, square is offset temperature [47]

associated with the mass reduction of sintered Ag paste when heated in an air or nitrogen environment [47]. Air atmosphere is more oxidative in decomposing these organics from the sintered nano-Ag paste than nitrogen atmosphere [39, 47]. This difference also translated to higher mass loss because of oxidative decomposition instead of pure desorption under inert atmosphere [48]. Hence, it is common to sinter the Ag paste in nitrogen atmosphere at higher sintering temperature than air atmosphere to compensate for the lower mass loss in nitrogen atmosphere.

However, it is fallacious to note that removal of organics is associated to the only exothermic event, as it can also involve endothermic peak when binders are removed from Ag paste, as shown in Fig. 20. This ambiguity is again attributed to the different reaction experienced by these binders under different sintering environment. In an oxidizing atmosphere, the combustion of organics resulted in exothermic peak [39] compared to an endothermic peak associated with organic binders desorbed from the Ag paste when heated in an inert or reduced atmosphere [39, 49]. Similar exothermic peak was observed for Ag paste added with Ag_2O and triethylene glycol (TEG), at around 150 °C in the DSC peak, to signify the reduction of Ag_2O to Ag nanoparticles which bind the surrounding Ag fillers to form the Ag joint. The complementary information from TGA confirmed the contribution of TEG as the reducing agent in this sintering step.

Although removing the organic coating surrounding the Ag particle initiates the sintering step, the Ag particle size also plays a role in determining the minimum sintering temperature. Figure 21 compares the DSC profiles of as-received Ag nanoparticles before annealing and separately annealed at 100 °C, 150 °C, 200 °C, and 250 °C [50]. The sharp exothermic peak of the as-received Ag nanoparticles marked the initiation of Ag nanoparticle sintering. When the Ag nanoparticles were annealed, the exothermic peak and heat of reaction associated with this sintering step

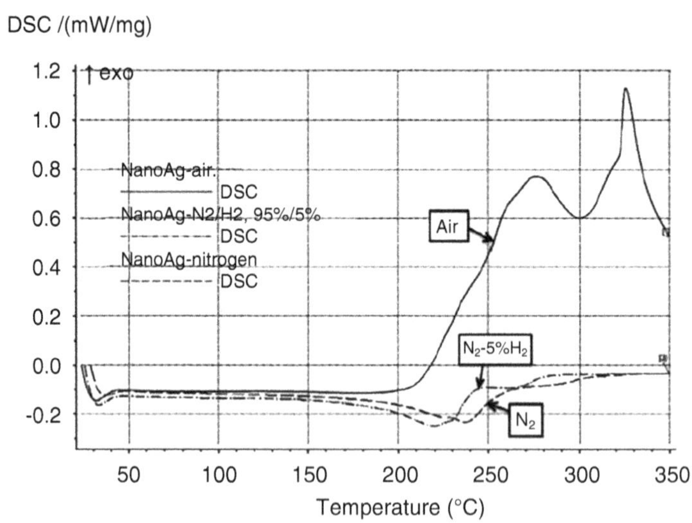

Fig. 20 DSC thermograms of nano-Ag pastes when sintered under air, N_2 and N_2-5%H2 atmospheres [39]

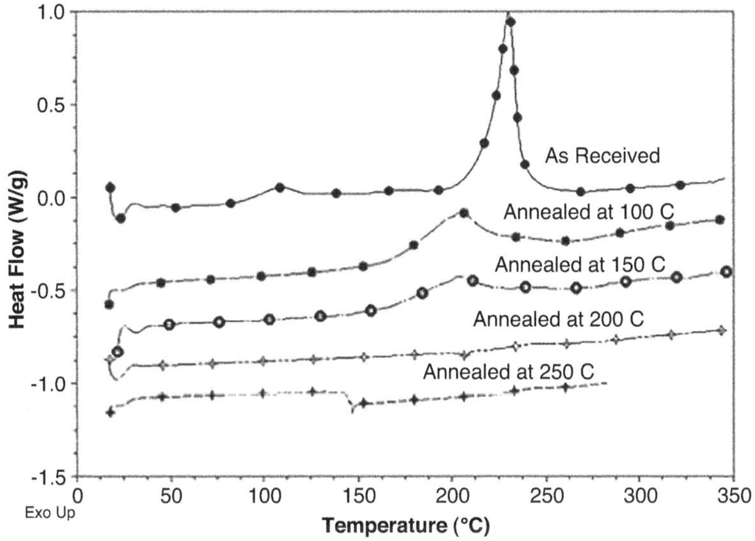

Fig. 21 DSC thermograms of Ag nanoparticles before and after annealed at different temperatures. The DSC heating was conducted in N_2 atmosphere to minimize exothermic contribution from combustion of impurities [50]

reduced and shifted to a lower temperature. Microstructural examination showed that the annealing increased the initial size of the Ag nanoparticles that reduced and shifted the exothermic peaks. The exothermic peak completely diminished as the annealing temperature of the Ag nanoparticles was increased to 200 °C to signify the absence of sintering step during the DSC analysis .

6.2 Thermomechanical Analyzer (TMA)

Silver paste with a high percentage of Ag fillers resulted in less dimensional change than those with the lower percentage of Ag fillers after the sintering process [14]. Hence, there is a possibility that the sintered Ag may have a different CTE from the bulk Ag, but experimental results reported similar CTE of 19 ppm/°C, as measured by TMA based on the changes in sintered Ag thickness with increase of temperature [51]. Others used dilatometer to confirm the lack of changes in CTEs (i.e. 19–21 ppm/oC) despite having relative density varying from 62% to 97% [53].

In spite of these similar CTEs, TMA analysis shows dimensional changes during the sintering of Ag particle. Figure 22 shows the TMA results of the Ag nanoparticle pellets sintered at different temperatures [50]. The initial increase in dimensions is attributed to thermal expansion of the Ag compacts. These dimensions began to decrease at 60 °C, 150 °C, and 250 °C for the initial Ag nanoparticle compacts and

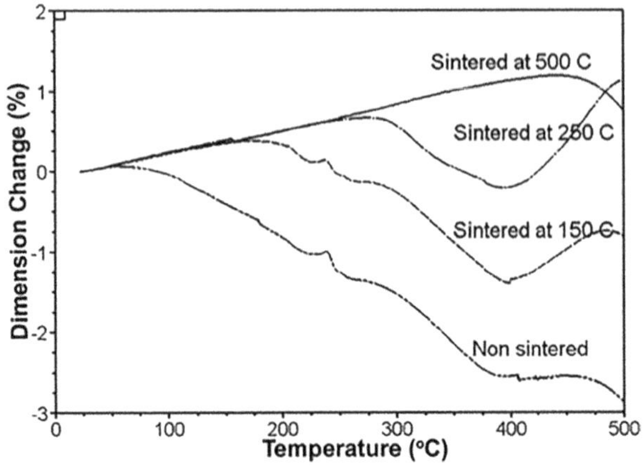

Fig. 22 TMAs thermograms of various Ag nanoparticle pellets of non-sintered and sintered at peak temperatures of 150 °C, 250 °C and 500 °C [50]

those compact sintered at 150 °C and 250 °C, respectively. The shrinkage was attributed to the sintering of nanoparticles after exceeding the sintering temperatures until the peak temperature of 400 °C. Similar to the exothermic reduction in Fig. 21, the higher sintering temperature experienced by the initial Ag nanoparticles increased the effective size of nanoparticles, resulting in lower surface energy for solid-state diffusion of atoms in the sintering step and lower dimensional change or shrinkage. In the case of Ag nanoparticles sintered at 500 °C, the dimension increase was attributed purely to thermal expansion effects with no shrinkage in the compact, suggesting complete sintering and shrinkage before the TMA test.

6.3 Scanning Electron Microscopy-Energy Dispersive Spectroscopy (SEM-EDS)

The main difference between soldered and sintered Ag DA in terms of die coverage is their wetting behavior at the die joint area. In the absence of contamination, the molten Sn-based solder wets the Cu or Ag substrate completely with the intermetallic formation and then self-aligns the dies in the joint area. On the other hand, the atomic Ag particles from the Ag paste merely wet these substrates at an atomic level via solid-state diffusion to form the bond, without going through solidification experienced in soldering. Hence, the interfacial and microstructural properties needed to be clearly examined to determine the Ag joint quality and reliability. For example, the sintered Ag layer showed separation from the Au back-metallized die after 24 h of thermal aging in 300 °C, while similarly sintered Ag layer showed no separation from the Ag back-metallized die [52]. The better adhesion for the Ag metallized die is attributed to the similar atomic structure and spacing between metallization and Ag paste. Such observation provides valuable information to the selection of metallization and reliability assessment for the sintered Ag applications.

In addition to the grain size and other microstructural properties, SEM also provides additional information on the porosity of the sintered Ag layer. This porosity reflects the joint density and influences the mechanical, thermal, and electrical properties of the sintered Ag joint [53]. Fortunately, this porosity is an adjustable property which depends on the pressure [54], temperature [55], time [55], and sintering atmosphere [41] during processing.

Although most SEM provides basic measuring feature, quantification of porosity often involved several other image processing steps with software such as Image J and a numerical-based computing tool, such as MATLAB, to accomplish the porosity quantification. As mentioned in Sect. 5.1, these image tools allow image filtering to remove the unwanted noises, followed by "binarizing" the image via setting the desired threshold in the gray scale, as shown in Fig. 23. Then, the software quantifies the porosity based on the area covered by either white or black within the region of interest.

The quantitative analysis of porosity can be characterized by total porosity area, pore size distribution, and pore shape factor. These porosity properties were influenced by the sintering parameters [55] and subsequent thermal aging during reliability test [9]. Figures 24 and 25 show that high sintering temperature, slow heating rates, and long holding time promote small and spherical pore sizes which are the preferred morphology to enhance the mechanical and compliance properties of sintered Ag joints [55]. Such information is valuable for the development engineer to optimize the microstructure properties of the sintered Ag.

Unlike solder joint, sintered Ag does not form any intermetallic formation at the joint area, whether the metallization is Au or Ag, for the as-sintered condition or upon further heat treatment. This absence of intermetallic layer is attributed to the solid solution between these metallic elements as shown in their respective phase diagrams. Figure 26 shows that further aging produces high frequency of small and spherical pore sizes that provide the desirable mechanical properties to the Ag joint. Hence, characterization of pore morphology for sintered Ag joints provides

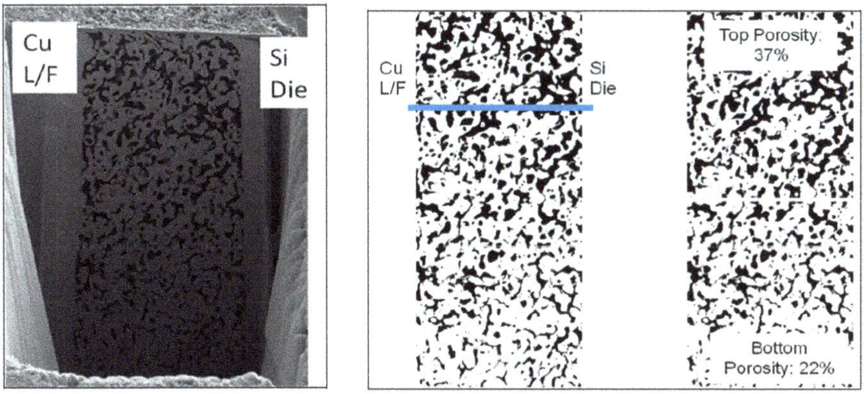

Fig. 23 SEM image of sintered Ag joint used for porosity quantification. Before image processing (left). After image processing ("binarizing") (right) [56]

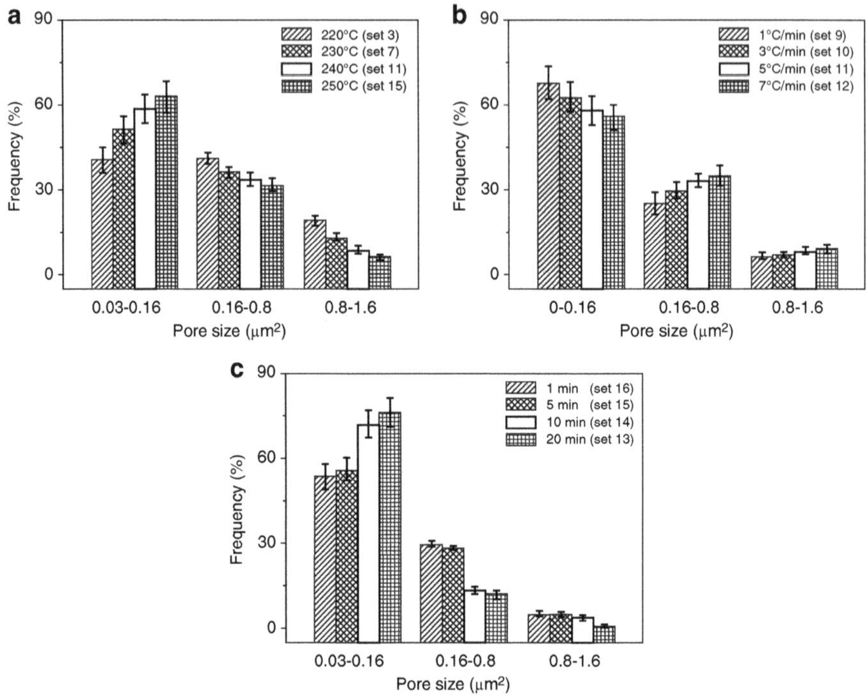

Fig. 24 Pore size distribution of sintered Ag nanoparticles with different (**a**) sintering temperature, (**b**) heating rates, and (**c**) holding times [55]

cumbersome but yet necessary and crucial information during this development process.

Fractography often provides insights on the causes of failure because it exposes the fracture mechanism and the weakest path of the materials or joint. In the case of die shear test, the die shear strength measures the highest strength of the weakest link within the DA layer. The variation of bonding strengths is closely related to the fracture morphologies. Figure 27 shows fracture surfaces of the sintered layer using pure Ag nanoparticles, pure Cu nanoparticles, and various mixtures of Ag and Cu nanoparticles [57]. Sintered Ag-Cu joint with more than 50 wt% of Ag fractured with elongated dimple structure to indicate ductile fracture that corresponds to high shear strength, as shown in Fig. 28. Conversely, the sintered Ag-Cu joint, with more than 50 wt% of Cu nanoparticles, fractured without much deformation and yielding. This intergranular-like fracture behavior indicated a brittle fracture that corresponds to low shear strength.

Similar to the function of flux in solder, Ag paste includes binder and solvent to remove oxides from Ag-metallized substrate or bare Cu substrate to provide effective sintering between the Ag paste and substrate. Together with the imaging from SEM, EDS is often used to measure the effectiveness of this oxide removal. For

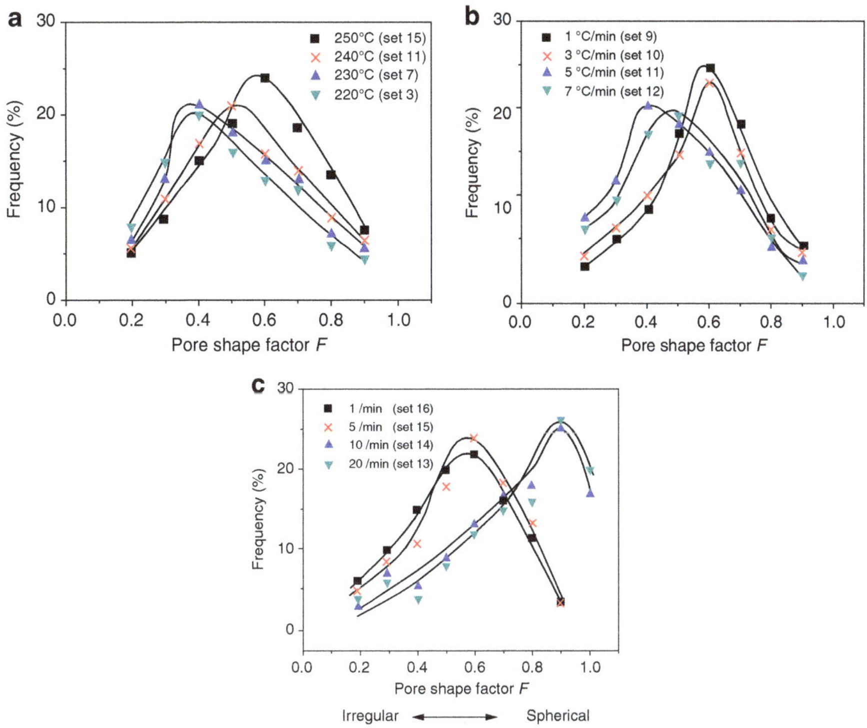

Fig. 25 Pore shape factor of sintered Ag nanoparticles with different (**a**) sintering temperature, (**b**) heating rates, and (**c**) holding times [55]

example, the point EDS result showed Cu and O elements on the fracture surface of the sintered Ag joint to determine the feasibility of sintering the Ag paste in the different atmospheric environment (Fig. 29). Copper oxide also indicates lack of metallic bonding at the Ag-substrate interface but instead a brittle fracture at the joint area that does not provide significant yielding but instead low die shear strength.

Although point EDS is useful to analyze the composition of a localized area, line EDS and elemental mapping are needed to study the diffusion activity at the interfacial area. For example, one research group analyzed the interfacial reaction between sintered Ag nanopores and ENIG using line EDS in a scanning TEM mode [58]. The dotted line from line EDS, shown in Fig. 30, confirmed that Ni and Au diffuse into the sintered Ag. However, the Ni layer successfully prevented the reverse diffusion of Ag and Au into Ni barrier metal. Hence, Ni(P) layer served as a good barrier layer to prevent Ag diffusion into the underlying substrate.

The EDS elemental mapping also provides diffusion of pores, elemental Ag and Au on the sintered Ag joint area. In one seminal study, Paknejad et al. demonstrated that Au diffuses rapidly into the grain boundaries of the sintered Ag layer, as shown in Fig. 31 [32]. The SEM image showed that sintered Ag occupies the Au

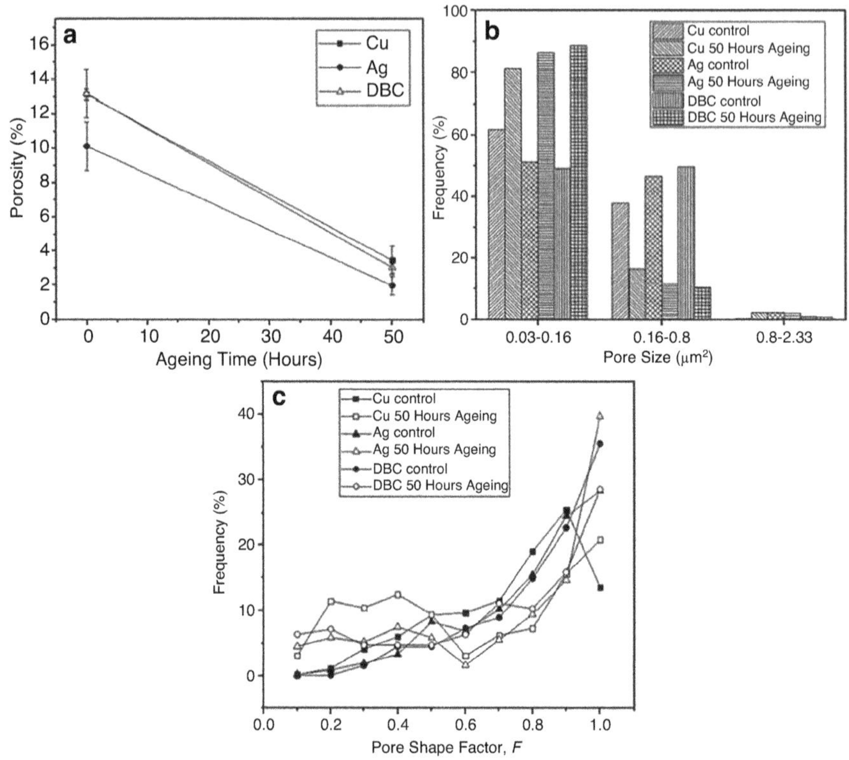

Fig. 26 Distribution of (**a**) porosity (%), (**b**) pore size (mm^2), and (**c**) pore shape factor, F in sintered nano-Ag joints on Cu, Ag-plated and DBC substrates during thermal aging at 300 °C for 50 h [9]

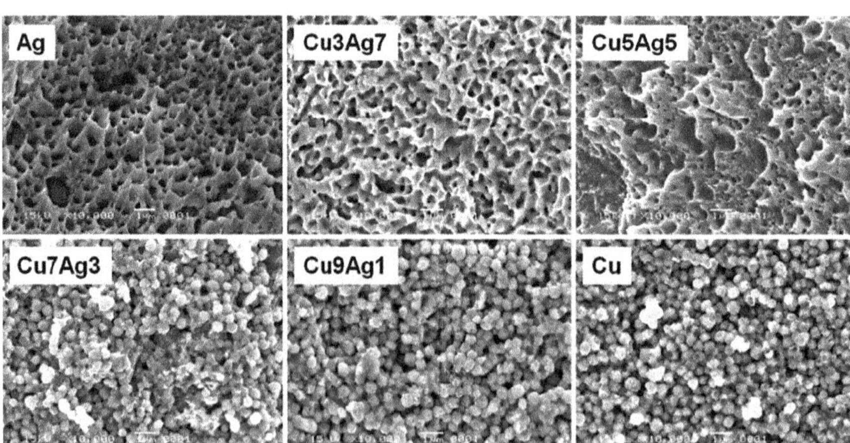

Fig. 27 SEM images of fracture surface originated from joints sintered with various mixtures of Ag and Cu nanoparticles. (Note: Cu3Ag7 means 30 wt% Cu and 70 wt% Ag in the metallic paste) [57]

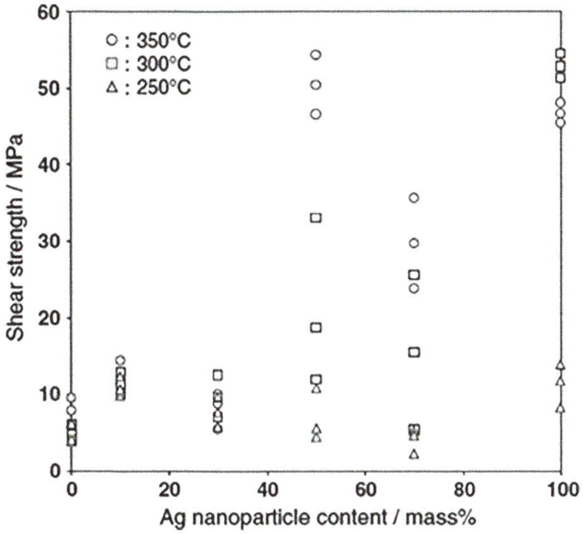

Fig. 28 The shear strength of sintered joints fabricated using various mixtures of Ag and Cu nanoparticles. The shear strength results correspond to the fracture surfaces in Fig. 27 [57]

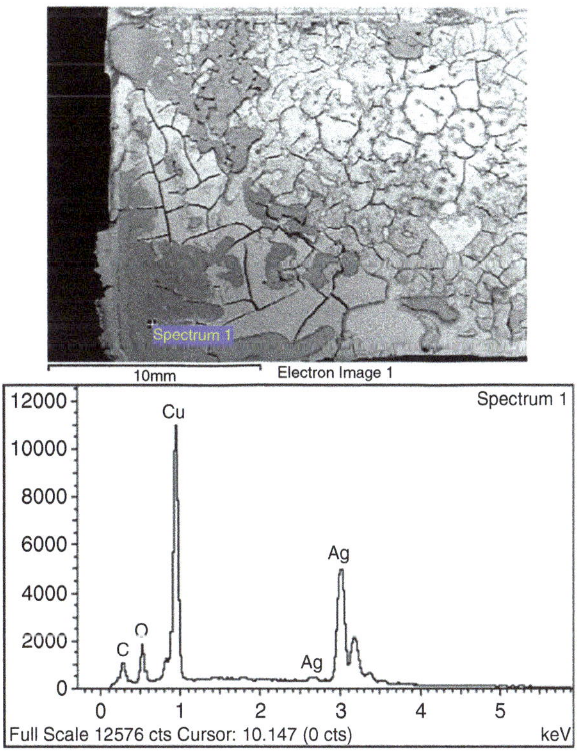

Fig. 29 The SEM image with location of point EDS on the fracture surface of sintered Ag nanoparticles in air. The corresponding EDS spectra of the point EDS located on the SEM image [39]

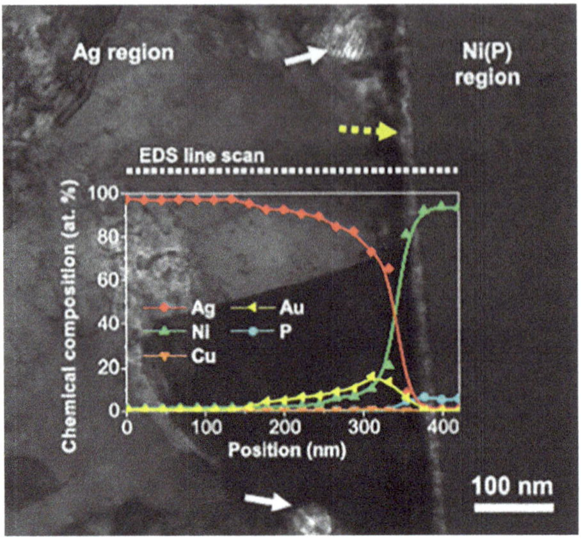

Fig. 30 The EDS line analysis across the interface of sintered Ag and ENIG layer [58]

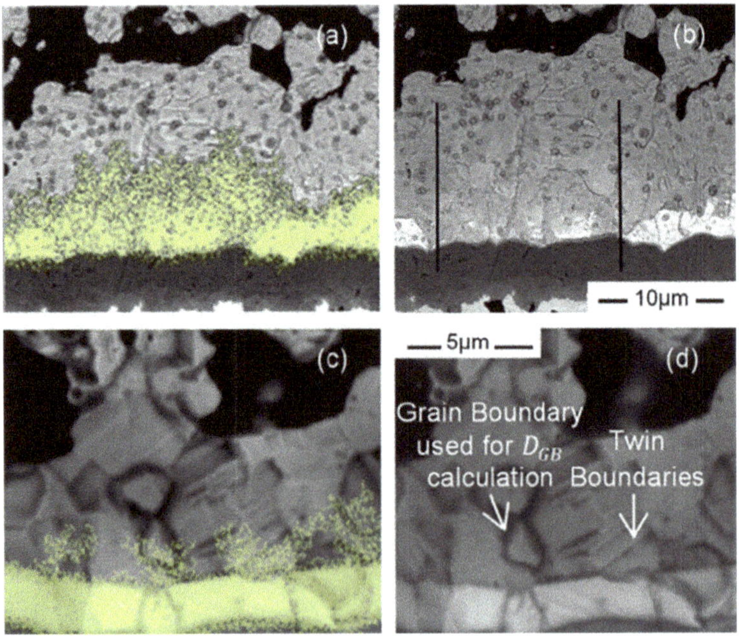

Fig. 31 EDS elemental mapping of Au. (**a**) Au distribution mapping superimposed on SEM image after 100 h of aging. (**b**) Figure 31a without Au mapping, portion of Au metallization seems to have dissolved. (**c**) Au elemental mapping showing Au concentrated at the grain boundaries of sintered Ag after 24 h of aging. (**d**) Figure 31c without Au mapping, lattice defects such as grain boundaries and twin boundaries are visible [32]

Fig. 32 An example of SEM image of a smeared sintered Ag sample during cross section analysis

metallization layer while the EDS elemental mapping suggested that the high porosity layer was attributed to the fast Ag migration into Au metallization.

Due to its softness, the sintered Ag could easily smear during mechanical grinding and polishing, as shown in Fig. 32. The smearing could hide the micro-structures, including the pores on the sintered Ag layer. As mentioned earlier in Sect. 5.1, some researchers suggested acid etching[2] to remove the smear after fine polishing, but such step is highly subjective and skill dependent. Others used FIB and ion milling machine to produce fine and smear-free microstructures, but at a much higher cost and lower throughput than mechanical polishing for the SEM imaging. However, mechanical polishing is the only viable approach to ascertain the bonding quality of the die-attached process because of the large bonding area, but final polishing by ion milling is often carried out to reveal the finer details of this sintered Ag microstructure.

6.4 *Transmission Electron Microscopy (TEM)*

The high resolution of TEM provides a detailed and useful microstructural analysis at the interfacial region, i.e., die back-metallization to sintered layer and sintered layer to substrate metallization. The TEM sample (i.e. an electron-transparent lamellae) is thinned with a highly precise milling and polishing plasma FIB machine [47, 59]. Subsequent TEM imaging is able to ascertain the bonding quality by inspecting any discontinuity at the interfacial area. Since delamination between sintered Ag and die can be as thin as few nanometers, this delamination is only visible under TEM [59]. An additional EDS line scan across the interface also confirms the purity of Ag content to detect any possibility of sulfur contamination within the joint area.

[2]Typical composition of this acid is 25% NH_4/distilled water/100% H_2O_2 = 11:10:16 for 2–3 s (Zhao et al. [30])

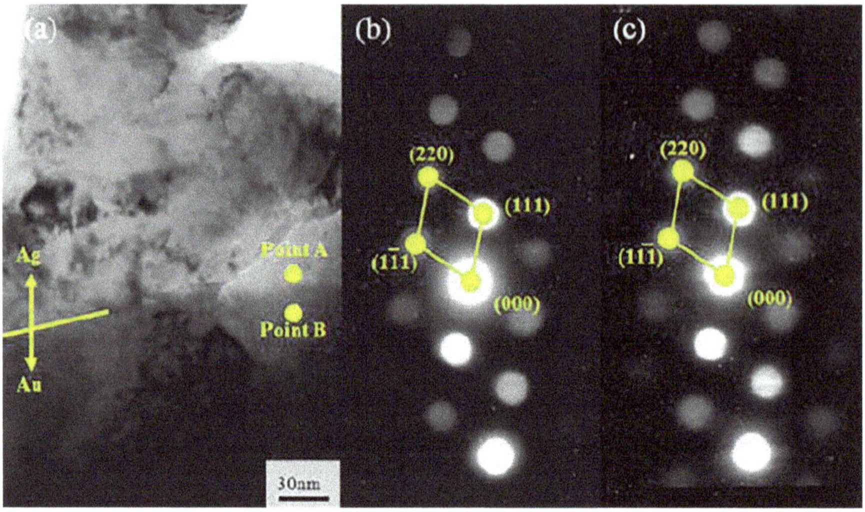

Fig. 33 (**a**) TEM image and the corresponding diffraction patterns of (**b**) point A (Ag) and (**c**) point B (Au). The sintered Ag bonding is produced by Ag_2O-TEG paste and heats up to 433 K with a pressure of 5 MPa [60]

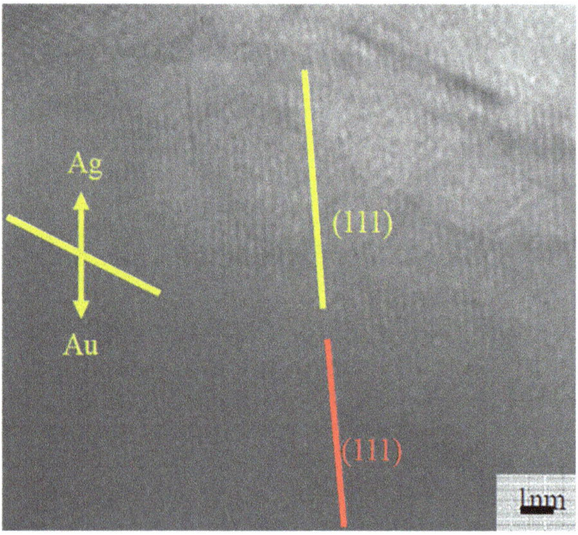

Fig. 34 HRTEM image showing the lattice structure of the sintered Ag-Au interface produced by the same paste and sintering parameter as per Fig. 33 [60]

Besides Ag metallization, the substrate has used metallization from, copper or aluminum that may produce a weak interface with the sintered Ag joint due to its dissimilar atomic spacing. Hence, lattice image with the HRTEM analysis is often used to elucidate the bonding mechanism at this interfacial region. In the case of Au-plated substrate, Figs. 33 and 34 show the identical diffraction patterns at the Au to Ag interface based on the corresponding HRTEM of the lattice

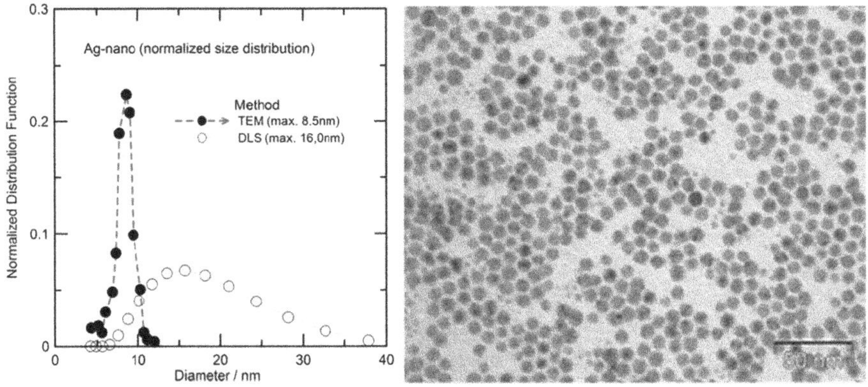

Fig. 35 Normalized size distribution of Ag nanoparticles (left). The average size of the Ag nanoparticles is measured at 8.5 nm. An example of TEM image of Ag nanoparticles used for size distribution quantification [62]

images [60]. Furthermore, the lattice image at the Au to Ag interface revealed a similar crystal orientation between Au and Ag. This similar orientation suggests an epitaxial growth of Ag on Au surface that is attributed to the identical atomic spacing of Au and Ag. This experimental result validated the molecular dynamic simulation of this Ag sintering on the Au substrate [60].

In addition, TEM can also provide the Ag particle size distribution in the Ag paste though such study is seldom carried out in the microelectronic packaging factory. Figure 35 shows Ag nanoparticle size distribution of a typical Ag paste from TEM image is narrower than those measured from dynamic light scattering (DLS) because of the different operating principles. TEM is based on physical calculation of a limited number of samples compared to DLS that depends on standard constants and assumptions, which may not bc truc. Silvcr pastc produccrs rcly on thcir particlc size distribution information to further optimize the Ag paste performance. If the Ag paste formulation is not optimized, Ag particles will aggregate and have much larger effective radius, thus losing low-temperature sinterability [61].

6.5 *Time-of-Flight Secondary Ion Mass Spectroscopy (TOF-SIMS)*

Compared to EDS that only provides elemental information, TOF-SIMS provides elemental fragments of molecules and compound-based composition in the DA joint. This capability enables the TOF-SIMS to determine the root cause of delamination at the interface of die back-metallization and sintered Ag layer [59]. In this said analysis, the initial TEM imaging of the die back-metallization revealed an amorphous layer of a few nanometer at the outer surface of Ag layer. Subsequent depth profiling analysis of TOF-SIMS confirmed silver sulfide (Ag_2S) present in this sintered Ag layer. A separate TOF-SIMS study in their process steps suggested that the die back-metallization is likely to have reacted with ambient sulfur to produce this silver sulfide; this study led to a change in process control to eliminate this risk of sulfur contamination in the sintered Ag joint.

TOF-SIMS technique also can deduce the carbon chain length of the fatty acids used in the development of sintered Ag paste [48]. A longer carbon chain translates to a higher decomposition temperature for the fatty acids. In the case of an organic mixture used in a Ag paste, the fatty acids with the longest carbon chain would determine the minimum sintering temperature necessary to decompose the compound to form a reliable sintered Ag joint. This result often complements the DSC and TGA results that can be highly dependent on the heating rate and sintering environment. Also, TOF-SIMS could be used to detect organic remnant on the sintered Ag joint since these organic residues surrounding the Ag particle could also inhibit the Ag sintering [63].

6.6 *Software Modeling and Simulation*

Although rarely used in a manufacturing environment, molecular dynamics (MD) and Monte Carlo-based simulation provided additional information on the molecule and microscopic interactions between particles during the sintering process [64]. In one study, Ogura et al. compared their MD model with the TEM results of the Ag_2O-derived silver nanoparticles bonding on the Au-plated substrate [60]. Figure 36 shows the simulated atomic arrangements of 4 nm-sized Ag particles sintered on the Au substrate with (011) orientation at 523 K. At time zero, Ag nanoparticles gathered closely on the Au substrate with nanoscale pores visible among the two nanoparticles and the Au substrate. Ag nanoparticles necked to reduce the pore at the beginning of the sintering step. The atom migration accelerated to minimize the large surface energy and grain boundary energy that further accelerated the pore shrinkage. This simulation confirmed that the bonding proceeds via epitaxial growth when the pores disappear within 10 ps, and the Ag atoms arranged and adhered on the Au substrate.

6.6.1 Finite Element Analysis (FEA)

The use of FEA and related results have been discussed as part of reliability and thermomechanical modeling studies of this book, but FEA is again included here within the context of design tools to implement sintered Ag joint in a manufacturing environment. The FEA methods have been used to solve numerous engineering problems involving stress-strain analysis, heat transfer, etc. Two different sets of parameters, namely, natural behaviors and disturbances, influence the behavior of any engineering system [65]. Natural behavior refers to the intrinsic physical properties of an engineering system such as elastic modulus, heat capacity, CTE,

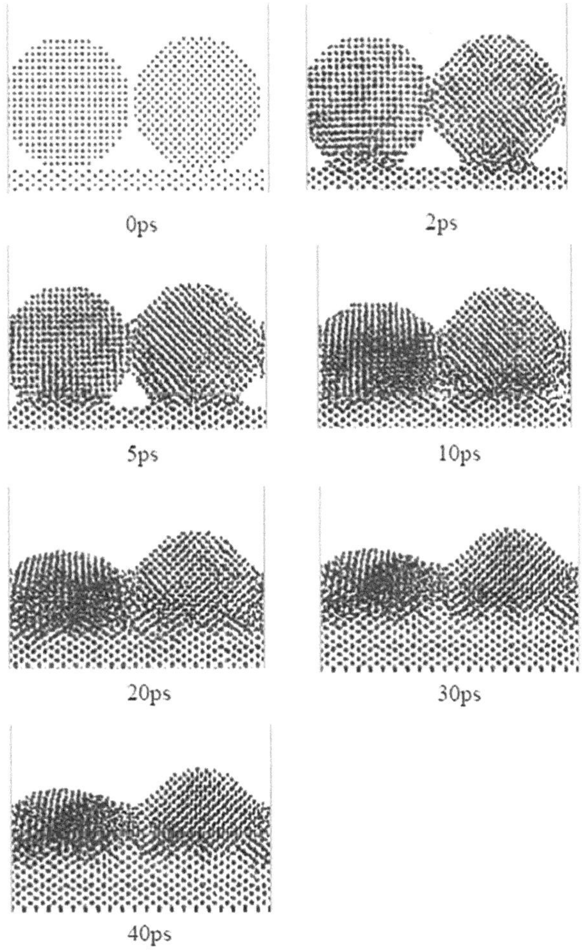

Fig. 36 Molecular dynamic simulation of two 4 nm Ag nanoparticles bonding on Au substrate with (011) orientation at 523 K [60]

etc., while disturbances refer to the external factors that exert on the engineering system such as differential forces, temperature, pressure, etc.

While the material properties of sintered Ag joints or bulk Ag properties are not exactly new, the values differ depending on the processing condition. Hence, caution should be exercised when selecting the appropriate properties for the sintered Ag. In the case of DA application, FEA can point the area of weakest interface based on the thermomechanical stresses and strains of the sintered Ag joints and dies. Further numerical values can also optimize the die-bonding layout of the package. Some researchers have used FEA to analyze the temperature and stress distribution of sintered Ag nanoparticle in both single- and double-sided cooling assembly using the material properties reported here; (elastic modulus = 6.28 GPa; Poisson ratio = 0.37; CTE (ppm/K) =19.6; thermal conductivity (W/mK) = 220 at the reference temperature of 298 K) [66] in [67]. Their results showed that the double-sided cooling assembly has lower peak temperature and narrower temperature distribution than single sided cooling resulting in greater thermal stresses on the double-sided assembly than the single-sided assembly [66]. Such information is useful in the design and layout of the package.

Besides package design, various studies also confirmed that sintered Ag, per se, has better thermal fatigue life than solder joint for DA application because of the lower elastic modulus (i.e. 9 GPa (69)) and resulting thermomechanical stress, as demonstrated in their FEA simulation [10, 68]. Separately, Bailey group presented a systematic flow of simulating thermal cycling stresses on the sintered Ag structure in a power electronic module after subjecting it to a tensile stress and temperature cycling profile shown in Fig. 37 [70].

In their study, they also used the accumulated plastic strain, caused by CTE mismatch, to model a Coffin-Manson fatigue life model, as per Eq. 2:

$$N_f = C(\Delta\varepsilon_{\mathrm{pl}})^{-k} \tag{2}$$

where $\Delta\varepsilon_{\mathrm{pl}}$ is the accumulated plastic strain during thermal cycle. The values of C and k for nanosilver-sintered joint were provided elsewhere [10]. Equation 3 calculates the accumulated plastic strain based on volume averaging technique for various thicknesses of sintered Ag layer:

$$\Delta\varepsilon_{\mathrm{pl}} = \frac{\sum_j \Delta\varepsilon_{\mathrm{pl}}^j V_j}{V_{\mathrm{total}}} \tag{3}$$

where V_{total} is the total volumes of all the elements, $\Delta\varepsilon^j{}_{\mathrm{pl}}$ is the accumulated plastic strain, while V_{j} is the volume of the jth element. The plastic strain distribution of the sintered Ag joints are plotted as per Fig. 38. The simulation result also suggested that the thicker sintered Ag layer reduced the plastic strain during the thermal cycling test.

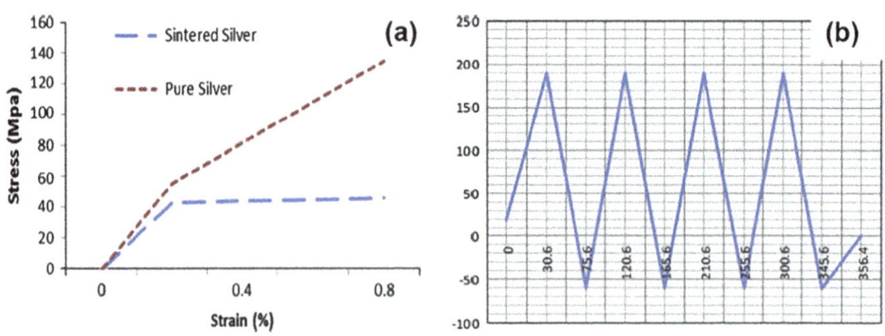

Fig. 37 (**a**) Tensile stress of both sintered Ag and bulk Ag. (**b**) Temperature profile of thermal cycling imposed on the finite element model [70]

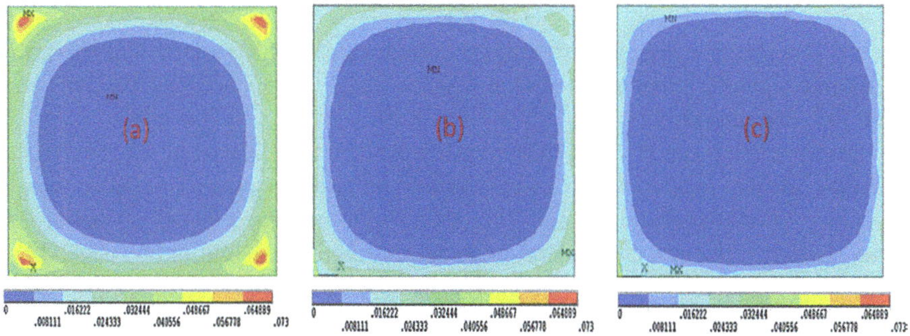

Fig. 38 Accumulated plastic strain on various sintered Ag thickness: (**a**) 20 μm, (**b**) 40 μm, and (**c**) 60 μm [70]

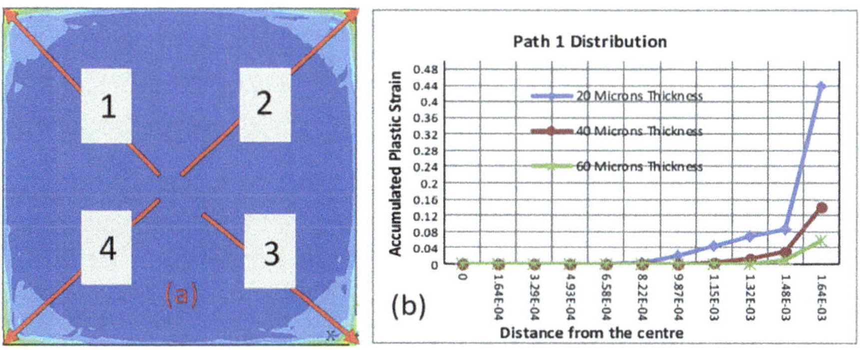

Fig. 39 (**a**) Accumulated plastic strain distribution of sintered Ag and (**b**) changes of accumulated plastic strain as a function of distance from die center to the corner [70]

Further analysis showed that the accumulated plastic strain concentrated at the corner and perimeter of the die (Fig. 39a) [70]. When a line was drawn from the center to the corner of the die, one can see an exponential growth of the plastic strain with respect to the distance from the center as plotted in Fig. 39b [70]. The plastic strain distribution from FEA coincided with experimental result that shows crack initiates at die edge sintered by Ag joints [10]. However, the benefits of thicker sintered Ag joint should be balanced with the adverse effect of an increase in thermal impedance and extreme shrinkage in the sintered Ag joint [49].

6.6.2 Thermal Impedance Analysis

Thermal impedance, Z_{th} of the DA layer, is often used to characterize different DA materials; lower thermal impedance enhances the overall device performance and reliability [71, 72]. Based on setup by Cao et al. [71], the nano-sintered Ag joint showed 12% less thermal impedance than other solder joints after 500 thermal cycles. The superior thermal performance of nano-sintered Ag joint could be attributed to its higher thermal conductivity of 1.56× and 1.73× more than SN100C and SAC305, respectively [71, 72].

In separate research, Sabbah et al. investigated the effect of type of Ag pastes (i.e., micron-scale and nanoscale Ag) and TIM (thermal interface materials) to the overall assembly thermal impedance [72]. The researchers concluded that the types of Ag paste did not influence the thermal impedance significantly, but an adhesive TIM performed better than grease and nonadhesive TIM because of the uncontrolled thickness of the grease and nonadhesive TIM deposit created discontinuity at the interfaces [72]. In addition, thermal impedance measurements have been found to be effective in detecting delamination during reliability test, as demonstrated by Rudzki et al. in their positive correlation between the delamination area and the changes of thermal impedance at the sintered Ag joint [73].

6.7 *Thermography*

In semiconductor packaging, thermography has been used to detect the heat distribution of the device in a working condition environment. The heat distribution study provides vital characteristics of power electronics and power modules that release abundant heat. In one study, thermography study confirmed that the lower maximum temperature experienced by the devices sintered by Ag corresponds to the high thermal conductivity of the sintered Ag [74]. At the package level, similar thermography study also showed that the high thermal conductivity of the sintered Ag joint leads to a lower thermal impedance for the whole assembly to improve the current carrying capability of the die [74].

Thermography has also proven to be effective in detecting delamination of sintered Ag joint. Based on the lock-in thermography technique, the localized heat of IGBT dies corresponded to the delamination area during biasing with short current pulses [75]. This technique also operates with heat sources that are periodically intensity-modulated at a specific lock-in frequency. Besides, thermograph is often used to monitor the heat energy emitted from the die during active temperature cycling [76]. When the thermal dissipation property was degraded by failure in the sintered Ag layer, the gate current lowered, and the time per cycle increased during the temperature cycling. This study also showed that the thermomechanical stresses shift to the die center due to the largest heat energy radiating from the center, compared to thermal cycling test where the heating was uniform across the die area [75].

6.8 X-Ray Radiography

X-ray inspection is a very popular tool for detecting assembly defects such as voids in the DA layer but suffered from poor sensitivity in detecting delamination [77]. When the DA material changes to Ag paste, the porosity of sintered Ag joint aggravates the image contrast of X-ray inspection. The thin BLT and porosity which leads to undefined absorption difference is the reason for low-contrast image [78]. In addition, the spinel structure of DBC in power module mimics and overlaps with the images produced by the sintered Ag porosity, and thus, confusing the analysis [78].

However, x-ray inspection still serves as a quick and a first inspection of the joint quality in leadframe packages with large voids because of its short cycle time and a contamination-free inspection.

6.9 C-Scanning Acoustic Microscopy (C-SAM)

Acoustic microscopy refers to image production using high-frequency ultrasonic inspection techniques. Ultrasonic energy can detect discontinuity due to material change; internal flaws such as cracks, voids, and delaminations; and inclusions that interfere with the ultrasonic transmission and/or reflection. Similar to X-ray inspection, acoustic microscopy could provide the internal examination of samples that are not accessible to light.

However, the pores of the sintered Ag joint may not be visible to acoustic inspection if they are smaller than the wavelength of sound. C-SAM is unable to resolve the porosity of the sintered Ag layer since porosity is micron- to nano-sized and typical wavelength of ultrasound is larger by a few orders of magnitude [79]. Based on Brand et al. study on C-SAM use in sintered Ag joint, it is not recommended to scan the device from the die side to avoid the shading artifacts from the pattern surface on the die and wire bond [79]. Brand et al. have also conducted a systematic study on optimizing various inspection parameters of the C-SAM to acquire best defect detection and image quality [79]. The focal length of the transducer is affecting the final acoustic image; a shorter focal length transducer yields a clearer and sharper acoustic image. However, a lower-frequency and higher focal length transducer is often employed when scanning from substrate side because the lower-frequency transducer translates to higher penetration depth through the substrate to reach the DA layer. This imaging characteristic is important during transducer selection since the substrate layer is usually thicker than the die.

In terms of niche application, C-SAM is more suitable to detect crack failure in the power devices/modules due to its high sensitivity in phase difference [59]. These die or DBC cracks from pressure sintering are usually only a few micron in width and barely visible under X-ray inspection. Apart from its role as an in-line instrument to monitor product quality, C-SAM also analyzes the quality of the sintered joints during and after reliability tests to differentiate the voids caused by non-optimized sintering parameters or thermomechanical stresses during reliability test. Common industry practice uses 20% un-bonded area as the failing criteria for die-attached layer after the reliability test.

Besides its primary function to detect flaws within the assembly layers, SAM machine is also used to estimate the thickness of DA layer [79]. Based on their parametric images of sintered Ag layer thickness, the color can be used to reflect the thickness variation by analyzing the difference in phase of die echoes. The thickness is computed by inserting the material sound velocity. However, such analysis must consider the selection of entrance and exiting echoes of sinter layers because the coupling of the thin sintered Ag joints and long ultrasonic pulse will result in overlapping of entrance and exiting echoes and thus confounding this analysis.

7 Conclusions and Outlook

The use of Ag sintering, be it with or without pressure, will be a mature technology for mass manufacturing in the years to come. The development of large-scale manufacturing equipment and commercialization of the sintered Ag material by several manufacturers lends a strong reason to support such a statement. The increasingly stringent requirements for power semiconductor package reliability, such as in automotive and aerospace applications, mean that sintered Ag has a niche market to fill. In essence, the process control methodology and tools for sintered Ag joints are very similar to other die-attach materials. However, due to its physical structure in how the joint is formed, additional care is needed to assess characteristics of the joint. This approach may bring about new methods for failure analysis on the package joint.

Acknowledgments We greatly acknowledge the feedback and support from Marco Koelink (Boschman Technologies BV), Giulio Locatelli (Locatelli Meccanica S.r.l.), Eric Kuah (ASM Technology) colleagues, managers, and friends involved in bringing this sintered Ag technology to the mainstream market. KSS also acknowledges Universiti Kebangsaan Malaysia Research Grants (GUP-2017-055 "Production of Metallic Conducting Nanowires for Industrial Applications") for this work.

References

1. H.S. Chin, K.Y. Cheong, A.B. Ismail, Metall. Mater. Trans. B Process Metall. Mater. Process. Sci. **41**, 824–832 (2010)
2. V.R. Manikam, C. Kuan Yew, IEEE Trans. Compon. Packag. Manuf. Technol. **1**, 457–478 (2011)
3. B.J. Baliga, IEEE Electron Device Lett. **10**, 455–457 (1989)
4. Infineon, Die Attach -5 Project (2018), https://www.infineon.com/dgdl/DA5_customer_presentation_1612016.pdf?fileId. Accessed 22 Feb 2018
5. K.S. Siow, Y.T. Lin, J. Electron. Packag. **138**, 020804-1–020804-13 (2016)

6. K.S. Siow, M. Eugénie, in *2016 I.E. 37th International Electronic Manufacturing Technology (IEMT) & 18th Electronics Materials and Packaging (EMAP) Conference*, 2016. pp. 1–6
7. K.S. Siow, J. Electron. Mater. **43**, 947–961 (2014)
8. F. Yu, J. Cui, Z. Zhou, K. Fang, R. Johnson, M. Hamilton, IEEE Trans. Power Electron. **32**, 7083–7095 (2017)
9. S.T. Chua, K.S. Siow, J. Alloys Compd. **687**, 486–498 (2016)
10. M. Knoerr, S. Kraft, A. Schletz, *12th IEEE Electronics Packaging Technology Conference*, Singapore, 2010. pp. 56–61
11. V.A. Ivensen, Densification of metal powders during sintering, Consultants Bureau, 1973
12. J.K. Mackenzie, R. Shuttleworth, Proc. Phys. Soc. Sect. B **62**, 833 (1949)
13. M.B.G. Koelink (Boschman Technologies), Private Communication, 2017
14. T. Wang, M. Zhao, X. Chen, G.Q. Lu, K. Ngo, S. Luo, J. Electron. Mater. **41**, 2543–2552 (2012)
15. W. Ng, K. Kumagai, K. Sweatman, T. Nishimura, *17th IEEE Electronics Packaging and Technology Conference*, Singapore, 2015, pp 1–6
16. H. Mutoh, N. Moriyama, K. Kaneko, Y. Miyazawa, H. Kida, in *IOP Conference Series: Materials Science and Engineering, Vol. 61*, Institute of Physics Publishing, 2014
17. E.N. Kablov, V.I. Lukin, N V S. Ryl, A.F. Cherkasov, E.K.A.N. Afanas, Method of drying coating of silver-coating paste, Federal Noe Gup Vrnii Aviat Materialov Fgup Viam, RU2564518, 2014
18. O. Khaselev, E. Boschman, Method for die and chip attachment, Alpha Assembly & Adv Packaging Ctr B.V., WO2016100470, 2016
19. Heraeus, Heraeus Sinter Materials (2016), http://www.heraeus.com/en/het/products_and_solutions_het/sinter_materials/sinter_materials___page.aspx. Accessed on 24 Oct 2016
20. R. Bayerer, O. Hohlfield, Method for producing a composite and a power semiconductor module, Infineon Tech, US20130203218A1, 2012
21. W. Knapp, Method for mounting electronic components on substrates, ABB Res. US6935556B2, 2005
22. L. Viswanathan, L.M. Mahalingam, D. Abdo, J. Molla, Packaged semiconductor devices and methods of their fabrication, Freescale Semiconductor, US9099567B1, 2015
23. H. Hauenstein, Sintering utilizing non-mechanical pressure, International Rectifier Corporation, US20140224409A1, 2014
24. L. Wang, *16th IEEE International Conference Electronic Packaging Technology*, Changsha, China, 2015, pp. 1317–1320
25. Boschman, Advanced molding and sintering systems: sinterstar innovate F XL (2016), http://www.boschman.nl/index.php/sintering-systemplatforms/sinterstar-innovate-f-xl.html. Accessed on 9 Mar 2016
26. C. Gobl, J. Faltenbacher, in *6th IEEE International Conference Integrated Power Electron System*, Nuremberg, 2010. p. 1
27. Y. Zhao, Y. Wu, K. Evans, J. Swingler, S. Jones, X. Dai, *15th IEEE International Conference on Electronic Packaging Technology*, ed. by K. Bi, Z. Tian, Z. Xu. 2014, pp. 200–204
28. Smart Equipment Technology, FC150 Automated Die/Flip Chip Bonder (2016), http://www.set-sas.fr/en/cat422408%2D%2DFC150.html?Cookie=set
29. Datacon 2200 Evo plus BESI (2017), https://www.besi.com/products-technology/product-details/product/datacon-2200-evo/. Accessed on 24 Feb 2018
30. S.Y. Zhao, X. Li, Y.H. Mei, G.Q. Lu, Microelectron. Reliab. **55**, 2524–2531 (2015)
31. S. Chen, G. Fan, X. Yan, C. LaBarbera, L. Kresge, N. C. Lee, *16th IEEE International Conference on Electronic Packaging Technology*, 2015, pp. 367–374
32. S.A. Paknejad, G. Dumas, G. West, G. Lewis, S.H. Mannan, J. Alloys Compd. **617**, 994–1001 (2014)
33. C. Buttay, A. Masson, J. Li, M. Johnson, M. Lazar, C. Raynaud, H. Morel, *IMAPS International Conference on High Temperature Electronics Network*, Oxford, 2011, pp. 84–90
34. R.Y. Agustin, J.M. Jucar, J.S. Talledo, *23rd ASEMEP National Technical Symposium* 2015, pp. 1–6

35. J. Weidler, R. Newman, C. J. Zhai, *52nd IEEE Electronic Components and Technology Conference 2002. (Cat. No.02CH37345)*, 2002, pp. 1172–1177
36. T. Adams, Inverted Acoustic System Cuts IGBT Failures (2011), http://www.powerelectronics.com/power-electronics-systems/invertedacoustic-system-cuts-igbt-failures, accessed on 24 Feb 2018
37. G.Q. Lu, J.N. Calata, Z. Zhang, J.G. Bai, *6th IEEE Conference High Density Microsystems Design Packaging Component Failure Analysis*, Shanghai, 2004, pp. 42–46
38. V. R. Manikam, S. Paing, A. Ang, *15th Electronics Packaging Technology Conference Singapore*, 2013, pp. 152–155
39. S.T. Chua, K. S. Siow, A.Jalar, 36th IEEE International Electronics Manufacturing Technology, Johor Bahru, 2014. pp. 1–6
40. V.R. Manikam, E.N. Tolentino, *16th IEEE Electronics Packaging Technology Conference*, Singapore, 2014, pp. 1–5
41. H. Zheng, D. Berry, K.D.T. Ngo, G.Q. Lu, IEEE Trans. Compon. Packag. Manuf. Technol. **4**, 377–384 (2014)
42. Y. Mei, G.Q. Lu, X. Chen, S. Luo, D. Ibitayo, IEEE Trans. Device Mater. Reliab. **11**, 312–315 (2011)
43. F. Cosiansi, E. Mattiuzzo, M. Turnaturi, P.F. Candido, *9th IEEE International Conference on Integrated Power Electronics Systems*, 2016, pp. 1–5
44. Japan Electronics and Information Technology Industries Association (JEITA) Technical Standardization Center, https://home.jeita.or.jp/tsc/downloadE.html. Accessed on 24th Feb 2018
45. H. Ogura, M. Maruyama, R. Matsubayashi, T. Ogawa, S. Nakamura, T. Komatsu, H. Nagasawa, A. Ichimura, S. Isoda, J. Electron. Mater. **39**, 1233–1240 (2010)
46. S. Takata, T. Ogura, E. Ide, T. Morita, A. Hirose, J. Electron. Mater. **42**, 507–515 (2013)
47. E. Ide, S. Angata, A. Hirose, K.F. Kobayashi, Acta Mater. **53**, 2385–2393 (2005)
48. C. Fruh, M. Gunther, M. Rittner, A. Fix, M. Nowottnick, *3rd IEEE Electronic Systems Integrated Technology Conference*, Berlin, 2010. pp. 1–5
49. W. Schmitt, *6th International Conference on Integrated Power Electronics Systems*, 2010, pp. 1–6
50. K.-S. Moon, H. Dong, R. Maric, S. Pothukuchi, A. Hunt, Y. Li, C.P. Wong, J. Electron. Mater. **34**, 168–175 (2005)
51. J.G. Bai, Z.Z. Zhang, J.N. Calata, G.Q. Lu, *8th IEEE High Density Microsystem Design Packaging and Component Failure Analysis* 2005, Shanghai, 2006
52. Y. Fang, R.W. Johnson, M.C. Hamilton, IEEE Trans. Compon. Packag. Manuf. Technol. **5**, 1258–1264 (2015)
53. A.A. Wereszczak, D.J. Vuono, H. Wang, M.K. Ferber, Z. Liang, Oak Ridge National Laboratory Technical Report ORNL/TM-2012/130, 2012, https://www.osti.gov/biblio/1041433, Accessed on 24 Feb 2018
54. S. Wolfgang, T. Krebs, *9th IEEE International Conference on Integrated Power Electronics Systems*, 2016, pp. 1–7
55. S. Fu, Y. Mei, X. Li, P. Ning, G.-Q. Lu, J. Electron. Mater. **44**, 3973–3984 (2015)
56. K. S. Siow, *35th IEEE International Electronics Manufacturing Technology*, Ipoh, 2012, pp. 1–6
57. Y. Morisada, T. Nagaoka, M. Fukusumi, Y. Kashiwagi, M. Yamamoto, M. Nakamoto, J. Electron. Mater. **39**, 1283–1288 (2010)
58. M.S. Kim, H. Nishikawa, Scr. Mater. 92, 43–46 (2014)
59. B. Boettge, B. Maerz, J. Schischka, S. Klengel, M. Petzold, *8th IEEE International Conference on Integrated Power Electronics Systems*, 2014, pp. 1–7
60. T. Ogura, M. Nishimura, H. Tatsumi, N. Takeda, W. Takahara, A. Hirose, Open Surf. Sci. J. **33**, 55–59 (2011)
61. Z. Zhang, Processing and Characterization of Micro-scale and Nanscale Silver Paste for Power Semiconductor Device Attachment, Virginia Polytechnic Institute and State University PhD thesis, 2005

62. Z. Pešina, V. Vykoukal, M. Palcut, J. Sopoušek, Electron. Mater. Lett. **10**, 293–298 (2014)
63. M.A. Asoro, D. Kovar, P.J. Ferreira, Chem. Commun. **50**, 4835–4838 (2014)
64. M.P. Allen, Introduction to molecular dynamics simulation, in *Computational Soft Matter: From Synthetic Polymers to Proteins*, ed. by N. Attig, K. Binder, H. Grubmüller, K. Kremer (Eds), vol. 23, (2004), pp. 1–28
65. S. Moaveni, Finite Element Analysis Theory and Application with ANSYS, Upper Saddle River, N.J (Pearson Education, 2008)
66. J. Lian, Y. Mei, X. Chen, X. Li, G. Chen, K. Zhou, *13th International Conference on Electronic Packaging Technology & High Density Packaging*, 2012. pp. 232–237
67. J.G. Bai, J.N. Calata, G.Q. Lu, *19th IEEE Annual Applied Power Electronics Conference and Exposition*, 2004, pp. 1240–1246
68. G. Chen, L. Yu, Y.H. Mei, X. Li, X. Chen, G.Q. Lu, J. *Mater. Process. Technol.* **214**, 1900–1908 (2014)
69. J.G. Bai, J.N. Calata, G.Q. Lu, IEEE Trans. Electron. Packag. Manuf. **30**, 241–245 (2007)
70. P. Rajaguru, H. Lu, C. Bailey, Microelectron. Reliab. **55**, 919–930 (2015)
71. X. Cao, T. Wang, K.D.T. Ngo, G.Q. Lu, IEEE Trans. Compon. Pack. Manuf. Tech. **1**, 495–501 (2011)
72. W. Sabbah, R. Riva, S. Hascoet, C. Buttay, S. Azzopardi, E. Woirgard, D. Planson, B. Allard, R. Meuret, *7th International Conference on Integrated Power Systems*, Nuremberg, Germany, 2012. p. 1
73. J. Rudzki, L. Jensen, M. Poech, L. Schmidt, F. Osterwald, *7th IEEE International Conference on Integrated Power Electronics Systems*, 2012, pp. 1–6
74. S. Duch, T. Krebs, W. Schmitt, *9th IEEE International Conference on Integrated Power Electronics Systems*, 2016. pp. 1–6
75. C. Schmidt, F. Altmann, O. Breitenstein, Mater. Sci. Eng. B **177**, 1261–1267 (2012)
76. M. Hutter, C. Weber, C. Ehrhardt, K. D. Lang, *9th IEEE International Conference on Integrated Power Electronics Systems*, 2016. pp. 1–7
77. C. Weber, M. Hutter, H. Oppermann, K. D. Lang, *PCIM International Exhibition and Conference for Power Electronics, Intelligent Motion, Renewable Energy and Energy Management*, 2014. pp. 1–8
78. U. Sagebaum, *ECPE European Centre for Power Electronics Double Workshop, Munich*, Germany, 2014
79. S. Brand, F. Naumann, S. Tismer, B. Boettge, J. Rudzki, F. Osterwald, M. Petzold, *9th IEEE International Conference on Integrated Power Electronics Systems*, 2016. pp. 1–6

Thermomechanical Modeling of High-Temperature Bonded Interface Materials

P. P. Paret, D. J. DeVoto, and S. V. J. Narumanchi

1 Introduction

Optimal performance and accurate lifetime estimations of power electronics components are vital for different renewable energy and energy-efficiency technologies such as wind turbines, concentrating solar power systems, solar photovoltaics, and electric-drive vehicles. While numerous advancements related to the design and development of semiconductor devices, their testing and manufacturing technology, and the surrounding packaging have been made in the past few decades, the device material has by and large remained as silicon (Si) in the power electronics industry. However, the industry trend toward miniaturizing power electronics packages with increased power density demands for better performance, and cost-effective solutions are gradually pushing Si to its limits. To overcome these new challenges facing the power electronics industry, wide bandgap (WBG) devices, such as silicon carbide (SiC) and gallium nitride (GaN), were proposed as alternatives, and significant progress has been made in the application of these devices [1]. The ability of WBG devices to operate at higher temperatures, higher voltages, and faster switching frequencies enables a significant reduction in the footprint of packages/ components incorporating these devices, resulting in compact, high-performance power electronics packages with increased power density. These compact electronics packages will find applications in multiple platforms, offering significant energy and cost savings that will drive the market penetration of advanced clean energy technologies in many sectors such as automotive, military, commercial, and residential buildings, industrial applications, and grid integration. Rapid adoption and implementation of high performance, compact, reliable, and energy-efficient electronic technologies such as WBG devices and packaging are becoming critical [2].

P. P. Paret (✉) · D. J. DeVoto · S. V. J. Narumanchi
National Renewable Energy Laboratory (NREL), Golden, CO, USA
e-mail: Paul.Paret@nrel.gov; Douglas.DeVoto@nrel.gov; Sreekant.Narumanchi@nrel.gov

K. S. Siow (ed.), *Die-Attach Materials for High Temperature Applications in Microelectronics Packaging*, https://doi.org/10.1007/978-3-319-99256-3_4

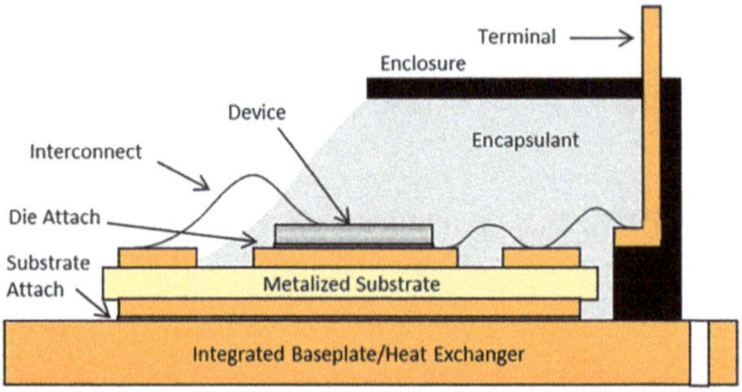

Fig. 1 A traditional power electronics package

As with any novel technology, numerous challenges need to be addressed before WBG devices can be implemented in electronics packages and become widely adopted. WBG devices are not a drop-in replacement for Si devices; a redesign of the entire package and other components becomes necessary. For example, a power electronics package/module in an automotive inverter, as shown in Fig. 1, will need to be redesigned with novel materials if it is to withstand the thermal and mechanical challenges introduced by the use of WBG devices in elevated temperature operating conditions. Materials that can reliably operate at higher temperatures (>200 °C) need to be used. Also, other power components positioned nearby, such as capacitors and bus bars, need to be manufactured with high-temperature-rated materials, or aggressive cooling mechanisms need to be utilized to keep the components under safe operating conditions. An important aspect that requires special attention is the set of bonded materials deployed at the die-attach and substrate-attach level, as shown in Fig. 1.

In a package, a significant amount of heat is generated by the semiconductor device during its operation that needs to be removed to keep temperatures within limits. However, due to the asperities on the component surfaces in contact to each other, the air gaps that are formed between the surfaces [3, 4] can cause a large resistance to heat transfer, which in turn results in large increases in temperatures in the package. Bonded materials ensure an efficient path for heat transfer at the interfaces by closing these air gaps. Hence, the semiconductor device is typically attached by a bonded material such as solder at the die-attach level to a metallized substrate. The substrate is composed of a ceramic bounded by copper layers on either side and provides electrical isolation. This substrate is then mounted onto a baseplate or directly to a heat exchanger, typically made of copper or aluminum, via the substrate attach.

Previously, lead solders had predominantly been used as bonded materials in the electronics packaging industry; however, the challenges introduced by the high-temperature operating conditions with the advent of WBG devices coupled with the Restriction of Hazardous Substances (RoHS) Directive in Europe banning lead-based formulations have pushed the electronics industry to shift its focus to lead-free, high-temperature-compatible bonded interface solutions. In addition to different formulations of gold (Au) solders, sintered silver and transient liquid-phase diffusion bonds are promising alternatives for high-temperature applications.

Irrespective of the bonded material used, a critical aspect that needs to be characterized is its reliability—the capacity of the material to operate successfully or without any significant degradation issues under regular or extreme loading conditions for its intended period of use. In other words, a non-reliable bonded material acts as a major bottleneck to the smooth flow of heat from the device to the baseplate/heat spreader, and the resulting increase in temperature can cause catastrophic failure of the device and the entire package.

An important aspect of the reliability characterization of a bonded material is its lifetime prediction model [5, 6], which is usually developed with inputs from an array of experimental tests and computational modeling. The experimental part involves subjecting the bonded material in a test sample configuration to accelerated tests, mainly thermal cycling, thermal shock, power cycling, corrosion or humidity tests, and vibration tests. A bonded material when subjected to these tests fails faster than in real-use conditions, and, depending on the test type, failure data (type of failure, cycles to failure) can be obtained in a matter of a few weeks or months. On the other hand, in computational models, the accelerated testing condition is simulated on the bonded material, and theoretical parameters such as stress, strain, or strain energy density are computed. Finally, these parameters are correlated with the experimentally obtained failure data to form the lifetime prediction model. It should be noted that a lifetime prediction model developed for a bonded material in a certain configuration (e.g., ball-grid array) could yield erroneous results when applied to the same material tested in a different configuration (e.g., flat bonds) because the failure data from experimental tests are dependent on the geometry of the sample.

A quick survey of the literature on lifetime prediction models of bonded materials reveals that thermal cycling and power cycling are the most studied and reported accelerated tests in the electronics industry [7]. In thermal cycling, while a certain temperature profile is directly applied to the bonded material in an electronics package, the objective of power cycling is to create a more real-life temperature distribution within the package. Also, power cycling is mostly tested on die-attach bonded materials, whereas relevant failure data of substrate-attach materials (larger area) are obtained from thermal cycling. Under thermal cycling, the component layers are made up of different materials in an electronics package and expand and contract at different rates due to their differences in the coefficient of thermal expansion (CTE). This CTE mismatch induces mechanical stresses mainly on the bonded material, causing its failure. To capture sufficient insights on this fatigue-induced failure on a continuum level and to assist in the reliability evaluation of bonded materials, thermomechanical computational models are developed. Thermomechanical modeling of bonded materials within an electronics package mainly serves three purposes:

1. It helps to locate those regions within the bonded material that are most prone to failure. Through a detailed assessment of stress, strain, or deformation plots, parts of the material where these parameters have exceeded a certain threshold limit can be identified.

2. The reliability potential offered by different package designs can be compared and quantified. Often, there is an initial baseline package design which later undergoes optimization to improve its electrical and thermal performance and reliability. Modeling can guide the design optimization in the right direction and thus save significant amount of time and effort that would otherwise be expended on experimental characterization.
3. It provides the required inputs to formulate a predictive lifetime model of the bonded interface material layer. Different types of predictive lifetime models, such as stress-based, strain-based, strain energy density-based, and fracture mechanics-based, have been developed in the past, and all these empirical lifetime models require at least a key modeling parameter as an input.

In this chapter, we study and compare the thermomechanical performance of three different bonded materials, SAC305 solder, sintered silver, and 95Pb5Sn solder, under a thermal cyclic load of −40 °C to 200 °C through modeling. First, we report strain energy density/cycle values of these materials computed using a nonlinear finite element method (FEM) approach in the model and demonstrate the evolution of strain energy density in a thermal cycle by plotting their hysteresis loops. Next, we describe an alternative thermomechanical modeling approach that is based on fracture mechanics theory but is relatively less adopted. However, this fracture mechanics-based approach could result in more accurate lifetime predictions as it takes into consideration the failure mechanisms in the bonded material, thus accounting for a more realistic depiction of the material failure.

2 Thermomechanical Modeling

Reliability evaluation of bonded interface materials and formulation of their lifetime prediction or fatigue models is not a nascent area but is still an indispensable research objective for the electronics industry as novel materials are synthesized and innovative package designs continue to be developed. The role and influence of thermomechanical modeling in this field have been extensively studied and published [8–10], even as it continues to get refined for improved accuracy and to suit different packaging configurations. A wide range of literature exists reporting the thermomechanical analyses conducted on different solder interface materials [11–16] at temperatures below 150 °C. The main objective is to develop a high-fidelity, nonlinear FEM model from which a suitable theoretical parameter that adequately represents the creep and fatigue deformation of the solder on a macroscale level can be extracted. Although no failure mechanisms such as cracks or voids are considered in the FEM model, the computed parameter, when linked with experimental data such as cycles to failure, would result in a reasonably accurate lifetime prediction model, provided the right set of material properties and mesh elements are deployed. Also, no microstructural representation of the material is needed as the objective of the FEM model is to capture the macroscopic, continuum-level displacement and stress fields of the material in response to an applied load.

The few different types of bonded materials that are used in the electronics packaging industry for high-temperature applications (>150 °C) include, but are not limited to, lead-free solders (Sn-Ag-Cu, Sn-Ag, Sn-Au), high-lead-content solders (Pb-Sn, Pb-Sn-Ag), and sintered silver [17]. Joints based on transient liquid phase bonding are being explored in the research space, but to the best of our knowledge, they are yet to be implemented in a commercial package. In the following sections, we compare and report the thermomechanical performance of three bonded materials—SAC305, sintered silver, and 95Pb5Sn—using the thermomechanical modeling approach as described in the previous paragraph. SAC305 solder has a melting point between 217 °C and 220 °C, the melting point of 95Pb5Sn solder is between 305 °C and 312 °C, and the melting point of sintered silver is 962 °C. While SAC305 solder has a relatively lower melting point and has a typical operating temperature less than 150 °C, we included it in this study as a baseline for comparative analysis. One of the challenges in the thermomechanical modeling of these bonded materials lies in capturing their deformation behavior at higher temperatures, under different loading variations of ramp rate and dwell or soak time. There is little information in the literature on their reliability or lifetime prediction models at temperatures above 150 °C. In the next section on FEM analysis, we report in detail the various steps that we considered in our modeling approach to illustrate the thermomechanical behavior of the three bonded materials. A study of their strain energy density values and hysteresis loops is presented.

2.1 Material Properties

In a thermomechanical FEM analysis of an electronics package under any loading condition, it is critical to choose the right set of material properties for each material in the model to obtain accurate stress and displacement fields. The required material properties for a thermomechanical simulation are elastic modulus, Poisson's ratio, and CTE. In addition, and more importantly, a suitable constitutive model that accounts for the complex deformation behavior of the bonded material must be provided as input. One particular constitutive model that the electronics industry has widely used for different bonded materials to account for both the rate-dependent (creep) and rate-independent (plasticity) deformation nature is the Anand model [18]. Originally developed to characterize the deformation nature of metals under hot-working, the Anand model has become popular partly because it is available as a built-in model in most of the commercial FEM packages. In this model, a single scalar quantity in which both plasticity and creep are embedded is introduced to represent the isotropic resistance to inelastic deformation. Thus, separate analyses for plastic and creep deformation are avoided. The various material properties of all materials, including the Anand model parameters of the three bonded materials used in this study [19–21], are listed in Tables 1 and 2.

Table 1 Material properties

Material	Elastic modulus (GPa)	Poisson's ratio	Coefficient of thermal expansion ($\times 10^{-6}$/°C)
Copper (Cu)	112	0.34	16.5
Silicon nitride (Si_3N_4)	300	0.28	2.8
SAC305	53	0.36	21.6
Sintered silver	7	0.37	20
95Pb5Sn	24	0.4	28.7

Table 2 Anand model parameters

Material parameters	Parameter description	SAC305	Sintered silver	95Pb5Sn
A (s^{-1})	Pre-exponential factor	3501	9.81	3.25×10^{12}
Q/R (°K)	Activation energy/universal gas constant	9320	5706	15,583
ξ	Multiplier of stress	4	11	7
m	Strain rate sensitivity of stress	0.25	0.66	0.14
$\hat{s}$ (MPa)	Coefficient for deformation resistance saturation	30.2	67.3	72.7
n	Deformation resistance	0.01	0.00326	0.0043
h_0 (MPa)	Hardening/softening constant	1.8×10^5	1.58×10^4	1787
a	Strain rate sensitivity of hardening or softening	1.78	1	3.73
s_0 (MPa)	Initial deformation resistance	21	2.77	15.09

2.2 Model Setup

Various factors to consider in setting up an effective FEM model include, but are not limited to, creating a geometric model that is closely representative of the actual sample, selecting a robust mesh and ensuring mesh-independent results, and assigning the right set of material properties and solver parameters. We chose the ANSYS Workbench platform to create the FEM model and performed the simulations in parallel mode on Peregrine, the National Renewable Energy Laboratory's supercomputing cluster. For our simulations, we created a geometric model of roughly 50-mm × 50-mm cross-sectional footprint, in which a 5-mm-thick Cu baseplate was attached to a 0.72-mm-thick active-metal-bonded substrate (0.32-mm-thick Si_3N_4 with 0.2-mm-thick Cu foil on either side of Si_3N_4) using the bonded material. A schematic of the quarter-symmetric model is shown in Fig. 2. The substrate attach was considered to be the main region of interest in this study. Material layers such as die-attach, devices, wire bonds, and encapsulants were omitted in the model for the ease of computations.

Under cyclic loads, it is to be expected that both shear and bending deformation would occur in the model, especially given its layered structure, as shown in Fig. 2.

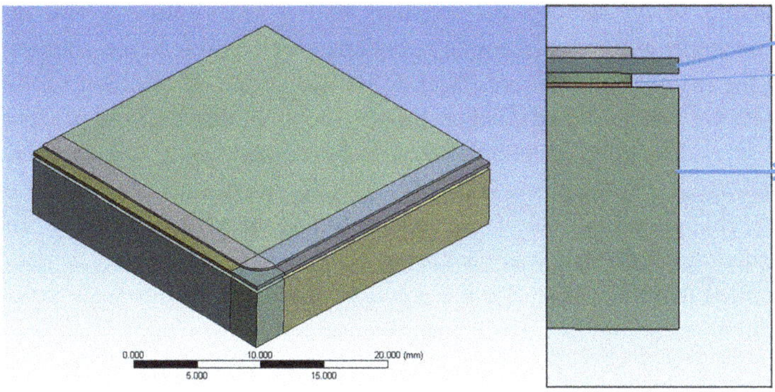

Fig. 2 50-mm × 50-mm sample model

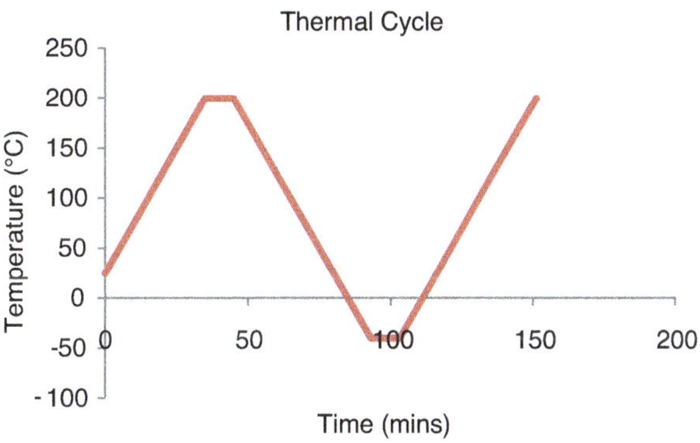

Fig. 3 Thermal cycling profile used in this study. Ramp rate is 5 °C/min; dwell time at temperature extremes is 10 mins

Choosing elements with a higher-order formulation is preferred in such cases as elements with only translational degrees of freedom could lead to issues such as shear locking, in which the elements become overly stiff and result in spurious deformation values. In this study, the meshing procedure consisted of utilizing the multizone technique and various body and element sizing features in ANSYS to create a hexahedral mesh for the entire model, with an increased mesh density at the bonded material layer. The results presented in this chapter are mesh-independent and are accurate to within 5%.

A thermal cycling profile selected from the JEDEC standard [22] or a modified version is typically used in the electronics packaging industry to study the thermomechanical behavior of a bonded material. While such a loading profile may not be the best representation of the actual temperature variations in a package in its field conditions, it encompasses the different possible temperature-related worst-case scenarios that a package could get subjected to in its lifetime. A thermal cycle (Fig. 3) with temperature extremes of −40 °C and 200 °C was applied on the entire model as the loading condition in this study. We chose a ramp rate of 5 °C/min and a dwell time of 10 min at both temperature extremes. For each load step of the thermal cycle, we ensured that a sufficient number of sub-steps were given as input to apply the loading in an incremental manner, thereby facilitating a smooth convergence of the solution. Lack of enough sub-steps would still lead the solution to converge, but the results obtained are likely to be inaccurate. In addition to the thermal cyclic load, we constrained the center point node in all directions to prevent rigid body motion and applied symmetric boundary conditions to the quarter-symmetric model. Both the displacement and force convergence criteria were specified, and the changing stiffness of the model due to the change in shape of the elements as they get deformed was taken into account. We performed simulations for each bonded material, and after the solution phase, the results were post-processed to calculate the strain energy density per cycle values.

2.3 Solvers

In general, the two most common types of solvers used for a nonlinear FEM analysis are the direct solver and the iterative solver. Both these solvers take different approaches with respect to each other, and the preference of one solver over the other depends on the amount of computational resources available for the solution phase. If there is sufficient physical memory available to hold all the solution matrices, then a direct solver will be more computationally efficient than an iterative solver. However, when the mesh element count is very large, demanding a larger amount of physical memory than is available, the solution will transfer matrices to and from the hard memory, thereby degrading the solver performance. Under such circumstances, choosing an iterative solver is desirable.

2.4 Strain Energy Density Results

Strain energy density at a particular node in the FEM model is the time integral of the product of stresses and incremental strains in all six directions at that node. Research in the past [12, 13] has found strain energy density to be a good indicator of material fatigue, and is recommended over other parameters such as stress and inelastic strain in formulating lifetime models. For all three bonded materials under study, the maximum value of strain energy density was found to be at the corner regions. To avoid any singularity issues, we calculated a weighted volume average of the strain energy density values per cycle over the fillet region at the corner and considered it as the metric for comparison between the different bonded materials. Figure 4 shows a contour plot of the strain energy density over the SAC305 solder and the corner fillet region that was selected as the region of interest. Figure 5 shows a comparison of strain energy density/cycle of SAC305, sintered silver, and 95Pb5Sn solder.

Among the three interface materials selected in this study, SAC305 exhibited the highest value of strain energy density per cycle, followed by sintered silver and 95Pb5Sn in order. In terms of experimental conclusions, a higher value of strain

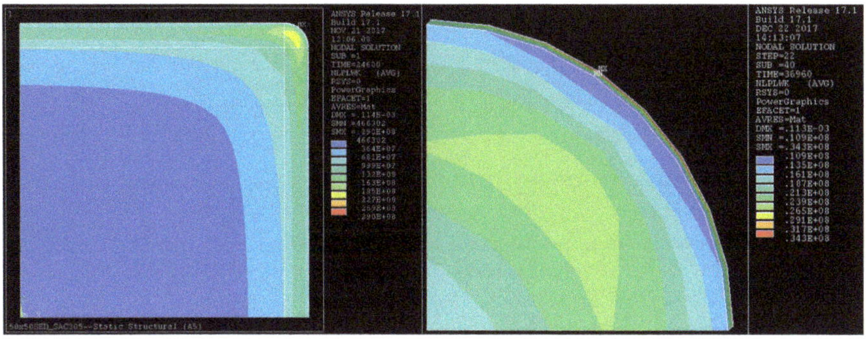

Fig. 4 Strain energy density contour plot of SAC305 (left) and corner fillet region (right)

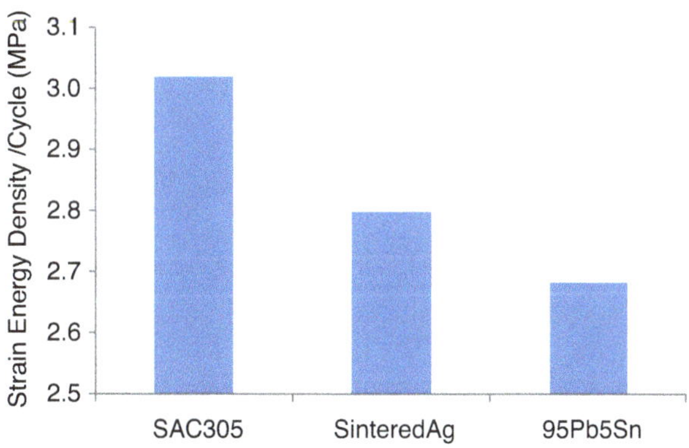

Fig. 5 Comparison of strain energy density/cycle of SAC305, sintered silver, and 95Pb5Sn solder

energy density translates to shorter cycles to failure. In other words, if test samples with the different bonded materials based on the simulated geometry were to be subjected to experimental accelerated thermal cycling, SAC305 samples would fail faster than the other two sample types. The homologous ratio (operating temperature/melting point in °K) is quite high for SAC305 under the simulated thermal cycle, particularly at the elevated temperatures, which means the creep deformation that occurs within the material would be significant enough to cause imminent failure. For this reason, in real applications, SAC305 is used only up to a maximum operating temperature of around 125 °C.

These modeling results predict that 95Pb5Sn solder possesses a superior reliability over sintered silver under thermal cycling. Nevertheless, sintered silver has an advantage over 95Pb5Sn in that it is a lead-free material and does not pose any significant risk to the environment. The recent power electronics industry pushes toward WBG devices (with ~200–250 °C operating junction temperature), and the RoHS regulations make sintered silver a strong candidate for high-temperature applications. A brief note on sintered silver is provided at the end of this chapter.

In addition to strain energy density per cycle values, stress-strain graphs reveal a significant amount of valuable information regarding the deformation behavior of

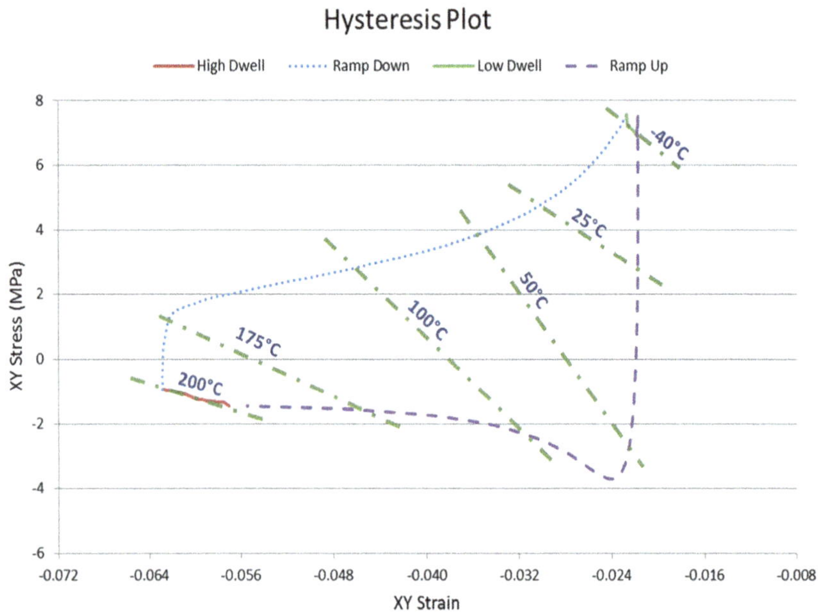

Fig. 6 Shear stress-shear strain hysteresis plot of SAC305 solder

the bonded material. The impact of each loading condition within a thermal cycle on the strain energy density value of the bonded material can be assessed from a stress-strain plot. A plot of any translational or shear components of stress versus strain results would essentially result in a hysteresis loop, the area of which is in fact the strain energy density value in that particular direction. Figure 6 shows a stress-strain plot of SAC305 solder in the XY plane. These values were obtained at a single node in the corner region, and the XY plane (in-plane direction) was arbitrarily chosen. The dashed lines connect the points in the loop that correspond to the same temperature points in the cycle. Four distinct phases are identified in the thermal cycle: high dwell, ramp down, low dwell, and ramp up. It is clear from Fig. 6 that the ramping loads, due to their longer durations, contribute substantially more to the strain energy density formation than the dwell loads. Creep strain is considered to be the predominant strain in these temperatures, although a separate analysis of creep and plastic strain is not possible with the Anand viscoplastic constitutive model. An analysis of the hysteresis loop in the clockwise direction is presented here. Starting with the high-dwell phase (solid line at bottom left) when the temperature is held constant at 200 °C for 10 min, strain continues to build up in the solder joint while a slight drop in stress can be observed. As the temperature starts to ramp down to −40 °C (dotted line), strain remains constant initially, but at about 175 °C, it begins to drop, whereas stress now increases in the opposite direction. At −40 °C, stress reaches the maximum; however, the low-dwell phase (solid line at top right) contributes the least to strain energy density as minimal changes in strain occur in this phase. In the subsequent ramp-up phase (dashed line), strain initially remains constant but for a longer duration than in the ramp-down phase, followed by an increase in value until the temperature reaches 200 °C. Stress reverses in direction, reaches a peak value, and thereafter remains almost a constant as a significant amount of creep strain is generated in the solder joint.

Plotting stress-strain data of sintered silver and 95Pb5Sn solder will result in similar hysteresis loops, and the corresponding loop areas would be indicative of the amount of strain energy density developed in these materials.

3 A Fracture Mechanics Approach in Thermomechanical Modeling

A fracture mechanics-based modeling approach essentially involves computation of parameters such as stress intensity factor or J-integral near a crack tip singularity that is manually embedded in the geometric model. Similar to strain energy density, these parameters can be correlated to experimental fatigue data to formulate lifetime prediction models of bonded interface materials. The rationale behind adopting a fracture mechanics-based approach is that the failure mechanism observed for most bonded materials under thermal cycling is crack initiation and propagation [5]. As a result, a parameter that can adequately describe the deformation and stress field around a crack front in the material could, in principle, offer improved accuracy in predicting its lifetime; however, this methodology is still unproven in its effectiveness to predict lifetime accurately. In addition, no established practices that describe the modeling approach in an FEM model to calculate a fracture mechanics parameter under accelerated test loads were found in the literature. In this section, we outline a simple modeling strategy to compute and process J-integral values in an FEM setting for use in a lifetime prediction model. To explain the calculation of the J-integral in ANSYS, we resort to the same geometric model (Fig. 2) and similar loading conditions as described in the previous section, with sintered silver as the bonded material.

3.1 *Elastic-Plastic Fracture Mechanics for Cyclic Loading*

Although lifetime prediction models with a basis in fracture mechanics theory are not strictly a novel research area, they are far less attempted compared to the traditional approach, possibly due to the complexity involved in deriving a meaningful parameter that can relate to material failure data. Introduction of crack modeling and solution of crack-related parameters in commercial FEM packages may solve the issue to an extent, but the lack of conclusive data in the literature that prove an improvement of accuracy and versatility over the traditional strain-based or energy-based models may be another factor hindering the wider implementation of

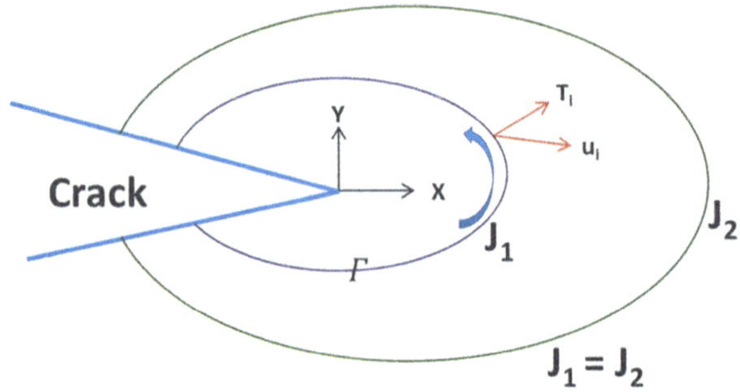

Fig. 7 J-integral path independence

fracture mechanics concepts in developing a lifetime prediction model. Also, the plastic and creep deformation nature of bonded materials is beyond the scope of linear elastic fracture mechanics and requires a comprehensive analysis using elastic-plastic fracture mechanics theory.

The J-integral is an elastic-plastic fracture mechanics parameter that was developed to represent the stress and displacement fields around a crack tip or crack front in a nonlinear elastic body. It denotes the energy release rate in a cracked body for a unit increase in crack area and can be computed as a contour integral along a counterclockwise path around the crack front. The path-independent nature of the J-integral (illustrated in Fig. 7) makes it a convenient parameter for crack tip studies [23]. However, it must be noted that the path independence of the J-integral for elastic-plastic materials holds only under certain circumstances, such as when the material behavior can be described based on a deformation theory of plasticity or when the material is subjected to purely monotonic loading [24]. According to the deformation theory of plasticity, the total strain in a material is purely a function of the total stress and is history independent. Under such circumstances, elastic-plastic materials can be idealized as nonlinear elastic materials, and the J-integral effectively characterizes the crack tip stress and displacement field. For a contour surrounding the crack tip or front, the J-integral is given by the equation:

$$J = \int_{\Gamma} (W.dy - T_i.\left(\frac{\partial u_i}{\partial x}\right).ds) \tag{1}$$

where

Γ is the contour surrounding the crack front
W is the strain energy density
T_i is the traction vector
u_i is the displacement vector

Most of the constitutive models that were developed to describe material deformation are based on an incremental theory of plasticity. In these models, material plasticity is defined in the form of a strain rate tensor as a function of other parameters such as time, temperature, and stress. A few researchers have reported that the incremental theory of plasticity mimics the deformation theory of plasticity in cases when the applied loading is proportional in all directions and the J-integral remains path-independent [25]. Nevertheless, even if the loading is nonproportional under the incremental theory of plasticity or in cases of cyclic loading, the path-independent nature of the J-integral does not completely break down [26]. In fact, it exhibits a converging trend with successive contours around a crack front. This converged value can potentially be considered as a valid parameter to input in the lifetime prediction models. However, the relevance of the J-integral obtained using this technique can be justified only after developing a lifetime prediction model incorporating experimental results and determining the accuracy of that model when applied to multiple loading scenarios.

3.2 Computing J-Integral in ANSYS

The modeling approach to calculating the J-integral in ANSYS remains the same as the approach outlined in the previous sections except for the introduction of a crack feature at the desired location in the model. For easier numerical implementation, the contour integral form of the J-integral as shown in Eq. (1) is converted to a volume integral in 3-D models using the domain integral method proposed by Shih et al. [27]. FEM simulations of sintered silver in ANSYS showed that higher values of strain energy density occur in the corner fillet regions where the cracks are likely to initiate. In a simulation on the 50-mm × 50-mm sample (Fig. 2), we manually inserted a crack feature in the corner region, as shown in Fig. 8. The semielliptical red curve denotes the manually inserted crack front. The most recent versions of ANSYS offer insertion of crack features of any arbitrary size. Six contours, shown as circles that run through the entire crack front, can be seen at both ends of the crack. We discretized these features to obtain convergence in the computed

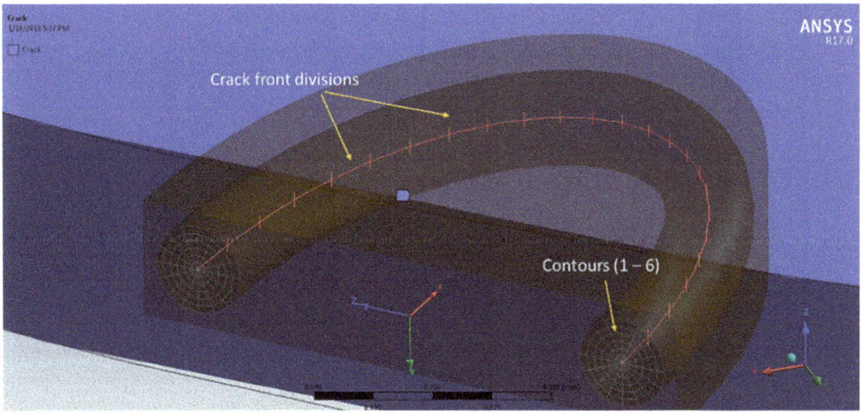

Fig. 8 Crack feature inserted in the model

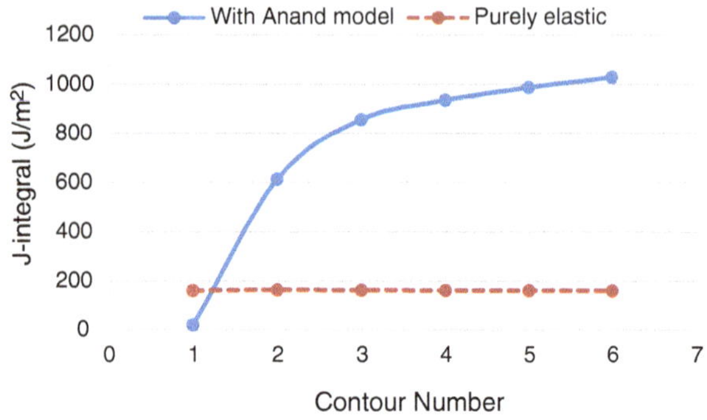

Fig. 9 Averaged J-integral values for each contour surrounding the crack front

J-integral values. This includes dividing the crack front into smaller divisions and splitting the circular contour lines into shorter segments. It is recommended that multiple simulations with a higher number of contours and increased crack front divisions be conducted as the level of discretization may vary depending on the material type and loading conditions.

In the post-processing phase, J-integral values were obtained for each contour surrounding all the nodal points along the crack front. To avoid any singularity issues, these values were then averaged along the crack front. A plot of the averaged J-integral values of each contour (at any given time step), as shown in Fig. 9, leads to the conclusion that the path dependence becomes less pronounced with each successive contour and is within 5% after six contours in this case. Also, if the FEM model included only purely elastic properties, i.e., without any elastic-plastic or viscoplastic constitutive model as input, then it would be sufficient to have just one contour around the crack front to calculate the J-integral, as proved by the straight dashed line plot in Fig. 9.

Finally, because the J-integral is a cumulative parameter, its value over a thermal cycle range in the model was obtained and taken as the final output for use in lifetime prediction models. Efforts are currently underway at the National Renewable Energy Laboratory to develop a lifetime prediction model for sintered silver by correlating J-integral/cycle values with experimental accelerated test results.

3.3 Other Fracture Mechanics Parameters

In view of the theoretical limitations of the J-integral—originally developed for nonlinear elastic materials and that can be extended to elastic-plastic materials only if defined by the deformation theory of plasticity and when subjected to monotonic loading—a few researchers have proposed either modifications to the J-integral equation or other parameters for applications in plastic materials under cyclic loading. Simha et al. [28] used the concept of configurational forces to derive an incremental plasticity J-integral (J_{ep}) that can be applied under nonproportional loading conditions or cyclic loading. While J_{ep} is path-dependent, its far-field value is considered to be an appropriate parameter to characterize the thermodynamic driving force for both crack growth and plastic deformation. A cyclic form of the J-integral was derived by Lamba [29] for calculating the strain concentration at a notch root and was found to obtain similar results as the Neuber rule, which is widely used in low-cycle fatigue analyses. Yet another parameter that was theoretically derived by Atluri and Nishioka [30] for use in incremental plasticity is the T_p integral; however, numerical implementation of this integral for a practical case study has not been carried out.

4 A Brief Note on Sintered Silver

Sintered silver has proven to be a promising candidate for use as a bonded material in electronics packages. As opposed to certain solders, it can potentially operate at higher temperatures without significant creep issues while still efficiently conducting heat away from the devices. Although the choice of sintered silver in electronics packages is not fully without any drawbacks due to the requirement of higher application pressures (~40 MPa) for bonding, research over time has led several manufacturers to develop sintered silver materials that require lower (3–5 MPa) or no bonding pressures. Pressure-bonded sintered silver has even been commercially deployed as a die-attach material [31] in inverters/converters, but large-area applications of pressure-less sintered silver material and its reliability evaluation are still in the research and development phase.

Compared to solders, the existing body of literature on sintered silver is limited mainly because it was introduced into the electronics market as a bonded material much later. However, given the current market trend toward high-temperature packaging applications, there is significant focus on developing sintered silver formulations that require simpler and less severe synthesis procedures and can still meet the desired reliability targets. An important area of research is the development and reliability analysis of nanoparticle-based sintered silver. It has been observed that switching to nanoscale formulations of sintered silver can increase the interparticle bonding force due to larger surface area and enable sintering at lower temperatures. Furthermore, only lower pressures (<3 MPa) may be required to obtain a uniform, defect-free bond. A detailed summary of the status of sintered silver as a bonded material in the microelectronics packaging industry is provided by Siow in [32]. Different types of sintered silver formulations and methods of application are discussed, and its long-term reliability prospects based on elevated temperature storage tests are commented on.

From a modeling perspective, an aspect of sintered silver that requires further research is the identification or development of a suitable constitutive model for use in FEM simulations. Chen et al. [33] studied the ratcheting behavior of a sintered nanosilver-based sample assembly in the context of power electronics packaging. In this study, two viscoplastic constitutive models—one based on the Ohno-Wang and

Armstrong-Frederick (OW-AF) nonlinear kinematic hardening rule and the other one based on the Anand model [20]—were investigated for their accuracy and effectiveness in predicting the cyclic ratcheting effects of nanosilver joints. The OW-AF model results were found to be in better agreement with experimental results than the Anand model. In another study on high-temperature applications of sintered silver, Croteau et al. [34] conducted uniaxial tensile tests on filmlike specimens at different stress rates, strain rates, and temperatures to obtain the required stress-strain curves from which the desired constitutive model parameters can be extracted. In addition, they conducted die shear tests and thermal cycling tests and reported on the dependence of shear strength results on temperature and bond line thickness. In addition to constitutive models, lifetime prediction models of sintered silver, encompassing suitable modeling parameters, such as strain energy density or J-integral, and accelerated experimental data, must be formulated especially given the promise of sintered silver as a replacement for solders in WBG-device-based electronic applications [35, 36].

5 Summary

Thermomechanical modeling plays a critical role in the reliability evaluation and lifetime prediction of bonded materials used in electronics packages, particularly for high-temperature applications. In this chapter, we studied and compared the reliability of three different bonded materials—SAC305, sintered silver, and 95Pb5Sn—under an accelerated thermal cyclic loading condition. Volume-averaged strain energy density values were obtained for each of these materials through FEM modeling. In addition to the FEM simulations to calculate strain energy density, a modeling strategy based on fracture mechanics theory to compute the J-integral around a crack front is presented. The path-dependent nature of the J-integral is analyzed with sintered silver as the bonded material under a purely elastic case and using Anand viscoplastic model as the constitutive model input. The fracture mechanics approach is relatively new and less employed in the electronics industry; however, the potential advantage of this approach over the traditional method is that it directly addresses the failure mechanisms commonly observed in bonded materials and could result in improved accuracy for lifetime prediction models. More research and validation studies are required before the fracture mechanics approach can be considered as superior to the strain energy density approach. Finally, a summary on the status and challenges of sintered silver in high-temperature packaging applications is reported.

Acknowledgments We acknowledge the financial support for the work provided by Susan Rogers, the technology manager for the Electric Drive Technologies Program, Vehicle Technologies Office, US Department of Energy Office of Energy Efficiency and Renewable Energy (EERE). This work was supported by the US Department of Energy under Contract No. DE-AC36-08GO28308 with Alliance for Sustainable Energy, LLC, the manager and operator of the National

Renewable Energy Laboratory. We would also like to thank Andrew Wereszczak (Oak Ridge National Laboratory) for his guidance and insightful comments on fracture mechanics-based modeling of sintered silver. The U.S. Government retains and the publisher, by accepting the article for publication, acknowledges that the US Government retains a nonexclusive, paid-up, irrevocable, worldwide license to publish or reproduce the published form of this work or allow others to do so, for U.S. Government purposes.

References

1. J. Millán, P. Godignon, X. Perpiñà, A. Pérez-Tomàs, J. Rebollo, A survey of wide bandgap power semiconductor devices. IEEE Trans. Power Electron. **29**(5), 2155–2163 (2014)
2. U.S. Drive, Electrical and electronics technical team roadmap, in *Partnership Plan, Roadmaps, and Other Documents*, Oct. 2017. Available: https://energy.gov/sites/prod/files/2017/11/f39/EETT%20Roadmap%2010-27-17.pdf
3. M.M. Yovanovich, E.E. Marotta, Thermal spreading and contact resistances, in *Heat Transfer Handbook*, ed. by A. Bejan, A.D. Kraus (Wiley, Hoboken, 2003), pp. 261–395
4. R. Prasher, Thermal interface materials: historical perspective, status and future directions. Proc. IEEE **94**(8), 1571–1586 (2006)
5. P. Paret, D. DeVoto, S. Narumanchi, Reliability of emerging bonded interface materials for large-area attachments. IEEE Trans. Compon. Packag. Manuf. Technol. **6**(1), 40–49 (2016)
6. H. Lu, C. Bailey, C. Yin, Design for reliability of power electronic modules. *Microelectron. Reliab.* **49**, 1250–1255 (2009)
7. B. Ji, X. Song, E. Sciberras, W. Cao, Y. Hu, V. Pickert, Multiobjective design optimization of IGBT power modules considering power cycling and thermal cycling. IEEE Trans. Power Electron. **30**(5), 2493–2504 (2015)
8. X. Fan, M. Pei, P. Bhatti, Effect of finite element modeling techniques on solder joint fatigue life prediction of flip-chip BGA packages, in *IEEE ECTC*, San Diego, CA, 2006
9. J. Pang, T. Tan, S. Sitaraman, Thermo-mechanical analysis of solder joint fatigue and creep in a flip chip on board package subjected to temperature cycling loading, in *IEEE ECTC*, Seattle, WA, 1998
10. S. Wen, G.Q. Lu, Finite-element modeling of thermal and thermomechanical behavior for three-dimensional packaging of power electronics modules, in *IEEE ITherm*, Las Vegas, NV, 2000
11. A. Schubert, R. Dudek, E. Auerswald, A. Gollhardt, B. Michel, H. Reichl, Fatigue life models for SnAgCu and SnPb solder joints evaluated by experiments and simulation, in *IEEE ECTC*, New Orleans, LA, 2003, pp. 603–610
12. W. Lee, L. Nguyen, G. Selvaduray, Solder joint fatigue models: review and applicability to chip scale packages. Microelectron. Reliab. **40**, 231–244 (2000)
13. A. Syed, Accumulated creep strain and energy density based thermal fatigue life prediction models for SnAgCu solder joints, in *IEEE ECTC*, Las Vegas, NV, 2004
14. A. Lajimi, J. Cugnoni, J. Botsis, Reliability analysis of lead-free solders, in *WCECS*, San Francisco, CA, 2008
15. S. Ridout, C. Bailey, Review of methods to predict solder joint reliability under thermo-mechanical cycling. Fatigue Fract. Eng. Mater. Struct. **30**, 400–412 (2006)
16. R. Darveaux, Effect of simulation methodology on solder joint crack growth correlation. J. Electron. Packag. **124**, 147–154 (2002)
17. R. Khazaka, L. Menizabal, D. Henry, R. Hanna, Survey of high temperature reliability of power electronics packaging components. IEEE Trans. Power Electron. **30**(5), 2456–2464 (2015)
18. L. Anand, Constitutive equations for hot working of metals. Int. J. Plast. **29**(2), 213–231 (1985)
19. M. Motalab, Z. Cai, J.C. Suhling, P. Lall, Determination of Anand constants for SAC solders using stress-strain or creep data, in *IEEE ITherm*, San Diego, CA, 2012

20. D. Yu, X. Chen, G. Chen, G.Q. Lu, Z.Q. Wang, Applying Anand model to low-temperature sintered nanoscale silver paste chip attachment. J. Mater. Des. **30**, 4574–4579 (2009)
21. G. Wang, Z. Cheng, K. Becker, J. Wilde, Applying Anand model to represent the viscoplastic deformation of solder alloys. J. Electron. Packag. **123**, 247–253 (2001)
22. JEDEC Standard Temperature Cycling, *JESD22-A104D*, Mar. 2009
23. J.R. Rice, A path independent integral and the approximate analysis of strain concentration by notches and cracks. J. Appl. Mech. **35**, 379–386 (1968)
24. T.L. Anderson, *Fracture Mechanics: Fundamentals and Applications*, 3rd edn. (CRC Press, Boca Raton, 1991)
25. J.R. Rice, Mathematical analysis in the mechanics of fracture, in *Fracture: An Advanced Treatise*, vol. II, ed. by H. Liebowitz (Academic Press, New York, 1968), pp. 191–311
26. W. Brocks, I. Scheider, *Numerical Aspects of the Path-Dependence of the J-Integral in Incremental Plasticity – How to Calculate Reliable J-Values in FE Analyses*. Internal Report, Institute für Werkstofforschung, Köln-Porz, Oct. 2001
27. C. Shih, B. Moran, T. Nakamura, Energy release rate along a three-dimensional crack front in a thermally stressed body. Int. J. Fract. **30**(2), 79–102 (1986)
28. N.K. Simha, F.D. Fischer, G.X. Shan, C.R. Chen, O. Kolednik, J-integral and crack driving force in elastic-plastic materials. J. Mech. Phys. Solids **56**(9), 2876–2895 (2008)
29. H.S. Lamba, The J-integral applied to cyclic loading. Eng. Fract. Mech. **7**, 693–703 (1975)
30. S.N. Atluri, T. Nishioka, Incremental path-independent integrals in inelastic and dynamic fracture mechanics. Eng. Fract. Mech. **20**(2), 209–244 (1984)
31. C. Gobl, Low temperature sinter technology die attachment for power electronic applications, in *Proceedings of CIPS*, Nuremberg 2010, p. 327
32. K.S. Siow, Are sintered silver joints ready for use as interconnect material in microelectronic packaging? J. Electron. Mater. **43**(4), 947–961 (2014)
33. G. Chen, Z. Zhang, Y. Mei, X. Li, G. Lu, X. Chen, Ratcheting behavior of sandwiched assembly joined by sintered nanosilver for power electronics packaging. Microelectron. Reliab. **53**, 645–651 (2013)
34. P. Croteau, S. Seal, R. Witherell, M. Glover, S. Krishnamurthy, A. Mantooth, Test results of sintered nanosilver paste die attach for high-temperature applications. J. Microelectron. Electron. Packag. **13**, 6–16 (2016)
35. J. Heilmann, I. Nikitin, U. Zschenderlein, D. May, K. Pressel, B. Wunderle, Advances and challenges of experimental reliability investigations for lifetime modelling of sintered silver based interconnections, in *Proceedings of 17th Eurosime Conference*, 2016
36. F. Henaff, S. Azzopardi, E. Woirgard, T. Youssef, S. Bontemps, J. Joguet, Lifetime evaluation of nanoscale silver sintered power modules for automotive application based on experiments and finite-element modeling. IEEE Trans. Device Mater. Reliab. **15**(3), 326–334 (2015)

Reliability and Failure Mechanisms of Sintered Silver as Die Attach Joint

Y. H. Mei, Z. Wang, and K. S. Siow

1 Introduction

The search for a high-temperature die-attach materials is driven by the need to replace leaded solder with environment-friendly alternatives [1] and to cater the advent of wide bandgap (WBG) semiconductor devices (i.e., silicon carbide (SiC) or gallium nitride (GaN)) which operates at the higher temperature of 300 °C or even 400 °C [2]. WBG devices operate at higher-power density and switching frequency than their traditional Si devices [3]. This higher operating temperature necessitates die-attach materials with higher melting temperature than conventional die-attach materials, to work reliably because typical operating condition needs to be less than 80% of their melting temperature [4]. Only a few Pb-free solders can meet this requirement, but they too face other limitations in implementation. For example, Au-rich solders, such as Au80-Sn20 and Au88-Ge12, suffered from high cost and poor processability due to high brittleness [5], while Zn-Al solders suffered from poor wettability and electrochemical corrosion [6]. Further details are available in Chaps. 8 and 10 of this book.

One possible solution for this elevated temperature operation is the sintered silver (Ag) joint which has a high working temperature resulting from the melting

Y. H. Mei (✉) · Z. Wang
Tianjin Key Laboratory of Advanced Joining Technology & Center of High-Temperature Electronic Packaging, School of Materials Science and Engineering, Tianjin University, Tianjin, China

Key Laboratory of Advanced Ceramics and Machining Technology, Ministry of Education, Tianjin, China
e-mail: yunhui@tju.edu.cn

K. S. Siow
Institute of Microengineering and Nanoelectronics, Universiti Kebangsaan Malaysia, Bangi, Malaysia

K. S. Siow (ed.), *Die-Attach Materials for High Temperature Applications in Microelectronics Packaging*, https://doi.org/10.1007/978-3-319-99256-3_5

temperature of the Ag at 961 °C [7] as well as high thermal and electrical conductivities compared to traditional solders [8, 9]. This sintered Ag joint as die-attach material was pioneered by Siemens in the late 1980s, to attach the silicon dies to molybdenum plates with micron-scale Ag particles [10]. High sintering pressure was used by the Siemens engineer to produce this die-attach joint reliably, but the high hydrostatic pressure may damage the devices directly or latently for the case of WBG semiconductor. An alternative approach is to reduce the Ag filler size to increase the surface reactivity and driving force to sinter the Ag paste [11]. At nanoscale dimension, it is possible to reduce or eliminate any applied pressure as well as reduce the sintering temperature to less than 300 °C [12].

However, the coefficient of thermal expansion (CTE) mismatch between the different components in power devices induced high thermomechanical stresses on the sintered Ag, resulting in instant catastrophic failure (when the stress exceeds the ultimate strength of sintered Ag and sintered Ag joints) or time-dependent crack growth leading to failure. The latter failed when the accumulated plastic strain and creep strain grew the cracks to decrease the effective bonding area and final failure. These crack initiation and growth are the main failure modes for sintered Ag joints as the die-attach application in power modules [13]. These different failure conditions motivated us to analyze and review the reliability of the sintered Ag joints under these static and cyclic conditions.

Although previous reviews had already summarized the mechanical properties of sintered Ag joint, this chapter began with mechanical properties, such as elastic modulus, strength (yield and ultimate tensile strength), creep strength, fatigue strength, and ratcheting of the sintered Ag joints [14]. The second section summarized the effect of temperature cycling, temperature aging, power cycling, electrochemical migration (ECM) on the reliability of sintered Ag joints, and related failure mechanisms. Finally, the chapter concluded with trends and challenges facing sintered Ag joints as die-attach materials in the next few years.

2 Mechanical Properties

2.1 Elastic Modulus

Elastic modulus or Young's modulus is calculated by dividing the applied stress by the strain of the material in the elastic region during tensile loading. A typical stress-strain curve of a sintered Ag material produced an apparent elastic modulus of 9 GPa, based on the steepest slope in this stress-strain curve [12] or elevated to 50 GPa for the dynamic elastic modulus [15]. Compared to a typical elastic modulus of 80 GPa for bulk Ag, this reduced elastic modulus for the sintered Ag can be attributed to the presence of pores in bulk sintered material [16] or the weaker bonding interface between the nanoscale Ag particles within the Ag joint [17].

Porosity is closely related to the relative density in terms of its influence on the elastic modulus of the sintered Ag. Caccuri et al. [18] studied the effect of density on

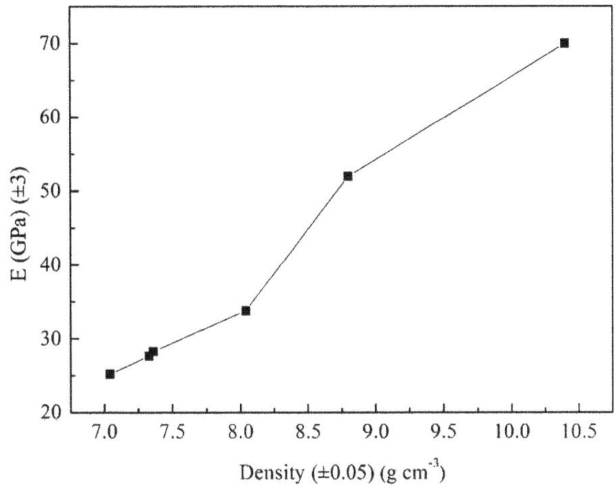

Fig. 1 Elastic modulus of sintered Ag against density at room temperature [18]

the elastic modulus of bulk sintered Ag at room temperature; the elastic modulus increased with the density increase of the bulk sintered Ag (Fig. 1).

Milhet et al. [15] validated the elastic modulus of the bulk sintered Ag experimentally using Eqs. (1) and (2) which were based on the Ramakrishna model described herein [19]:

$$E = \frac{(1-p)^2}{1+p(kvm)} E_{\mathrm{m}} \tag{1}$$

$$v = \frac{v_{\mathrm{m}} + \frac{3}{4}p - \frac{7}{4}pv_{\mathrm{m}}}{1+2p-3pv_{\mathrm{m}}} \tag{2}$$

(k) is between 2 and 3, (p) is pore volume, E_{m} is elastic modulus, v_{m} is Poisson's ratio of the fully dense bulk Ag. Another possible relationship to calculate the elastic modulus of the sintered Ag is based on Eq. (3) [20]:

$$E_{\text{porous layer}} = E_{\mathrm{Ag}} \frac{3(3-5P)(1-P)}{9-P\left(9.5-5.5v_{\mathrm{Ag}}\right)} \tag{3}$$

P refers to the porosity of the bulk sintered Ag. v_{Ag} is Poisson's ratio of sintered Ag with less than 40% porosity and E_{Ag} of 80 GPa. Equation (1) was developed to cater a random distribution of spherical pores of larger range, while Eq. (3) was applied on the special case with a porosity of less than 40%.

Besides porosity, testing temperature also influences the elastic modulus of the sintered Ag. This chapter replotted data from reference [21] to show that the elastic modulus decreases almost linearly with the increase of testing temperature (Fig. 2).

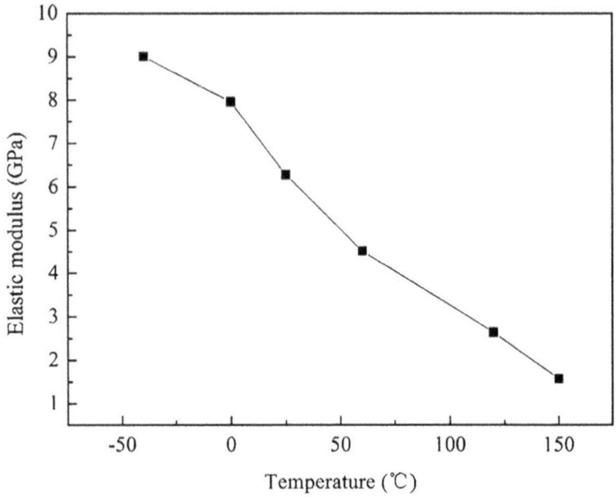

Fig. 2 Elastic modulus versus testing temperature for sintered nanoscale Ag films [21]

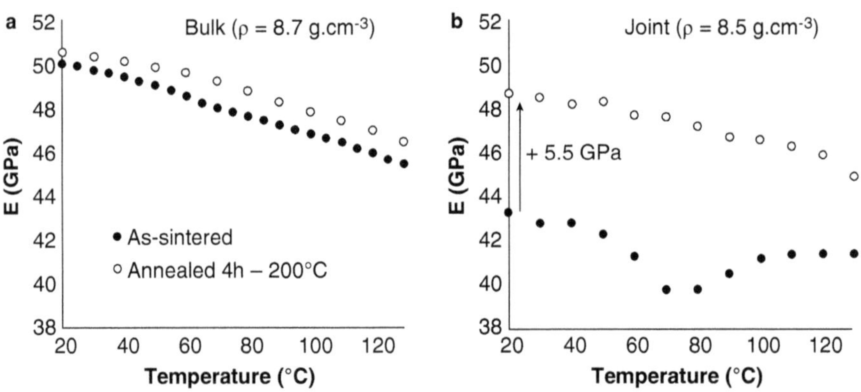

Fig. 3 Elastic modulus changes with testing temperature: (**a**) bulk Ag and (**b**) sintered Ag joint [15]

Others compared the effect of temperature on the elastic modulus of a bulk sintered Ag sample and a sintered Ag joint to arrive at a similar conclusion for bulk sintered Ag, but the sintered Ag joint showed at inflection in elastic modulus at around 70 °C (Fig. 3) [15]. This inflection was caused by microstructural instability in the sintered Ag joint (not experienced by the bulk Ag), which could be alleviated by annealing at 200 °C for 4 h [15].

Since the elastic modulus of sintered Ag was lower than Pb-Sn and Pb-free solders, the sintered Ag could serve as a buffer layer to release stress more effectively than the latter during elastic deformation [22]. Elsewhere, it has been shown that the sintered Ag joint with an approximate density of 80% will achieve the highest thermal fatigue life [23]. This result demonstrated the importance of controlling the porosity size and distribution to obtain the desired elastic modulus for maximum reliability.

2.2 *Strength*

2.2.1 Tensile Strength

The ultimate tensile strength of a bulk sintered Ag is approximately 43 MPa [12] which is comparable to those reported for Pb-Sn and Pb-free solders [22]. In the case of sintered Ag film (15 mm × 5 mm × 0.1 mm), the tensile strength decreased with increasing the testing temperature or decreasing strain rate in spite of its low homologous temperature [21]. Fractography was not provided in those literature, but increased mobility was likely to play a role in this strength reduction at high-temperature testing.

A similar trend on sintering temperature was provided by Akada et al. who sintered the Ag pastes with two cylindrical copper (Cu) specimens plated with (out) Ni/Au [24] (Fig. 4). Higher sintering temperature produced sintered Ag joints of higher density (less porosity) to support the stress-strain of the sintered Ag joint. Similar to the measured elastic modulus, bulk sintered Ag of higher density produced joint with higher tensile strength because of the reduction in porosity within the joint which supported a higher loading capacity (Fig. 5) [18].

Microstructurally, sintered Ag joints fractured cohesively, with intergranular fracture and apparent plastic deformation in the sintered Ag joint during a tensile test, but the fracture strength and strain of sintered Ag were usually lower than those of bulk Ag. This lower fracture stress and strain of sintered Ag was attributed to the notch effect of numerous pores present in the sintered Ag joint. Localized deformation around the location of highest pores density could fracture the sintered Ag joint prematurely [25].

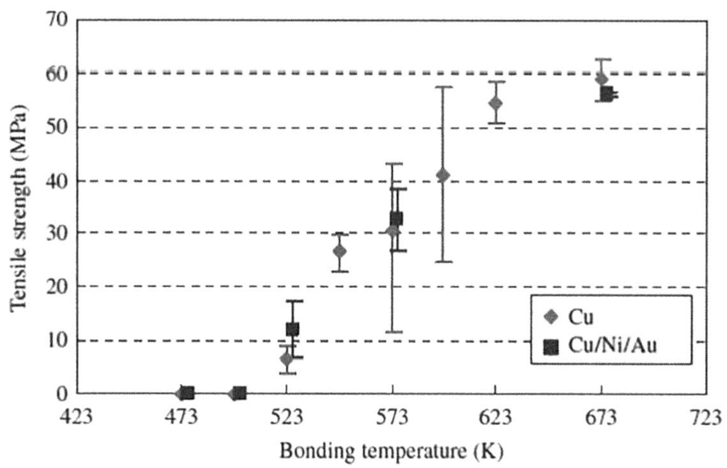

Fig. 4 Relationship between sintering temperature and tensile strength of Cu-Cu joints and Cu/Ni/Au-Cu/Ni/Au joints [24]

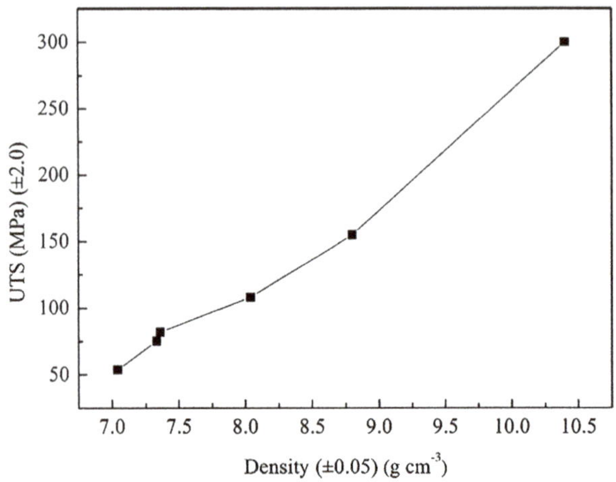

Fig. 5 Ultimate tensile strength against density of sintered Ag joint [18]

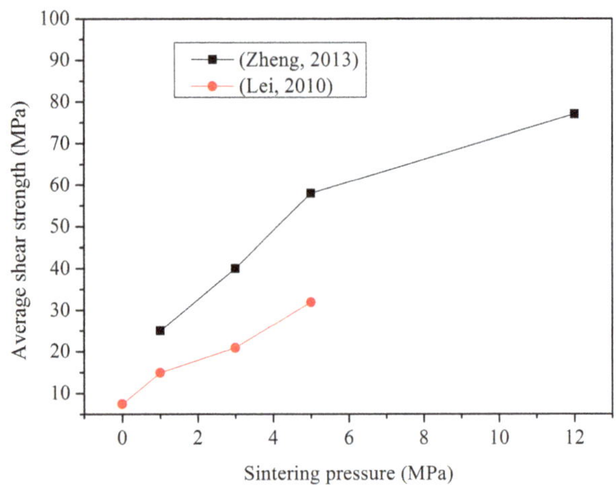

Fig. 6 Average shear strength of sintered Ag joints against sintering pressure [26, 27]

2.2.2 Shear Strength

Shear strength is the most widely reported parameter to compare different bonding techniques and conditions because of its simplicity and relevance to the stress state of the die-attach joint. Factors affecting the shear strength of the sintered Ag joint are sintering pressure, sintering temperature, dwell time at peak sintering temperature, bonding area, and substrate or die back metallization.

In terms of pressure, pressureless sintered Ag joint is the "holy grail" of die-attach joint as a drop-in solution, but it lacks the reliability required in mass production for demanding applications. Hence, pressure-assisted sintering is still the de facto standard in the power module industry, especially for application involving die sizes larger than 100 mm^2. Figure 6 shows that the higher sintering pressure resulted in higher die shear strength [26, 27]. This increased shear strength could be attributed to the closer contact between Ag particles and the

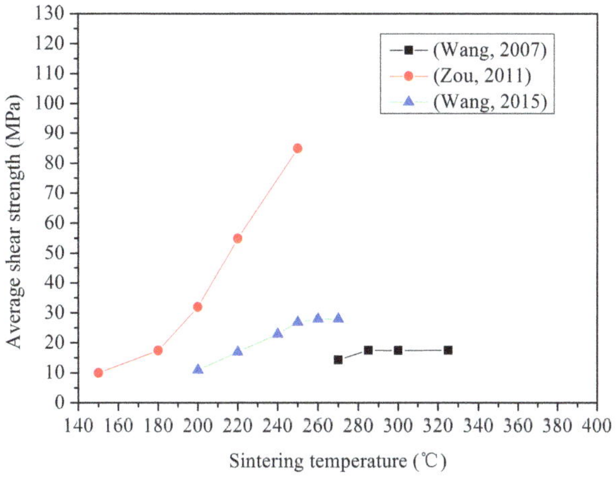

Fig. 7 Average shear strength of sintered Ag joints against sintering temperature [30–32]

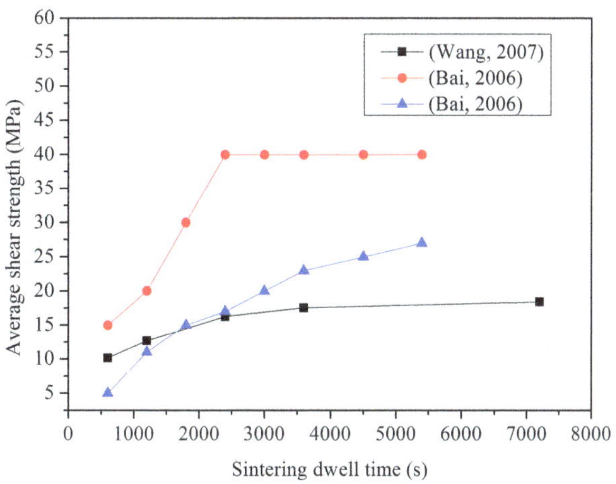

Fig. 8 Average shear strength of sintered Ag joints against sintering dwell time [30, 33]

substrate surface, and with one another when the pressure was increased, leading to increased interdiffusion and strong bond at the faying interfaces [28, 29].

Similar to sintering pressure, the increase of peak sintering temperature increases the sintered Ag shear strength (Fig. 7) [30–32]. Higher sintering temperature promotes the decomposition of dispersant in the paste, allowing the Ag particles to coalesce and increase the die shear strength up to a plateau where additional pressure is needed to improve the strength again.

Similar to the sintering temperature, the shear strength also increases with the increase of dwell time at the peak sintering temperature up to a plateau (Fig. 8) [30, 33]. Longer sintering time allows the sintering kinetics of Ag particles to take effect and coalesce among themselves. Generally, sintering pressure, temperature, and dwell time form the “iron” triangle to control the ultimate die shear strength of the sintered Ag joint; a reduction of one parameter increases the other two parameters [34].

While there are some disagreements over the exact influence of bonding area, large bond area, in the absence of applied pressure, will result in low shear strength because of the difficulty in evaporating the solvents and decomposed binders from the die center area to surrounding atmosphere [35]. Oxygen is also needed to diffuse and fully decompose the organics and binders present in the Ag paste located under the die [28, 36].

In the case of substrate metallization, Ide et al. [37] compared the effect of Al, Ti, Cu, and Ag metallization on the bonding strength of the Ag joints which were sintered at 300 °C for 300 s under an applied pressure of 5 MPa. Sintered Ag joints with Al and Ti substrates suffered from low shear strength because their stable oxides prevented strong bonding between Ag particles and metallized substrates. The same study also showed that Cu, Ag, and Au possess the necessary surface characteristics to be sintered with Ag paste. In comparison between Ag and Au, sintered Ag joint on Ag-metallized substrate produced higher die shear strength than Au-metallized substrate because of the faster self-diffusion of Ag nanoparticles into the Ag substrate than the interdiffusion between Ag nanoparticles and Au-metallized substrate [33]. In other words, additional thermal energy was needed to activate the interdiffusion of Ag atoms into dissimilar materials with different lattice constants.

However, the most preferred surface to be sintered among the power device manufacturers is bare Cu because of its ubiquitous presence as the underlying substrate, and further Ag or Au metallization will incur an additional plating cost [26]. Bare Cu suffered from oxidation, and often Ag pastes were formulated to deoxidize the Cu oxides during the sintering step [37]. Microscopically, the bonding interface was disconnected in several localized positions between the sintered Ag and Cu, indicating the poor bonding of sintered Ag on Cu [38]. This localized bonding was attributed to Cu oxides that prevented intimate contact between the Cu and the sintered Ag paste. These oxides were not entirely deoxidized by the binders/solvents during the sintering step. Pre-diluted hydrochloric acid had been used to remove the Cu oxides on the Cu substrates to ensure proper bonding and sintering although such practice was only restricted to laboratory experiments [26, 37].

2.3 Creep

Creep failure is a time-dependent failure that deforms under mechanisms such as dislocation glide, dislocation creep, diffusional flow (Nabarro creep), and finally crack under permanent strain at elevated temperature. Creep crack growth is a common failure mechanism for die attachment because of the high-temperature operation in the power module application [39]. However, the creep fatigue of sintered Ag joint is not as widely documented nor researched compared to the solder joint.

In one of the earlier studies, Kariya et al. [17] found that creep occurs at lower homologous temperature for sintered nano-Ag joint ($0.24T_m$) than bulk Ag which creeps at $0.4T_m$ and the activation energy for the former was only half of the latter. This observation was attributed to the grain boundary of sintered Ag joints that were porous and of poor crystalline quality. This porous grain boundary also resulted in an intergranular fracture, plastic deformation, and cracks, especially at the triple points of the grain boundaries [40]; the triple point was the weakest point due to its stress concentrations [41, 42].

Other researchers also studied the creep behavior of sintered nano-Ag joints by varying their temperature and tensile stress [43]. Similar to creep failure in other structural materials, bulk sintered Ag [43] and Ag joints [44] also showed three stages in creep failure, namely, transient creep, steady-state creep, and accelerating creep stage; the steady-state stage being the most dominant part of the process. The creep lifetime of the bulk sintered Ag has been modeled with the Monkman-Grant relationship, namely [43]:

$$t_r \dot{\varepsilon}^{\mathrm{m}} = C_{\mathrm{MG}} \tag{4}$$

where the values for m and C_{MG} are 0.7836 and 8.1158, respectively, while t_r and $\dot{\varepsilon}^{\mathrm{m}}$ represent the lifetime (s) and creep strain rate (s^{-1}), respectively. In the case of sintered Ag joint, Li et al. [44] proposed a modified Arrhenius power law model to describe their creep properties based on the following model, namely:

$$\dot{\varepsilon}^{\mathrm{C}} = A' \tau^{n_0 + \frac{\beta}{T}} \exp\left(-\frac{Q_0 - \alpha \ln \tau}{RT} \right) \tag{5}$$

where $\dot{\varepsilon}^{\mathrm{C}}$ is the steady-state creep shear strain rate, τ is the applied shear stress, T is absolute temperature, n is the stress exponent, Q is the activation energy, R is Boltzmann's constant, and A is a material constant. This model described the creep strain evolution of sintered lap shear joints at 225 °C and 5 MPa (Fig. 9); the creep strain of the sintered lap joints increased with the increase of the shear stress and temperature. Such early creep models did not include fatigue stresses that were experienced simultaneously by the sintered Ag joints in power modules during operations. Instead, such composite fatigue-creep models could be borrowed from the widely-published solder joint related studies (Sect. 2.4).

In terms of fracture mechanism, the fracture surface of sintered nano-Ag lap joints creeping at 225 °C and 325 °C can be seen in Fig. 10. Tearing ridges were observed on the fracture surfaces of sintered Ag at 225 °C and 325 °C, but more extensive plastic deformation was also visible in the shear loading direction with the temperature increase [44]. Microscale cracks were also generated during the organics outgassing in the Ag sintering step [39]. At such high operating temperature, the creep stress strained the joint locally at the adjacent of those cracks. These cracks grew and merged with one another during operation, to reduce the effective bonding area of sintered Ag joints, resulting in the joint failure [39, 45].

Fig. 9 Creep strain evolution of nano-silver sintered lap shear joint at 225 °C, 5 MPa [44]

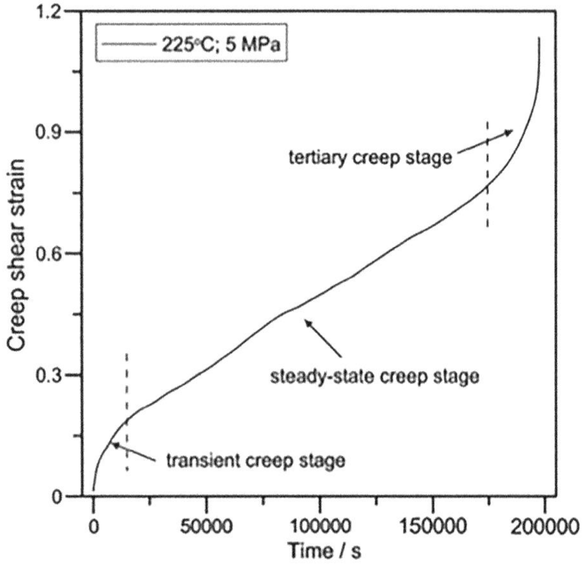

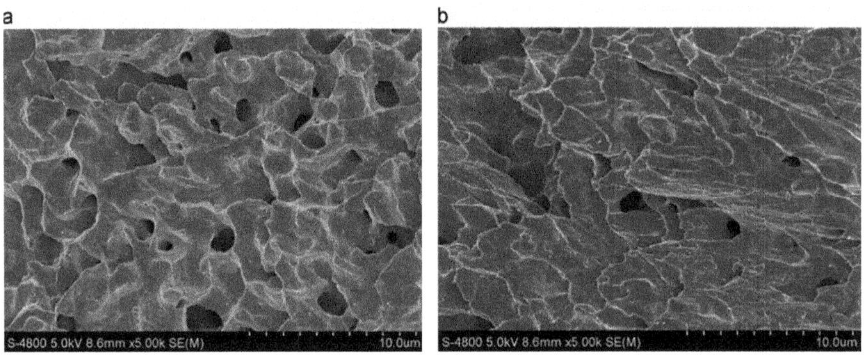

Fig. 10 Micrographs of fracture sections of sintered Ag joints due to creep at (**a**) 225 °C and (**b**) 325 °C [44]

2.4 *Fatigue and Ratcheting*

The operation environment of power devices generates temperature cycles that induce thermomechanical stresses between the electronic components and many interconnection joints due to the CTE mismatches between them [46]. The cyclic stress-strain conditions often produce fatigue failures in the sintered Ag joints.

During one low-cycle fatigue study for the sintered Ag, Shioda et al. observed that the fatigue life depended on the specimen shapes; the fatigue life increased with an increase of sintering temperature for notched specimens, while the increase of sintering temperature deteriorated the fatigue life for smooth specimens

[40]. This difference was attributed to the high-quality grain boundaries around the notch area of the specimen when the specimen was sintered at high temperature. On the other hand, smooth specimen induced localized spot of high stress that served as fatigue initiation sites to reduce the fatigue life.

Similar to other solder joints, the fatigue life of sintered Ag joints also decreased with an increase of stress amplitude and testing temperature [47]. The high testing temperature increased the dislocation mobility in the sintered Ag joints to increase the stress concentration at the interfaces of the Ag joints, resulting in low fatigue life. The Basquin model was used successfully to predict the fatigue life of sintered nano-Ag lap joint, as shown in the following equation [47]:

$$\frac{\Delta\tau}{2} = \sigma'_{\mathrm{ft}}(2N_{\mathrm{f}})^{b} \tag{6}$$

where $\Delta\tau/2$ is shear stress amplitude, N_{f} is reversals to failure, and 1 reversal is ½ cycle. Further studies confirmed that the fatigue life of sintered Ag joints reduced with the increase of stress amplitude while the increase of stress rate and stress ratio increased the fatigue life [48]. The influence of stress rate and ratio also depended on the testing temperature because high temperature would induce inelastic strain to reduce the fatigue life. Others also studied the dwell-fatigue behavior of the sintered nano-Ag lap joint [45, 47]. Their results also showed that the creep damage increased and dominated the failure with the increase of temperature and dwell time, instead of fatigue failure. A power law relationship between dwell-fatigue life N_{f} and mean shear strain rate $\dot{\gamma}_{\mathrm{m}}$ could be described as the following equation [45]:

$$N_{\mathrm{f}} = k_1\dot{\gamma}_{\mathrm{m}}^{k_2} \tag{7}$$

where k_1 and k_2 are constants fitting from testing data.

Besides fatigue failure, progressive accumulation of inelastic deformation resulted in ratcheting failure when the sintered Ag joints were subjected to cyclic loading with a non-zero mean stress higher than the material's yield strength [49, 50]. Since ratcheting behavior can lead to catastrophic failure of the sintered Ag joint, several studies have been conducted to elucidate the ratcheting mechanism [48, 51–55]. Based on the uniaxial ratcheting of sintered Ag, Wang et al. discovered that the ratcheting behavior of sintered Ag depends on the stress amplitude, mean stress, stress rate, temperature, holding time at peak stress, loading history [51], and peak stress [54]. Strain accumulated significantly with the increase of holding time at peak stress especially at high-temperature application. The ratcheting strain accumulated with high stress amplitudes or mean stresses but would reduce if stress cycling with low stress amplitudes or mean stresses was applied before loading [51], and also reduced with increasing loading rate [55]. The influence of force amplitude, mean force, and peak force dwell time became increasingly dominant with an increase of testing temperature because of the increased dislocation mobility [52].

In terms of modeling, Chen et al. proposed the composite viscoplastic constitutive model based on Ohno-Wang and Armstrong-Frederick (OW-AF) nonlinear kinematic hardening rule and the Anand model to simulate the ratcheting behavior of sintered Ag [53, 54]. Their result showed that the combined OW-AF model provides a better prediction for the ratcheting stress and strain than the traditional Anand model. The same study showed that the sintered nano-Ag joint has higher resistance to ratcheting failure than solder joints SAC305 under similar high testing conditions [56].

3 Reliability Assessment of Sintered Ag Joints

Accelerated reliability tests, such as thermal aging test, thermal cycling test, power cycling tests, and ECM tests, were used to assess the reliability of sintered Ag joints in their operation environment.

3.1 Thermal Aging

A thermal aging test measures changes in characteristic parameters, such as shear strength [57–59], porosity of sintered Ag [60], and void size [60], after the sintered Ag joints are subjected to a constant high-temperature environment for a fixed duration. These characteristic parameters serve as proxies for the evolution of the bonding quality in sintered Ag joint during the aging process. Among these parameters, shear strength is widely reported in the literature because of ease of obtaining and comparing results from various sources.

Figure 11 summarizes an increase of shear strength in the beginning before a subsequent decrease in shear strength with an increase of thermal aging time [57, 61–63]. In the beginning of the thermal aging, Ag atoms from the sintered Ag joint diffused into the substrate or wafer backside metallization to strengthen the joint strength. Then, the sintered Ag joint densified and strengthened with further thermal aging. Subsequent thermal aging decomposed any residual organics in the pores of the sintered Ag, resulting in swelling, delamination, and cracking to reduce the shear strength. At the later stage of thermal aging, sintered Ag joint recrystallized and suffered from grain growth, resulting in a further drop of shear strength. Other factors such as aging temperature, metallization type, and aging atmosphere could also affect the rate of shear strength reduction.

In terms of metallization finishes, microstructural differences were detected between sintered Ag joints on Ag and Au-plated substrates during thermal aging at 300 °C [60]. The effective bonding area between the plated Ag and the sintered Ag increased with aging, to further densify the sintered Ag joint. As mentioned in earlier paragraphs, the Ag atoms diffused from the Ag finish to the sintered Ag, leading to

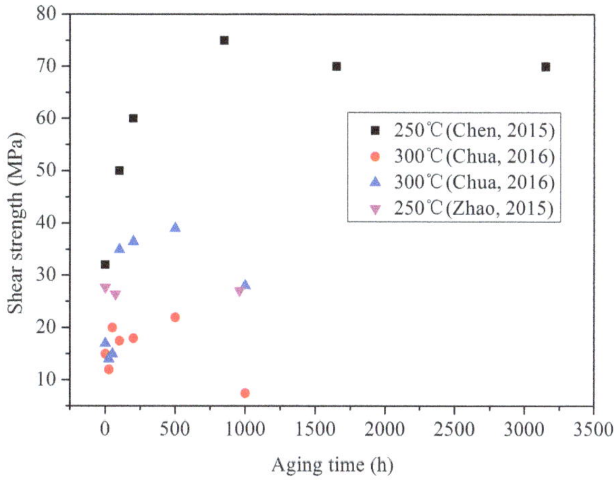

Fig. 11 Effect of thermal aging time on shear strength of sintered Ag [57–59]

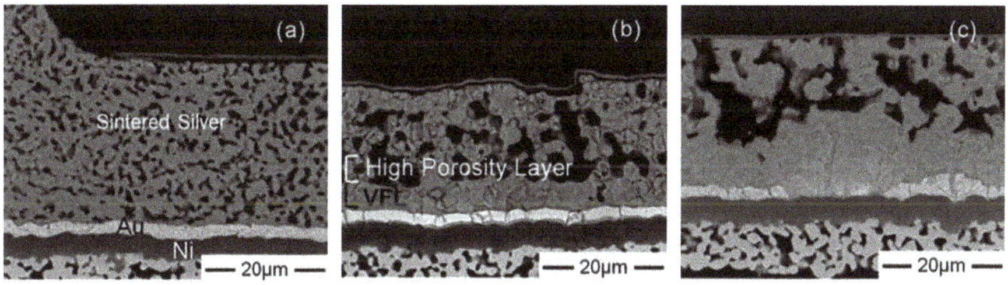

Fig. 12 SEM images of the evolution of interfacial microstructure during 300 °C aging for sintered Ag on Au substrates: (**a**) control sample, (**b**) 24 h, and (**c**) 100 h [60]

an increase in effective bonding area and lower porosity. In the case of the substrates metallized with Au, a void-free layer was formed at the interface between the sintered Ag and Au finish because of mutual diffusions between Ag atoms and Au atoms. The higher diffusivity of Ag atoms into the Au metallization at the aging temperature of 300 °C results in a high porosity zone emerging in the sintered Ag in adjacent to the die interface, as shown in Fig. 12.

3.2 *Thermal Cycling*

Besides high-temperature aging test, thermal cycling ranging from −40 °C to 125 °C (or 180 °C) has been used to study the reliability of the sintered Ag joints under accelerated condition. Similar to high-temperature aging test, several parameters, such as shear strength [64–67], thermal impedance [9], electrical resistance [68], and “life cycle to failure” [23, 69], were used to characterize the effects of these thermal cycles on the sintered Ag joints. The relationship of the shear strength of sintered Ag against the number of thermal cycles is presented in

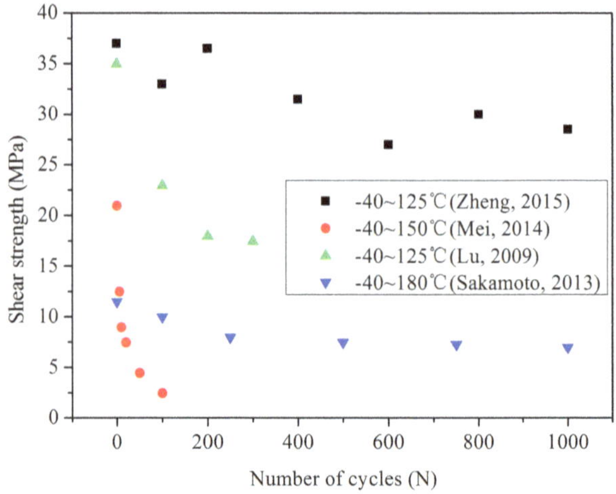

Fig. 13 Effect of the number of thermal cycles on shear strength of sintered Ag [64–67]

Fig. 13. Although the initial shear strength differed in different publications, the shear strength of sintered Ag decreased monotonically during the thermal cycling; this decrease was expected and attributed to the crack generated within the sintered Ag joint that reduced the shear strength.

This reduction is also affected by the interface metallization and joint configuration experienced in the package design. The shear strength of SiC die attachment on substrates with Au finish deteriorated less than that on substrates with Ag finish after the similar number of thermal cycles, though the initial shear strength of the Ag joint on Au finish was lower than that on substrates with Ag finish [33]. Unfortunately, no explanation was provided for this observation by the researchers, though it was likely caused by the microstructural stability related to the dissolution kinetic of the Au and Ag atoms in thermal cycle environment.

In the case of package design, the shear strength of the sintered Ag joint in a double-sided power module deteriorated more rapidly than that on a single-sided module because of the higher stress amplitude experienced in the double-sided power module [65]. Hence due to the ubiquity of double-sided design, it is common to introduce a buffer layer to relieve the stress in the die attachment region to enhance their reliability of sintered Ag joint.

In terms of stress distribution for single-sided module, the stress concentration at the die periphery produced cracks around the dies sintered by Ag paste, which grew to critical length to cause this time-dependent catastrophic failure [23, 69, 70]. This stress distribution has been demonstrated earlier by Rajaguru et al. when they modeled the accumulated strain around the package containing a SiC die sintered to the DBC substrate with Ag paste (Fig. 14) [71].

It is also possible to demonstrate this stress distribution experimentally by measuring the assembly curvature sandwiched by sintered Ag joints, as shown by Mei et al., with their optical measurement system [72, 73]. Their results showed that the residual curvature and stress decreased with the increase of the bond line

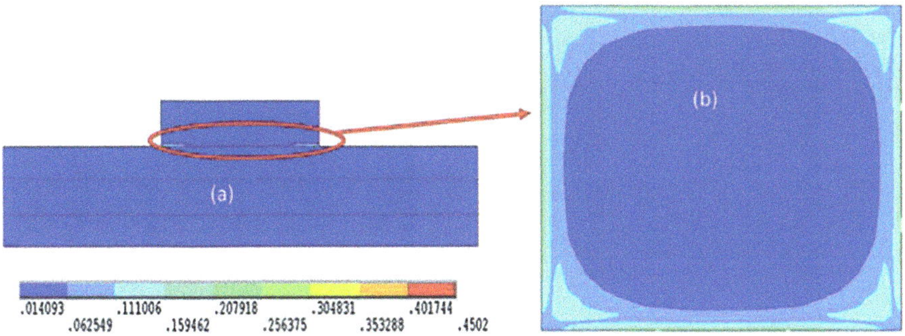

Fig. 14 Strain distribution in a sintered Ag joint [71]

thickness of the sintered Ag joint. Microscopically, Dudek et al. used a micromechanical cell model to simulate the stress distribution of sintered Ag joints in the initial porous microstructures as well as stress concentrations among the Ag particles [74]. Sintering necked at the Ag particle's contact points because of the stress concentration and inelastic strain among these Ag particles.

In terms of different Ag pastes, Sakamoto et al. [67] compared the reliability of Ag joints on Ag-plated substrates sintered with Ag nanoparticle paste and micron-scale Ag paste with the thermal cycling tests. The test result showed no visible cracks and voids in the sintered Ag joint or its interfaces during the thermal cycling. However, small voids appeared at the interface between the Cu substrate and the Ag plating layer, as shown in Fig. 15, possibly because of the high diffusivity of Cu into Ag, leading to some vacancies in the Cu substrate—Kirkendall effect. A diffusion barrier, such as nickel (Ni) plating, between the Cu substrate and the Ag plating layer was needed to reduce the formation of this void. Although cracks and voids were not observed in the sintered Ag joint as well as its interfaces during the thermal cycling, the interconnection of the sintered Ag could also fail by the additional voids appearing at the interface between the Cu substrate and the Ag plating layer that reduced the bonding strength.

However, the Ni plating also influences delamination risk for the sintered Ag joint. Figure 16 shows that delamination in the sintered Ag joint adjacent to the ENIG finish while no apparent defects could be found at the interface with electroplated Ni/Au substrate [75]. This delamination was attributed to the higher diffusivity of the Au atoms at the ENIG finish into the sintered Ag than those on the electroplated Ni/Au, caused by the different densities of grain boundaries and preferential orientation of the Au grains for the ENIG finishes. This delamination led to a reduced bonding strength and higher thermal resistance of the sintered Ag joints on the ENIG substrates.

Besides the influence of metallization finishes, Fig. 17 shows the increase of dwell-fatigue temperature will increase the grain size on the fracture surface [45]. At this higher temperature, the grain boundary and surface motion resulted in grain growth and coarsening in the sintered Ag that finally accelerated the fatigue failure.

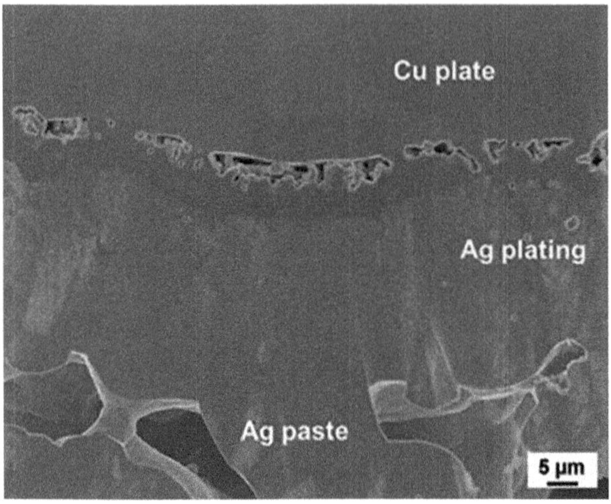

Fig. 15 Interface of Ag plating and Cu plate after thermal cycling up to 1000 cycles [67]

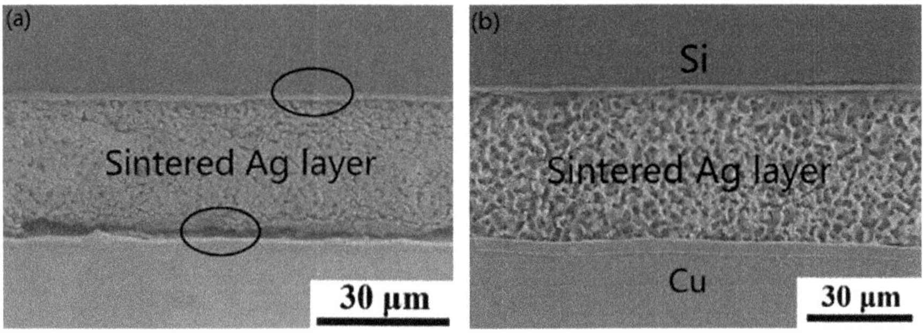

Fig. 16 SEM images of cross section of joint on (**a**) ENIG substrate and (**b**) Ni/Au substrate [75]

3.3 Power Cycling

Research on power cycling of modules packaged by sintered Ag is less widely available than the thermal aging or thermal cycling because power cycling test involved intricate test bench and analysis. Unlike the thermal cycling, the heat source of power cycling is the semiconductor dies of the power module rather than the external environment; the die is switched on and off to heat and cool the whole power module, respectively, during the power cycling,

A commonly used power cycling profile is shown in Fig. 18, but the failure criteria of power cycling test varied from one researcher to another when the die junction temperature T_j is cycled according to the changes in the voltage and current input [76]. For example, Bajwa et al. [77] took an increase of the forward voltage by 5–10% of the initial value as the failure criterion, while Weber et al. [78] took the occurrence of electric short as the failure criterion.

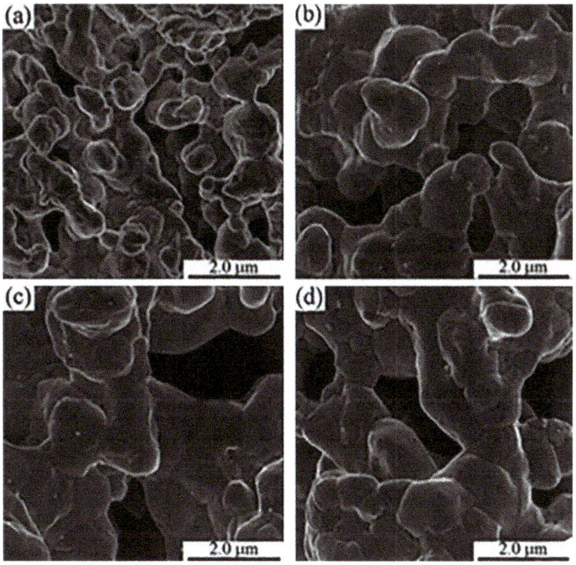

Fig. 17 High-magnification morphology of fracture surface of sintered Ag joints under different dwell-fatigue temperatures: (**a**) 125 °C, (**b**) 225 °C, (**c**) 275 °C, and (**d**) 325 °C [45]

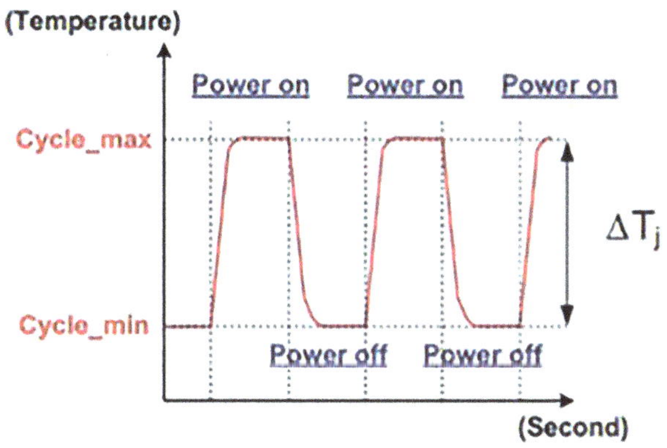

Fig. 18 A commonly used power cycling mode [76]

In general, the sintered Ag performed better than conventional solder or adhesives [78–80]. Bajwa et al. compared the power cycling results of power modules with SiC diodes die-attached by micron-Ag paste, Sn-Ag transient liquid phase (TLP) material, and electrically conductive adhesive [77]. The result showed that Ag-sintered samples survived 1.5 times longer than TLP-bonded samples and 6 times longer than adhesive-bonded samples. Others compared the multichip phase-leg IGBT modules bonded with pressureless nano-Ag paste with a commercial module bonded with Sn5Pb92.5Ag2.5 solder based on the second-level and minute-level power cycling tests [81]; the thermal impedance and electrical resistance of the Ag joints were lower than the solder joints under the same number of cycles.

In terms of package design, Amro et al. demonstrated that double-sided low-temperature joining technique (LTJT) bondings with sintering Ag paste could improve the power cycling reliability by a factor of 2 compared to the traditional solder connection [82, 83]. However, power cycling modeling showed that sintered Ag joints are more prone to brittle failure than those tested under the low-cycle fatigue test because of the higher stiffness for the Ag joints compared to the solder joint [84, 85]. The sintered Ag joint reliability could increase significantly by several hundred percent if brittle failure could be avoided using re-designed package layout.

3.4 Electrochemical Migration (ECM) of Sintered Ag

Electrochemical migration (ECM) is the formation of metallic dendrites between anode and cathode electrodes under an applied voltage across a nonmetallic medium in the moisture and oxygen-containing environment [86, 87]. Silver and, by extension, sintered Ag are the most susceptible element to ECM failure among the different metals [88].

Figure 19 presents the schematic diagram for the ECM mechanisms between two Ag electrodes; the Ag atoms of anode dissolved into Ag ions before depositing at the cathode in the form of Ag atoms to form a Ag conducting path, usually in a dendritic shape, between the anode and cathode. Formation of ECM dendrites between different electrodes of sintered Ag resulted in short circuit and damages to the semiconductor dies.

Factors which could influence the initiation and propagation of ECM in sintered Ag joints are bias voltage [89, 90], temperature [90], humidity [91], oxygen partial pressure [92], and spacing between electrodes [89]. Specifically, ECM of sintered

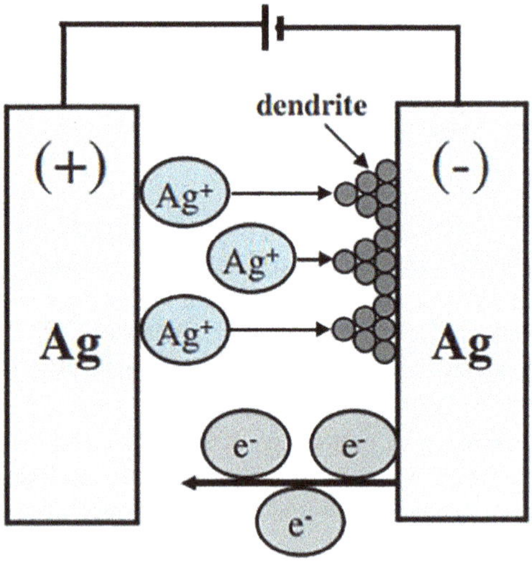

Fig. 19 Schematic diagram of ECM phenomenon between Ag electrodes [86]

Ag proliferated with an increase of bias voltage and temperature but delayed by the increase of electrode spacing [90, 93]. The ECM process sped up in a moist environment based on the following chemical equations:

$$Ag \rightarrow Ag^{+} + e^{-} \tag{8}$$

$$2Ag + 2OH^{-} \rightarrow 2AgOH \rightarrow Ag_2O + H_2O \tag{9}$$

$$Ag^{+} + e^{-} \rightarrow Ag \tag{10}$$

$$Ag_2O + 2H^{+} + 2e^{-} \rightarrow 2Ag + H_2O \tag{11}$$

Silver atoms at anode transformed to Ag^+ and Ag_2 based on Eqs. (8) and (9), which reduced to Ag atoms at cathode (based on Eqs. (10) and (11)). In the case of the dry atmosphere, oxygen decreased the onset time for significant leakage current to develop across the two electrodes exponentially [92]. Further research clarified the role of oxygen in the formation of ECM in the sintered Ag electrodes, as shown in the following equations [94]:

$$2Ag + 1/2O_2 \rightarrow Ag_2O \tag{12}$$

$$Ag_2O \rightarrow 2Ag^{+} + O^{2-} \tag{13}$$

At anode, Ag atoms oxidized to Ag_2O that dissociated into Ag cations and oxygen anions. The Ag cations migrated to the cathode to be reduced to dendritic silvers. Hence, an increase of partial pressure oxygen drives this Ag dendritic formation.

Based on the application of sintered Ag as die-attach material, it is not feasible to lower the bias voltage or temperature or increase the electrode spacing between electrodes because of the design requirement, limited real estate in the power module packaging in high-power density, and high working temperature environment. Hence, the only feasible technique to reduce the ECM formation is to separate the sintered Ag electrodes from water and oxygen. A parylene overcoat [93] or addition of epoxy into the formulation of Ag paste can achieve this protection and separation between moisture and sintered Ag joint.

In terms of sintering conditions, an increase of sintering temperature or sintering duration also produced a stable microstructure of sintered Ag joints to resist ECM in sintered Ag electrodes [86]. The sintered Ag resistance to ECM is also increased by adding Pd elements into the Ag paste which produced PdO to hinder the Ag dissolution and minimize the Ag migration steps in the electrodes [95–97]. Unfortunately, these PdO compounds also impeded the sintering step. Hence, others suggested Au [98] or Cu [99] to increase the resistance of Ag ion migration and ECM while maintaining the sinterability of the Ag paste; 5 vol% of Cu nanoparticles provided the optimum resistance to the Ag ion migration [99].

4 Conclusions and Outlook

In conclusion, reliable sintered Ag joints improve the robustness of the power device package. While sintered Ag joints can be produced reliably en masse, its long-term reliability depends on various factors discussed earlier (e.g., interfacial metallization and microstructural stability) in this chapter. In general, the elastic modulus and tensile and shear strength depend on the densification of sintered Ag controlled by sintering time, temperature, and pressure, while higher operating temperature reduces the elastic modulus due to increased mobility of various creep mechanisms at the higher temperature. The porous grain boundary of sintered Ag decreases the creep strength of the sintered Ag compared to bulk Ag while the fracture surface also shows intergranular failure with more considerable plastic deformation at higher operating temperature.

Various lifetime models predicted that the creep and fatigue-ratcheting life of sintered Ag joints are longer than the traditional die-attach joints. Initial thermal aging densified and improved the shear strength of sintered Ag before deterioration at later stage caused by delamination and cracking, while thermal cycling reduced their interfacial strength throughout the cyclic test. Substrate metallization can mitigate the various deterioration mechanisms due to the diffusivity of different elements at the interface; Ni plating reduces void formation at the Ag plated substrate to increase the strength but also creates delamination issue.

Besides this interfacial metallization approach, the reduction of CTE mismatch is the other strategy to improve the reliability of this Ag bonding technique; materials formulation and package design are the two main variables in reducing the CTE mismatch. While package design is determined by other requirements beyond the package engineers' influence, materials formulation is the key innovation in this area, as demonstrated by the high number of patents filed in this area [34]. Some researchers claimed that the addition of SiC or h-boron nitride (BN) powders into the nano-Ag paste could reduce the CTE mismatch between the sintered Ag and a die or a substrate by as much as 30% based on their finite element modeling [100]. An earlier discussion also showed that the modification of Ag paste formulation with other elements (e.g. Pd, Cu etc.) also reduced the mobility of Ag ions to decrease the ECM. Alternatively, the lower CTE value of aluminum nitride (AlN) substrate provided a better match than the Al_2O_3 substrate with sintered Ag joint to reduce the resulting thermomechanical stress [101].

However, such decision is also influenced by the cost structure of the package. At the same time, there are also constant efforts in developing the model to estimate this fatigue lifetime of these joints under different environments; such models are discussed in this chapter. Continuous research in this area will only enhance the confidence of power device community to adopt this sintered Ag joint as the die-attach joint of choice in the near future.

Acknowledgment SKS acknowledges Universiti Kebangsaan Malaysia Research Grants (GUP-2017-055 "Production of Metallic Conducting Nanowires for Industrial Applications") for this work.

References

1. O.A. Ogunseitan, Public health and environmental benefits of adopting lead-free solders. JOM **59**, 12 (2007)
2. C. Buttay, D. Planson, B. Allard, D. Bergogne, P. Bevilacqua, C. Joubert, et al., State of the art of high temperature power electronics. Mater. Sci. Eng. B **176**, 283–288 (2011)
3. J. Millán, P. Godignon, X. Perpiñà, A. Pérez-Tomás, J. Rebollo, A survey of wide bandgap power semiconductor devices. IEEE Trans. Power Electron. **29**, 2155–2163 (2014)
4. R. Khazaka, L. Mendizabal, D. Henry, R. Hanna, Survey of high-temperature reliability of power electronics packaging components. IEEE Trans. Power Electron. **30**, 2456–2464 (2015)
5. H.S. Chin, K.Y. Cheong, A.B. Ismail, A review on die attach materials for SiC-based high-temperature power devices. Metall. Mater. Trans. B **41**, 824–832 (2010)
6. G. Zeng, S. Mcdonald, K. Nogita, Development of high-temperature solders: Review. Microelectron. Reliab. **52**, 1306–1322 (2012)
7. C. Buttay, A. Masson, J. Li, M. Johnson, M. Lazar, C. Raynaud et al., Die attach of power devices using silver sintering – bonding process optimisation and characterization, in *Additional Conferences (Device Packaging, HiTEC, HiTEN, & CICMT)*, 2011, pp. 000084–000090
8. S.-Y. Zhao, X. Li, Y.-H. Mei, G.-Q. Lu, Effect of silver flakes in silver paste on the joining process and properties of sandwich power modules (IGBTs chip/silver paste/bare Cu). J. Electron. Mater. **45**, 5789–5799 (2016)
9. F.L. Henaff, S. Azzopardi, J.Y. Deletage, E. Woirgard, S. Bontemps, J. Joguet, A preliminary study on the thermal and mechanical performances of sintered nano-scale silver die-attach technology depending on the substrate metallization. Microelectron. Reliab. **52**, 2321–2325 (2012)
10. H. Schwarzbauer, R. Kuhnert, Novel large area joining technique for improved power device performance, in *Conference Record of the IEEE Industry Applications Society Annual Meeting*, vol. 2, 1989, pp. 1348–1351
11. C. Herring, Effect of change of scale on sintering phenomena. J. Appl. Phys. **21**, 301–303 (1950)
12. J.G. Bai, Z.Z. Zhang, J.N. Calata, G.Q. Lu, Low-temperature sintered nanoscale silver as a novel semiconductor device-metallized substrate interconnect material. IEEE Trans. Compon. Packag. Technol. **29**, 589–593 (2006)
13. F.L. Henaff, S. Azzopardi, E. Woirgard, T. Youssef, S. Bontemps, J. Joguet, Lifetime evaluation of nanoscale silver sintered power modules for automotive application based on experiments and finite-element modeling. IEEE Trans. Device Mater. Reliab. **15**, 326–334 (2015)
14. K.S. Siow, Mechanical properties of nano-silver joints as die attach materials. J. Alloys Compd. **514**, 6–19 (2012)
15. X. Milhet, P. Gadaud, V. Caccuri, D. Bertheau, D. Mellier, M. Gerland, Influence of the porous microstructure on the elastic properties of sintered Ag paste as replacement material for die attachment. J. Electron. Mater. **44**, 3948–3956 (2015)
16. V.A. Levin, V.V. Lokhin, K.M. Zingerman, Effective elastic properties of porous materials with randomly dispersed pores: finite deformation. J. Appl. Mech. **67**, 667–670 (2000)
17. Y. Kariya, H. Yamaguchi, M. Itako, N. Mizumura, K. Sasaki, Mechanical behavior of sintered nano-sized Ag particles. J. Smart Process. **2**, 160–165 (2013)
18. V. Caccuri, X. Milhet, P. Gadaud, D. Bertheau, M. Gerland, Mechanical properties of sintered Ag as a new material for die bonding: influence of the density. J. Electron. Mater. **43**, 4510–4514 (2014)
19. N. Ramakrishnan, V.S. Arunachalam, Effective elastic moduli of porous solids. J. Mater. Sci. **25**, 3930–3937 (1990)
20. J. Kähler, N. Heuck, G. Palm, A. Stranz, A. Waag, E. Peiner, Low-pressure sintering of silver micro-and nanoparticles for a high temperature stable Pick & Place die attach, in *Microelectronics and Packaging Conference (EMPC), 2011 18th European: IEEE*, 2011, pp. 1–7

21. D.J. Yu, X. Chen, G. Chen, G.Q. Lu, Z.Q. Wang, Applying Anand model to low-temperature sintered nanoscale silver paste chip attachment. Mater. Des. **30**, 4574–4579 (2009)
22. H. Ma, J.C. Suhling, A review of mechanical properties of lead-free solders for electronic packaging. J. Mater. Sci. **44**, 1141–1158 (2009)
23. M. Knoerr, S. Kraft, A. Schletz, Reliability assessment of sintered nano-silver die attachment for power semiconductors, in *2010 12th Electronics Packaging Technology Conference*, 2010, pp. 56–61
24. Y. Akada, H. Tatsumi, T. Yamaguchi, A. Hirose, T. Morita, E. Ide, Interfacial bonding mechanism using silver metallo-organic nanoparticles to bulk metals and observation of sintering behavior. Mater. Trans. **49**, 1537–1545 (2008)
25. C. Weber, H. Walter, M.V. Dijk, M. Hutter, O. Wittler, K.D. Lang, Combination of experimental and simulation methods for analysis of sintered Ag joints for high temperature applications, in *2016 I.E. 66th Electronic Components and Technology Conference (ECTC)*, 2016, pp. 1335–1341
26. H. Zheng, D. Berry, J.N. Calata, K.D.T. Ngo, S. Luo, G.Q. Lu, Low-pressure joining of large-area devices on copper using nanosilver paste. IEEE Trans. Compon. Packag. Manuf. Technol. **3**, 915–922 (2013)
27. T.G. Lei, J.N. Calata, G.Q. Lu, X. Chen, S. Luo, Low-temperature sintering of nanoscale silver paste for attaching large-area (>100 mm^2) chips. IEEE Trans. Compon. Packag. Technol. **33**, 98–104 (2010)
28. R. Khazaka, L. Mendizabal, D. Henry, Review on joint shear strength of nano-silver paste and its long-term high temperature reliability. J. Electron. Mater. **43**, 2459–2466 (2014)
29. Z. Zhang, L. Guo-Quan, Pressure-assisted low-temperature sintering of silver paste as an alternative die-attach solution to solder reflow. IEEE Trans. Electron. Packag. Manuf. **25**, 279–283 (2002)
30. T. Wang, X. Chen, G.-Q. Lu, G.-Y. Lei, Low-temperature sintering with nano-silver paste in die-attached interconnection. J. Electron. Mater. **36**, 1333–1340 (2007)
31. G. Zou, J. Yan, F. Mu, A. Wu, J. Ren, A. Hu, et al., Low temperature bonding of Cu metal through sintering of Ag nanoparticles for high temperature electronic application. Open Surf. Sci. J. **3**, 70–75 (2011)
32. M.Y. Wang, Y.H. Mei, X. Li, G.Q. Lu, Relationship between transient thermal impedance and shear strength of pressureless sintered silver as die attachment for power devices, in *2015 International Conference on Electronics Packaging and iMAPS All Asia Conference (ICEP-IAAC)*, 2015, pp. 559–564
33. J.G. Bai, G.Q. Lu, Thermomechanical reliability of low-temperature sintered silver die attached SiC power device assembly. IEEE Trans. Device Mater. Reliab. **6**, 436–441 (2006)
34. K.S. Siow, Y.T. Lin, Identifying the development state of sintered Ag as a bonding material in the microelectronic packaging via a patent landscape study. J. Electron. Packag. **138**, 020804 (2016)
35. S.A. Paknejad, S.H. Mannan, Review of silver nanoparticle based die attach materials for high power/temperature applications. Microelectron. Reliab. **70**, 1–11 (2017)
36. K. Qi, X. Chen, G.Q. Lu, Effect of interconnection area on shear strength of sintered joint with nano-silver paste. Soldering Surf. Mount Technol. **20**, 8–12 (2008)
37. E. Ide, S. Angata, A. Hirose, K.F. Kobayashi, Bonding of various metals using Ag metallo-organic nanoparticles: a novel bonding process using Ag metallo-organic nanoparticles. Mater. Sci. Forum **512**, 383–388 (2006)
38. B. Boettge, B. Maerz, J. Schischka, S. Klengel, M. Petzold, High resolution failure analysis of silver-sintered contact interfaces for power electronics, in *CIPS 2014; 8th International Conference on Integrated Power Electronics Systems*, 2014, pp. 1–7
39. Y. Tan, X. Li, G. Chen, Y. Mei, X. Chen, Three-dimensional visualization of the crack-growth behavior of nano-silver joints during shear creep. J. Electron. Mater. **44**, 761–769 (2015)
40. R. Shioda, Y. Kariya, N. Mizumura, K. Sasaki, Low-cycle fatigue life and fatigue crack propagation of sintered Ag nanoparticles. J. Electron. Mater. **46**, 1155–1162 (2017)

41. T. Herboth, M. Guenther, A. Fix, J. Wilde, Failure mechanisms of sintered silver interconnections for power electronic applications, in *2013 I.E. 63rd Electronic Components and Technology Conference*, 2013, pp. 1621–1627
42. J.G. Bai, J.N. Calata, L. Guangyin, L. Guo-Quan, Thermomechanical reliability of low-temperature sintered silver die-attachment, in *Thermal and Thermomechanical Proceedings 10th Intersociety Conference on Phenomena in Electronics Systems, 2006 ITherm 2006*, 2006, pp. 1126–1130
43. G. Chen, X.-H. Sun, P. Nie, Y.-H. Mei, G.-Q. Lu, X. Chen, High-temperature creep behavior of low-temperature-sintered nano-silver paste films. J. Electron. Mater. **41**, 782–790 (2012)
44. X. Li, G. Chen, L. Wang, Y.-H. Mei, X. Chen, G.-Q. Lu, Creep properties of low-temperature sintered nano-silver lap shear joints. Mater. Sci. Eng. A **579**, 108–113 (2013)
45. Y. Tan, X. Li, Y. Mei, G. Chen, X. Chen, Temperature-dependent dwell-fatigue behavior of nanosilver sintered lap shear joint. J. Electron. Packag. **138**, 021001–021008 (2016)
46. C. Kanchanomai, Y. Miyashita, Y. Mutoh, Low cycle fatigue behavior and mechanisms of a eutectic Sn–Pb solder 63Sn/37Pb. Int. J. Fatigue **24**, 671–683 (2002)
47. Y. Tan, X. Li, X. Chen, Fatigue and dwell-fatigue behavior of nano-silver sintered lap-shear joint at elevated temperature. Microelectron. Reliab. **54**, 648–653 (2014)
48. G. Chen, L. Yu, Y. Mei, X. Li, X. Chen, G.-Q. Lu, Uniaxial ratcheting behavior of sintered nanosilver joint for electronic packaging. Mater. Sci. Eng. A **591**, 121–129 (2014)
49. J. Ma, H. Gao, L. Gao, X. Chen, Uniaxial ratcheting behavior of anisotropic conductive adhesive film at elevated temperature. Polym. Test. **30**, 571–577 (2011)
50. M. Shariati, H. Hatami, H. Yarahmadi, H.R. Eipakchi, An experimental study on the ratcheting and fatigue behavior of polyacetal under uniaxial cyclic loading. Mater. Des. **34**, 302–312 (2012)
51. T. Wang, G. Chen, Y. Wang, X. Chen, G.-Q. Lu, Uniaxial ratcheting and fatigue behaviors of low-temperature sintered nano-scale silver paste at room and high temperatures. Mater. Sci. Eng. A **527**, 6714–6722 (2010)
52. X. Li, G. Chen, X. Chen, G.-Q. Lu, L. Wang, Y.-H. Mei, High temperature ratcheting behavior of nano-silver paste sintered lap shear joint under cyclic shear force. Microelectron. Reliab. **53**, 174–181 (2013)
53. G. Chen, Z.-S. Zhang, Y.-H. Mei, X. Li, D.-J. Yu, L. Wang, et al., Applying viscoplastic constitutive models to predict ratcheting behavior of sintered nanosilver lap-shear joint. Mech. Mater. **72**, 61–71 (2014)
54. G. Chen, Z.-S. Zhang, Y.-H. Mei, X. Li, G.-Q. Lu, X. Chen, Ratcheting behavior of sandwiched assembly joined by sintered nanosilver for power electronics packaging. Microelectron. Reliab. **53**, 645–651 (2013)
55. X. Chen, R. Li, K. Qi, G.-Q. Lu, Tensile behaviors and ratcheting effects of partially sintered chip-attachment films of a nanoscale silver paste. J. Electron. Mater. **37**, 1574 (2008)
56. G. Chen, L. Yu, Y.-H. Mei, X. Li, X. Chen, G.-Q. Lu, Reliability comparison between SAC305 joint and sintered nanosilver joint at high temperatures for power electronic packaging. J. Mater. Process. Technol. **214**, 1900–1908 (2014)
57. S.T. Chua, K.S. Siow, Microstructural studies and bonding strength of pressureless sintered nano-silver joints on silver, direct bond copper (DBC) and copper substrates aged at 300 °C. J. Alloys Compd. **687**, 486–498 (2016)
58. S. Chen, G. Fan, X. Yan, C. LaBarbera, L. Kresge, N.C. Lee, Achieving high reliability via pressureless sintering of nano Ag paste for die attach, in *2015 16th International Conference on Electronic Packaging Technology (ICEPT)*, 2015, pp. 367–374
59. S.-Y. Zhao, X. Li, Y.-H. Mei, G.-Q. Lu, Study on high temperature bonding reliability of sintered nano-silver joint on bare copper plate. Microelectron. Reliab. **55**, 2524–2531 (2015)
60. S.A. Paknejad, G. Dumas, G. West, G. Lewis, S.H. Mannan, Microstructure evolution during 300 °C storage of sintered Ag nanoparticles on Ag and Au substrates. J. Alloys Compd. **617**, 994–1001 (2014)

61. K.S. Siow, Are sintered silver joints ready for use as interconnect material in microelectronic packaging? J. Electron. Mater. **43**, 947–961 (2014)
62. S. Egelkraut, L. Frey, M. Knoerr, A. Schletz, Evolution of shear strength and microstructure of die bonding technologies for high temperature applications during thermal aging, in *2010 12th Electronics Packaging Technology Conference*, 2010, pp. 660–667
63. R. Kisiel, Z. Szczepański, P. Firek, J. Grochowski, M. Myśliwiec, M. Guziewicz, Silver micropowders as SiC die attach material for high temperature applications, in *2012 35th International Spring Seminar on Electronics Technology*, 2012, pp. 144–148
64. G.Q. Lu, M. Zhao, G. Lei, J.N. Calata, X. Chen, S. Luo, Emerging lead-free, high-temperature die-attach technology enabled by low-temperature sintering of nanoscale silver pastes, in *2009 International Conference on Electronic Packaging Technology & High Density Packaging*, 2009, pp. 461–466
65. Y.H. Mei, J.Y. Lian, X. Chen, G. Chen, X. Li, G.Q. Lu, Thermo-mechanical reliability of double-sided IGBT assembly bonded by sintered nanosilver. IEEE Trans. Device Mater. Reliab. **14**, 194–202 (2014)
66. H. Zheng, K.D.T. Ngo, G.Q. Lu, Temperature cycling reliability assessment of die attachment on bare copper by pressureless nanosilver sintering. IEEE Trans. Device Mater. Reliab. **15**, 214–219 (2015)
67. S. Sakamoto, T. Sugahara, K. Suganuma, Microstructural stability of Ag sinter joining in thermal cycling. J. Mater. Sci. Mater. Electron. **24**, 1332–1340 (2013)
68. F. Shancan, X. Yijing, M. Yunhui, Reliability of pressureless sintered nanosilver for attaching IGBT devices, in *2016 International Conference on Electronics Packaging (ICEP)*, 2016, pp. 382–385
69. T. Herboth, C. Früh, M. Günther, J. Wilde, Assessment of thermo-mechanical stresses in Low Temperature Joining Technology, in *2012 13th International Thermal, Mechanical and Multi-Physics Simulation and Experiments in Microelectronics and Microsystems*, 2012, pp. 1/7–7/7
70. D. Agata Skwarek, R. Dudek, P. Sommer, A. Fix, J. Trodler, S. Rzepka, et al., Reliability investigations for high temperature interconnects. Soldering Surf. Mount Technol. **26**, 27–36 (2014)
71. P. Rajaguru, H. Lu, C. Bailey, Sintered silver finite element modelling and reliability based design optimisation in power electronic module. Microelectron. Reliab. **55**, 919–930 (2015)
72. Y. Mei, G. Chen, X. Li, G.-Q. Lu, X. Chen, Evolution of curvature under thermal cycling in sandwich assembly bonded by sintered nanosilver paste. Soldering Surf. Mount Technol. **25**, 107–116 (2013)
73. Y. Mei, G. Chen, L. Guo-Quan, X. Chen, Effect of joint sizes of low-temperature sintered nano-silver on thermal residual curvature of sandwiched assembly. Int. J. Adhes. Adhes. **35**, 88–93 (2012)
74. R. Dudek, R. Döring, P. Sommer, B. Seiler, K. Kreyssig, H. Walter et al., Combined experimental- and FE-studies on sinter-Ag behaviour and effects on IGBT-module reliability, in *2014 15th International Conference on Thermal, Mechanical and Mulit-Physics Simulation and Experiments in Microelectronics and Microsystems (EuroSimE)*, 2014, pp. 1–9
75. Q. Xu, Y. Mei, X. Li, G.-Q. Lu, Correlation between interfacial microstructure and bonding strength of sintered nanosilver on ENIG and electroplated Ni/Au direct-bond-copper (DBC) substrates. J. Alloys Compd. **675**, 317–324 (2016)
76. T.Y. Hung, S.Y. Chiang, C.J. Huang, C.C. Lee, K.N. Chiang, Thermal–mechanical behavior of the bonding wire for a power module subjected to the power cycling test. Microelectron. Reliab. **51**, 1819–1823 (2011)
77. A.A. Bajwa, E. Möller, J. Wilde, Die-attachment technologies for high-temperature applications of Si and SiC-based power devices, in *2015 I.E. 65th Electronic Components and Technology Conference (ECTC)*, 2015, pp. 2168–2174
78. C. Weber, M. Hutter, S. Schmitz, K.D. Lang, Dependency of the porosity and the layer thickness on the reliability of Ag sintered joints during active power cycling, in *2015 I.E. 65th Electronic Components and Technology Conference (ECTC)*, 2015, pp. 1866–1873

79. S. Haumann, J. Rudzki, F. Osterwald, M. Becker, R. Eisele, Novel bonding and joining technology for power electronics - Enabler for improved lifetime, reliability, cost and power density, in *2013 Twenty-Eighth Annual IEEE Applied Power Electronics Conference and Exposition (APEC)*, 2013, pp. 622–626
80. N. Heuck, K. Guth, M. Thoben, A. Mueller, N. Oeschler, L. Boewer et al., Aging of new Interconnect-Technologies of Power-Modules during Power-Cycling, in *CIPS 2014; 8th International Conference on Integrated Power Electronics Systems*, 2014, pp. 1–6
81. S. Fu, Y. Mei, X. Li, C. Ma, G.Q. Lu, Reliability evaluation of multichip phase-leg IGBT modules using pressureless sintering of nanosilver paste by power cycling tests. IEEE Trans. Power Electron. **32**, 6049–6058 (2017)
82. R. Amro, J. Lutz, J. Rudzki, M. Thoben, A. Lindemann, Double-sided low-temperature joining technique for power cycling capability at high temperature, in *2005 European Conference on Power Electronics and Applications*, 2005, p. 10
83. R. Amro, J. Lutz, J. Rudzki, R. Sittig, M. Thoben, Power cycling at high temperature swings of modules with low temperature joining technique, in *2006 I.E. International Symposium on Power Semiconductor Devices and IC's*, 2006, pp. 1–4
84. R. Dudek, R. Döring, S. Rzepka, C. Ehrhardt, M. Günther, M. Haag, Electro-thermo-mechanical analyses on silver sintered IGBT-module reliability in power cycling, in *2015 16th International Conference on Thermal, Mechanical and Multi-Physics Simulation and Experiments in Microelectronics and Microsystems*, 2015, pp. 1–8
85. R. Dudek, R. Döring, S. Rzepka, C. Ehrhardt, M. Hutter, J. Rudzki et al., Investigations on power cycling induced fatigue failure of IGBTs with silver sintered interconnects, in *2015 European Microelectronics Packaging Conference (EMPC)*, 2015, pp. 1–8
86. B.-I. Noh, J.-W. Yoon, K.-S. Kim, S. Kang, S.-B. Jung, Electrochemical migration of directly printed Ag electrodes using Ag paste with epoxy binder. Microelectron. Eng. **103**, 1–6 (2013)
87. B.-I. Noh, J.-W. Yoon, W.-S. Hong, S.-B. Jung, Evaluation of electrochemical migration on flexible printed circuit boards with different surface finishes. J. Electron. Mater. **38**, 902–907 (2009)
88. J. Steppan, J. Roth, L. Hall, D. Jeannotte, S. Carbone, A review of corrosion failure mechanisms during accelerated tests electrolytic metal migration. J. Electrochem. Soc. **134**, 175–190 (1987)
89. G.Q. Lu, W. Yang, Y. Mei, X. Li, G. Chen, X. Chen, Effects of DC bias and spacing on migration of sintered nanosilver at high temperatures for power electronic packaging, in *2013 14th International Conference on Electronic Packaging Technology*, 2013, pp. 925–930
90. G.Q. Lu, W. Yang, Y.H. Mei, X. Li, G. Chen, X. Chen, Migration of sintered nanosilver on alumina and aluminum nitride substrates at high temperatures in dry air for electronic packaging. IEEE Trans. Device Mater. Reliab. **14**, 600–606 (2014)
91. C.-H. Tsou, K.-N. Liu, H.-T. Lin, F.-Y. Ouyang, Electrochemical migration of fine-pitch nanopaste Ag interconnects. J. Electron. Mater. **45**, 6123–6129 (2016)
92. Y. Mei, G.Q. Lu, X. Chen, S. Luo, D. Ibitayo, Effect of oxygen partial pressure on silver migration of low-temperature sintered nanosilver die-attach material. IEEE Trans. Device Mater. Reliab. **11**, 312–315 (2011)
93. R. Riva, C. Buttay, B. Allard, P. Bevilacqua, Migration issues in sintered-silver die attaches operating at high temperature. Microelectron. Reliab. **53**, 1592–1596 (2013)
94. G.Q. Lu, W. Yang, Y.H. Mei, X. Li, G. Chen, X. Chen, Mechanism of migration of sintered nanosilver at high temperatures in dry air for electronic packaging. IEEE Trans. Device Mater. Reliab. **14**, 311–317 (2014)
95. J.C. Lin, J.Y. Chan, On the resistance of silver migration in Ag-Pd conductive thick films under humid environment and applied d.c. field. Mater. Chem. Phys. **43**, 256–265 (1996)
96. J.C. Lin, J.Y. Chuang, Resistance to silver electrolytic migration for thick-film conductors prepared from mixed and alloyed powders of Ag-15Pd and Ag-30Pd. J. Electrochem. Soc. **144**, 1652–1659 (1997)

97. D. Wang, Y. Mei, K.S. Siow, X. Li, G. Lu, Roles of palladium particles in enhancing the electrochemical migration resistance of sintered nano-silver paste as a bonding material. Mater. Lett. **206**, 1–4 (2017)
98. T. Ito, T. Ogura, A. Hirose, Effects of Au and Pd additions on joint strength, electrical resistivity, and ion-migration tolerance in low-temperature sintering bonding using Ag_2O paste. J. Electron. Mater. **41**, 2573–2579 (2012)
99. M. Koh, K.-S. Kim, B.-G. Park, K.-H. Jung, C.S. Lee, Y.-H. Choa, et al., Electrical and electrochemical migration characteristics of Ag/Cu nanopaste patterns. J. Nanosci. Nanotechnol. **14**, 8915–8919 (2014)
100. N. Heuck, A. Langer, A. Stranz, G. Palm, R. Sittig, A. Bakin, et al., Analysis and modeling of thermomechanically improved silver-sintered die-attach layers modified by additives. IEEE Trans. Compon. Packag. Manuf. Technol. **1**, 1846–1855 (2011)
101. H. Sugihara, M. Yamagiwa, M. Fujita, T. Oshidari, Q. Yu, Thermal fatigue reliability of high-temperature-resistant joint for power devices, in *ASME 2009 InterPACK Conference*, 2009, pp. 937–943

Morphological Changes in Sintered Silver Due to Atomic Migration

S. Mannan, A. Paknejad, A. Mansourian, and K. Khtatba

1 Introduction

Sintered silver (Ag) is a material that is attractive as a high-temperature joining material because the processing can be accomplished at relatively low temperatures, typically 200–300 °C. Consequently, the use of sintered Ag as a die attach, its advantages and open questions have been reviewed by many authors [1–8]. The high melting point of Ag, 962 °C, however results in theoretical stability up to around 350 °C even if the extremely conservative criterion of homologous temperature equal to 0.5 is used. The use of submicron particles and nanoparticles not only facilitates the use of low-temperature processing but also allows the sintering to take place with reduced or zero applied sintering pressure, such that the range of pressures applied is typically 0–10 MPa. One advantage of applying pressure is to reduce porosity which lies in the 20–30% range for pressureless sintering and below 20% when pressure is applied. However there is not a clear consensus in the literature as to the desirability of lower porosity. On the one hand, increasing sintering pressure leads to lower porosity, higher interfacial contact area and hence higher initial shear strength. On the other hand, increased porosity results in lower Young's modulus and hence less stress transferred to the die during thermal cycling. The latter effect has led to reports that better reliability is achieved with higher porosity Ag during accelerated testing [8, 9]. This provides a stimulus for understanding the porous structure of sintered Ag rather than simply attempting to reduce porosity.

Simple estimates of porosity can be found by cross-sectioning a sintered Ag joint and using image processing to find the percentage of the image which is dark. Both optical and electron microscopy images can be used. Figure 1 shows a pressure-free

S. Mannan (✉) · A. Paknejad · A. Mansourian · K. Khtatba
Department of Physics, King's College London, London, UK
e-mail: samjid.mannan@kcl.ac.uk; sa.paknejad@kcl.ac.uk; ali.mansourian@kcl.ac.uk; khalid.khtatba@kcl.ac.uk

K. S. Siow (ed.), *Die-Attach Materials for High Temperature Applications in Microelectronics Packaging*, https://doi.org/10.1007/978-3-319-99256-3_6

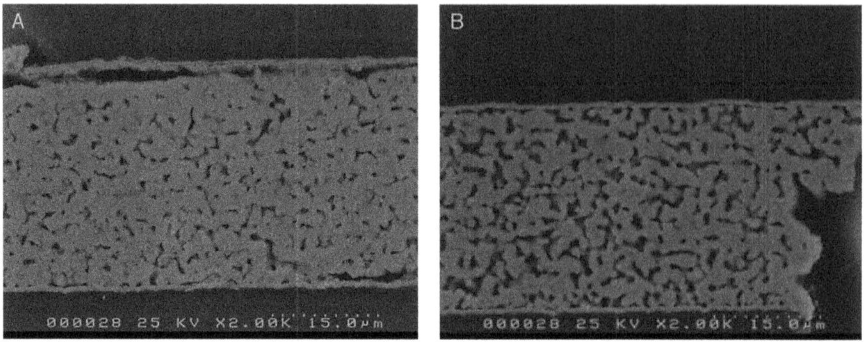

Fig. 1 (**a**) Interior and (**b**) edge of pressur-free sintered Ag sample

sintered sample where the porosity is calculated to be 25%. There is some evidence [10, 11] that the porosity in the interior is lower in the centre than at the edges of a die attach and that the coalescence process whereby larger voids grow at the expense of smaller ones also takes place more rapidly at the centre. This indicates that gaseous exchange between the paste and surrounding atmosphere plays an important role in Ag sintering as experiments involving different atmospheres confirm [12]. In particular, these results suggest that mechanical properties of the joints will vary along a cross section.

Using image processing techniques, it is also possible to determine eccentricity of the cross section of pores and other pore statistics such as pore surface area and homogeneity. However, 2-D data does not tell the whole story, and in particular it is desirable to understand what percentage of the pore network is open and what percentage is closed and the tortuosity of the pore network which enables calculations of gaseous diffusion through the pore network to be made. While X-ray tomography is a non-destructive technique to determine microstructure, resolutions are limited, and the tool of choice, serial block-face scanning electron microscopy (SBFSEM), allows high-resolution reconstruction using focussed ion beam (FIB) milling to re-create the 3-D structure from successive 2-D slices [13–15].

Figure 2 [13] shows the reconstructed volume of a sample sintered under pressure. The volume was constructed using SBFSEM, and changes in morphology were determined by analysing two samples which had similar starting densities; one was analysed straight after sintering and one after ageing. The samples were prepared using 10 MPa sintering pressure which resulted in an initial porosity of ~10%. The aged sample was aged for 1500 h at 125 °C, and this resulted in increased average pore size and pore sphericity, suggesting that minimisation of surface energy is the driving force for the changes. The ageing also results in the appearance of a higher proportion of isolated pores.

Figure 3 represents the work of Rmili et al. [14] in which varying sintering conditions were used resulting in variations in porosity. In this work the SBFSEM technique was again used, with porosity ranging from 56% to 16% and percentage of voids not connected to the open pore network ranging from 0.1% to 9.3% depending on sintering temperature, pressure and time. This work shows that using processing

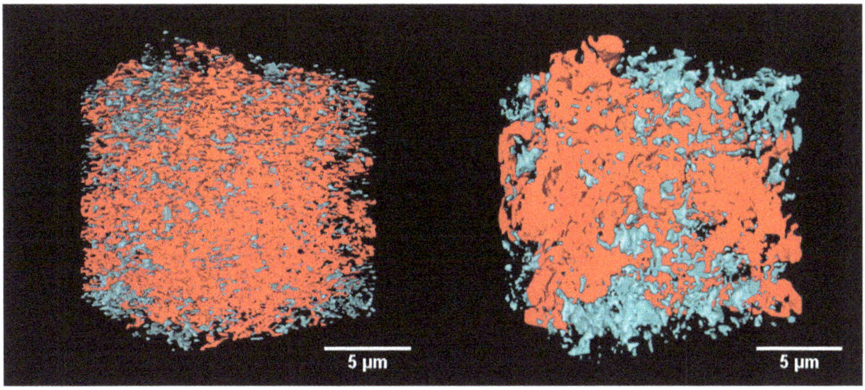

Fig. 2 As sintered (**a**) and aged (**b**) pore networks with single pore network shown as red [13]. Originally published at https://doi.org/10.1016/j.jmatprotec.2015.03.037 and reproduced here under https://creativecommons.org/licenses/by/4.0/

Fig. 3 3-D analysis of pore network as a function of porosity. Connected pores shown in yellow, representing 90.7% of volume. Originally published in [14] (https://doi.org/10.1007/s11664-015-4288-1)

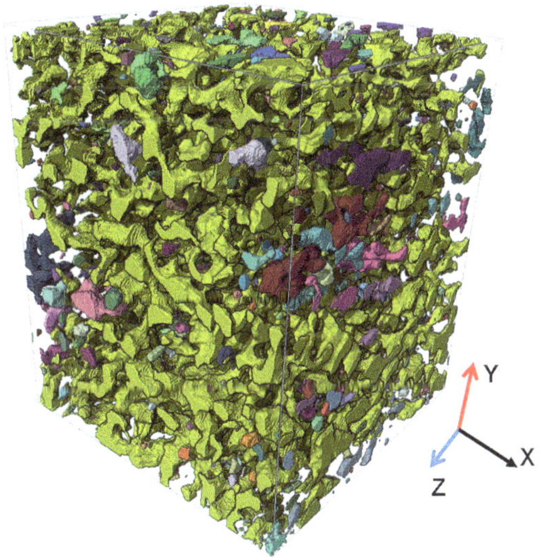

conditions with zero applied pressure results in a network of pores which are all effectively exposed to the ambient atmosphere after processing.

The presence of pores results in an internal free surface along which rapid atomic diffusion can take place. Typical values for surface diffusion on an oxide-free Ag surface are 0.91E−8 m^2/s m/s^2 [16]. Grain boundary diffusion coefficients depend on the nature of the grain boundary, but experimentally measured values for sintered Ag lie in the range 2×10^{-17} m^2/s [10]. Bulk diffusion inside a single crystal is characterised by extremely low diffusion coefficients of the order of 2×10^{-22} [17] and can essentially be neglected. Comparing these diffusion coefficients, we see that surface diffusion should be the dominant process for

Fig. 4 TEM image of sintered Ag grains

increasing the sphericity of pores and pore coalescence with the proviso that pores are connected. Once pores are isolated and closed off from the open pore network, then grain boundary diffusion and grain boundary migration will take over as the dominant process for further changes.

Details of grain boundaries are presented in Fig. 4 using transmission electron microscopy (TEM). Figure 4 shows the interface between grains and pores, grain boundaries and twin boundaries. The presence of a high density of twin boundaries has been reported as a general feature of nanoparticle sintered systems and may provide additional diffusion paths [10, 18]. The presence of rapid diffusion paths results in the possibility of rapid changes to microstructure under thermal and stress treatment, and these will be explored in Sects. 2 and 3 of this chapter.

2 Microstructure Evolution During Thermal Ageing

Ageing sintered Ag at elevated temperatures results in changes in microstructure. While changes have been reported even at temperatures as low as 125 °C [13], the changes accelerate above 200 °C, leading to coarsening of pores and grains with large pores growing at the expense of smaller pores [10, 13]. Coarser grain structure in sintered Ag has been reported as reducing fatigue lifetimes [19]. In contrast to homogenously dispersed small pores, heterogeneously dispersed larger pores could lead to local stress concentrations that explain the lower lifetime. Another effect that could reduce lifetime of sintered joints is the interdiffusion between sintered Ag and the contact metallisations on the die and the substrate. In particular it has been noted that sintered Ag in contact with electroless nickel immersion gold (ENIG) exhibits rapid loss of shear strength due to the formation of a high porosity layer in the Ag. Grain and pore coarsening are discussed in Sect. 2.1, while interdiffusion with contact layers is discussed in Sect. 2.2.

2.1 Microstructure Coarsening During Thermal Ageing

The variation in grain size as a function of storage time and temperature is depicted in Fig. 5. In this figure, the sintered Ag is viewed by an optical microscope through glass allowing individual grains and pore evolution to be tracked.

At first glance it might be considered odd to view the sintered Ag through glass, preventing the use of electron microscopy. Why not simply take a cross section such as that shown in Fig. 1 and age the sample? The answer is that after sectioning, no change in microstructure is seen even up to 400 °C. The result holds even if the sample is first perchlorate acid etched and heated under vacuum after sectioning. The morphology after sectioning remains frozen in its initial state. The obvious conclusion to draw from this is that oxidation occurs once the sample is exposed to air and that once the surface is oxidised, surface diffusion is restricted and the microstructure is stabilised. Indeed, the exterior surface of sintered Ag also does not exhibit obvious microstructural changes. In order to determine directly the presence and type of oxide on sintered Ag, samples were analysed using X-ray photoelectron spectroscopy (XPS). These analyses showed the presence of Ag_2O on Ag surfaces. Hydrocarbons were also detected but not identifiable with XPS. Ag_2O is known to

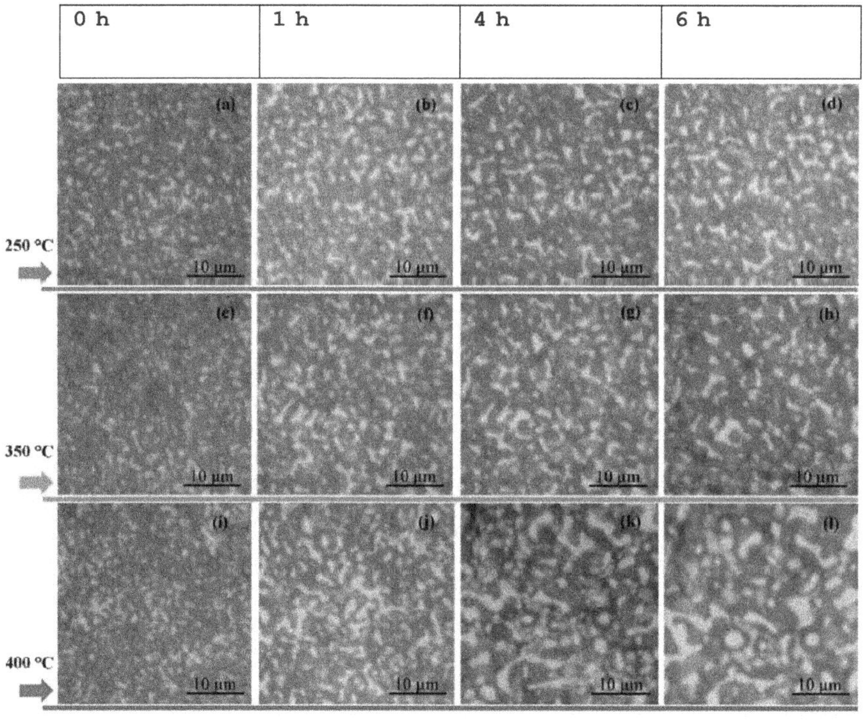

Fig. 5 Optical images of sintered silver under cover slip stored at high temperatures on top of a cartridge hearer. (**a–d**) 250 °C storage. (**e–h**) 350 °C storage. (**i–l**) 400 °C storage

decompose in air at 160 °C [20], but decomposition is complete only after 400 °C [21]. Above 400 °C there is a step change in the rate of morphological change, even on the exterior surface [22], providing further support to the hypothesis that the inner pore surfaces are initially largely oxide free, while exposure to air results in oxide formation and arrest of surface diffusion along the pores. Further support for this hypothesis lies in the differences in porosity observed in Fig. 1 between the centre and edges of the die. We can now interpret the higher porosity at the edges as a result of extensive oxidation of the Ag surfaces, slowing down surface diffusion and preventing pore coalescence. At the centre of the die, Ag surfaces may have some oxidation and residual organic contamination, but surface diffusion can occur more freely than at the edges.

Estimates of grain growth observed in Fig. 5 calculated using surface diffusivity values of pure Ag in vacuum are orders of magnitude too high, indicating that the initial surface of the pores contains a mix of Ag covered with residual hydrocarbons from the original paste and some oxides. An intriguing possibility now arises as to the possibility of deliberately arresting the surface diffusivity of Ag atoms on internal grain/pore boundaries by oxidation during or post-sintering. Oxygen should be present in an open pore network given that diffusion timescales are given by:

$$t \sim \frac{L^2}{D_{\mathrm{eff}}} \tag{1}$$

Here t is the timescale for diffusion, L is the typical length scale (~10^{-2} m) and D_{eff} is the effective diffusion coefficient for oxygen in air is 0.2×10^{-4} m^2 (clearly the pore atmosphere will not be air, but this assumption will suffice for an order of magnitude calculation). D_{eff} needs to take into account the tortuosity of the diffusion path, but again, for an order of magnitude calculation, tortuosity can be set to 1 in this relatively open network, leading to $D_{\mathrm{eff}} = 0.2 \times 10^{-4}$ m^2/s [23]. Substituting into Eq. (1) leads to penetration timescales of oxygen into the interior pores of seconds. The experimental results in Figs. 2 and 5 however show that even after several hours, oxide levels are not sufficient to freeze the microstructure. It may be that the residual hydrocarbons from the paste partially block diffusion or that some other mechanism prevents the surface from oxidising. After exposure to steam, however, the microstructure is effectively frozen as seen in Fig. 6. Figure 6 shows that after steaming the initial microstructure does not change for at least 24 h unless the temperature exceeds 400 °C. In the 300 °C ageing case, the lack of microstructural change has been verified for 600 h. Note that one grain out of approximately 7000 showed grain growth, as seen in Fig. 6d. This is believed to be a closed pore region where steam was unable to penetrate. In conclusion, it is possible to stabilise the microstructure by oxidising the inner pore surfaces by taking advantage of the fact that for pressure-free sintering, the pore network is open.

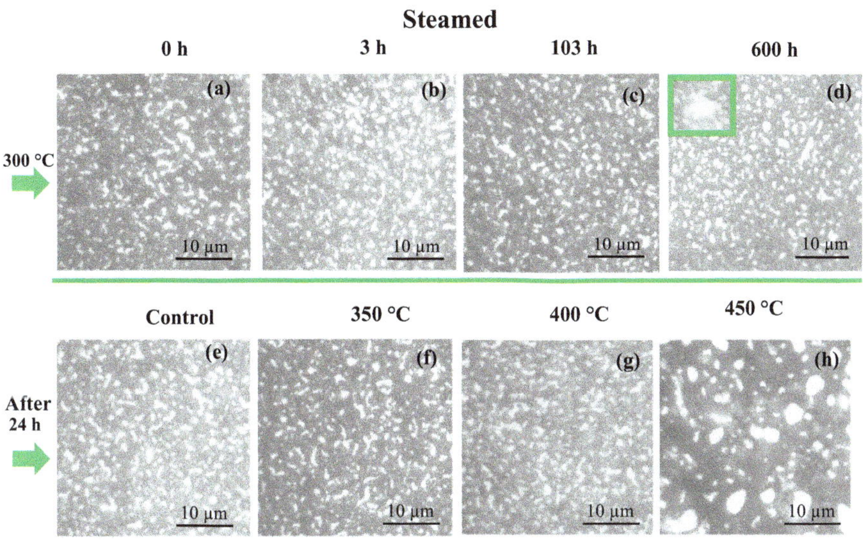

Fig. 6 Optical images of sintered Ag under cover slip. (**a–d**) Increasing time at constant temperature. (**e–h**) Increasing temperature at constant time

2.2 *Diffusion Phenomena with Metallic Contacts*

In the previous section, microstructural changes in pure sintered Ag were considered, and the key role of internal pore surfaces in facilitating mass transfer was explored. When the sintered Ag is in contact with the metallisations on the back of the die and on the substrate, the possibility of diffusion of Ag into the metallisation and from the metallisation into the Ag needs to be considered. Shear strengths for Ag metallisation to sintered Ag tend to increase after thermal ageing as the interfacial contact area increases due to diffusion at pore surfaces adjacent to the flat metallisation [10]. In the case of Ag metallisations, therefore, no additional complication arises with the proviso that oxidation of the underlying material by oxygen penetration through thin Ag layers is always a consideration, for example, oxidation of copper at 300 °C [24] has been reported to reduce shear strength.

Gold (Au) is a popular choice of metallisation for electronics. As a noble metal, oxidation is not an issue so that contacts can be stored indefinitely before joining takes place. When sintering to a Au surface, the initial shear strength is much higher than on Ag metallisations as a result of rapid interdiffusion. After thermal ageing for as little as 24 h, the strength deteriorates due to the same rapid diffusion phenomena which results in the formation of a pore-free layer immediately adjacent to the Au and a highly porous layer adjacent to that, as seen in Fig. 7 [10, 25]. In addition, the Au metallisation layer dissolves into the Ag.

In order to understand the diffusion mechanism, it is first necessary to understand that the free energy of an alloy of Ag and Au is minimised at a composition of 50%,

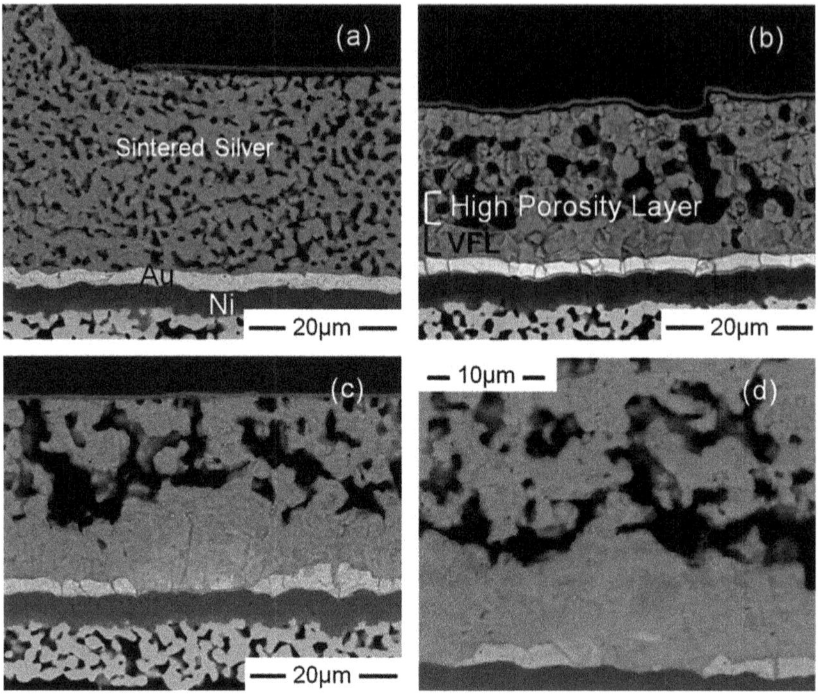

Fig. 7 Backscatter SEM images of (**a**) control. (**b**) 24 h (prepared using CSP). (**c**) 100 h. (**d**) 500 h sample (higher magnification). Originally published in [10] (https://doi.org/10.1016/j.jallcom.2014.08.062)

providing the thermodynamic driving force for the interdiffusion. Diffusion of Ag into the Au takes place along the grain boundaries of the Au. In the particular case of ENIG coatings, the grain boundaries in the Au continue on from the underlying nickel grain boundaries. Similarly, the diffusion of Au into the Ag takes place along the Ag grain boundaries. The presence of fast surface diffusion paths on the pores results in movement of the pores away from the Ag/Au interface as shown schematically in Fig. 8.

Recently, it has been shown that it is possible to convert the weakness of the pore-free/high-porosity structure into a strength by using a mesh insert between the die and the substrate as depicted conceptually in Fig. 9 [26]. If the mesh is Au, or Au-plated, then the pores will be driven outwards from the centre of each mesh cell, forming a single void at the centre of each cell. The 50Ag/50Au void-free layer will extend vertically, providing a thermodynamically stable system which has been shown to retain strength for at least 1000 h at 450 °C. The system still requires development with regard to increasing the initial joint strength and optimising the mesh design and silver porosity so as to take maximum advantage of the ability to locate the pores at desired locations, for example, at the periphery of the die.

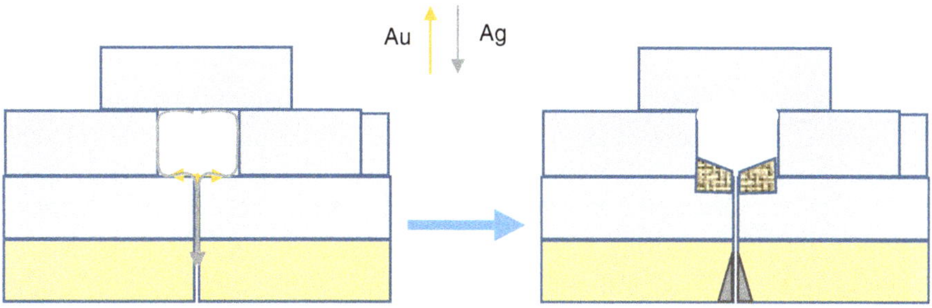

Fig. 8 Mass transfer leading to void movement away from interface

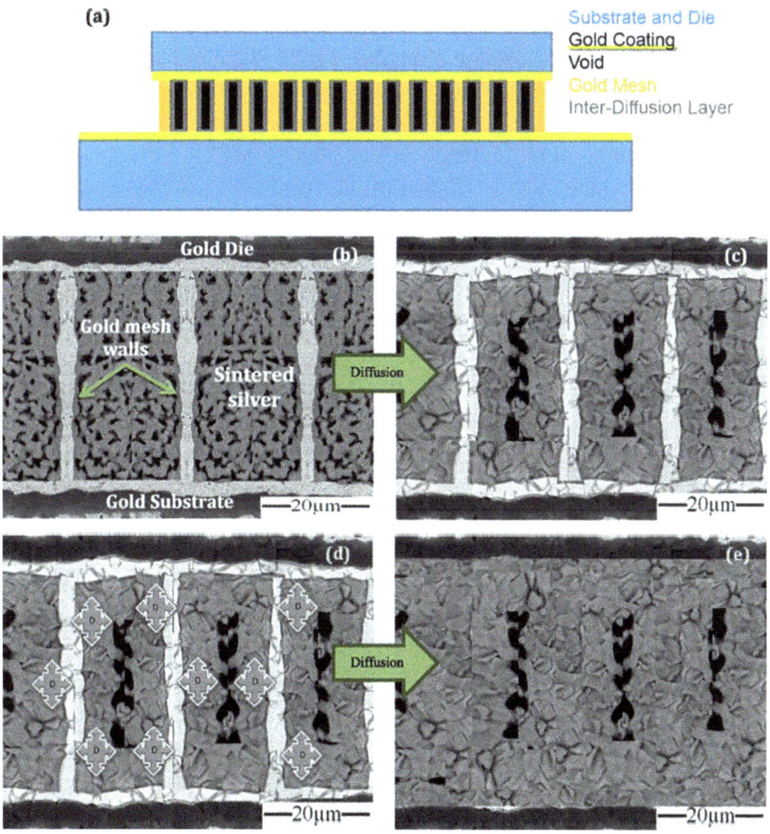

Fig. 9 Conceptual images of the die attach assembly. (**a**) Schematic cross section of a die attach assembly, where inserts (mesh) are used for formation of a continuous interdiffusion layer from die to substrate as the main part of the die attach. (**b**) Conceptual cross section of a die attach assembly showing the preformation of the continuous interdiffusion connection. (**c**) Conceptual cross section of the die attach assembly after storage of the assembly in (**b**) at high temperatures and forming the continuous interdiffusion layer. (**d**) Arrows labelled "D" showing schematic diffusion pathways of Au and Ag. (**e**) Showing the formation of the thermally stable structure after complete diffusion of Ag and Au. (Originally published in (https://doi.org/10.1016/j.matdes.2015.10.074)

3 Electromigration in Sintered Silver

Electromigration in this chapter is defined as the movement of metal ions along a conductor in the direction of electron flow due to momentum exchange between the conduction electrons and the lattice ions. Note that the term electromigration is sometimes used to describe movement of ions in the presence of an electric field, for example, on the insulating surface between two closely neighbouring conducting elements held at different potentials, which is also an important issue for Ag, but which is not explored further here. (The earlier phenomenon involving Ag ions is more aptly called electrochemical migration.) Instead, this section will explore the issues that can arise when sintered Ag is used not as a die attach, but as a joining material used to carry current and supply mechanical strength, replacing solder interconnections.

Electromigration has long been an issue for on-chip interconnects where the small dimensions of the current-carrying wires result in high current densities [27]. More recently, electromigration has become an issue for solder joints due to increasing miniaturisation leading to shrinking of joint dimensions. In general, mass transport induced by electromigration can result in hillocks or whiskers forming on the anode end of the conductor and material depletion and voids at the cathode end. The flux of ions (loosely referred to as the atomic flux) is given by [28]:

$$\vec{J}_{\mathrm{a}} = -D_{\mathrm{a}}\left[\nabla C_{\mathrm{a}} + \frac{|Z^{*}|\rho e}{k_{\mathrm{b}}T} C_{\mathrm{a}}\vec{J}_{\mathrm{e}} + \frac{Q^{*}}{k_{\mathrm{b}}T^{2}} C_{\mathrm{a}}\nabla T - \frac{f\Omega}{k_{\mathrm{b}}T} C_{\mathrm{a}}\nabla\sigma\right] \tag{2}$$

where C_{a} is the atomic concentration, D_{a} is the diffusivity of atoms, Z^{*} is the effective charge, ρ is the electrical resistivity, e is the elementary charge, k_{b} is Boltzmann's constant, T is the absolute temperature, $\vec{J}_{\mathrm{e}}$ is the current density, Q^{*} is the heat of transport, f is the atomic relaxation factor, Ω is the atomic volume and σ is the hydrostatic stress. The first term on the RHS (right hand side) of Eq. (2) represents self-diffusion, the second term electromigration, the third term thermomigration and the fourth term stress migration. While all terms may be significant in the general case, for electromigration in sintered Ag, the $\vec{J}_{\mathrm{e}}$ term dominates, and effects are seen in Ag when the order of magnitude of $\vec{J}_{\mathrm{e}}$ approaches 10^{8} A/m^{2}. Physical effects of electromigration are related not to the magnitude of the diffusion flux but rather to changes in atomic concentration, related to the divergence of $\vec{J}_{\mathrm{a}}$:

$$\mathrm{Div}\left(\vec{J}_{\mathrm{Em}}\right) = \left(\frac{E_{\mathrm{a}}}{kT^{2}} - \frac{1}{T}\right)\frac{C_{\mathrm{a}}Z^{*}e\rho}{kT} D_{0}\exp\left(-\frac{E_{\mathrm{a}}}{kT}\right)\vec{J}_{\mathrm{e}}\,.\nabla T \tag{3}$$

where D_{a} has been expressed by Arrhenius law as:

$$D_{\mathrm{a}} = D_{0}\exp\left(-\frac{E_{\mathrm{a}}}{k_{\mathrm{b}}T}\right) \tag{4}$$

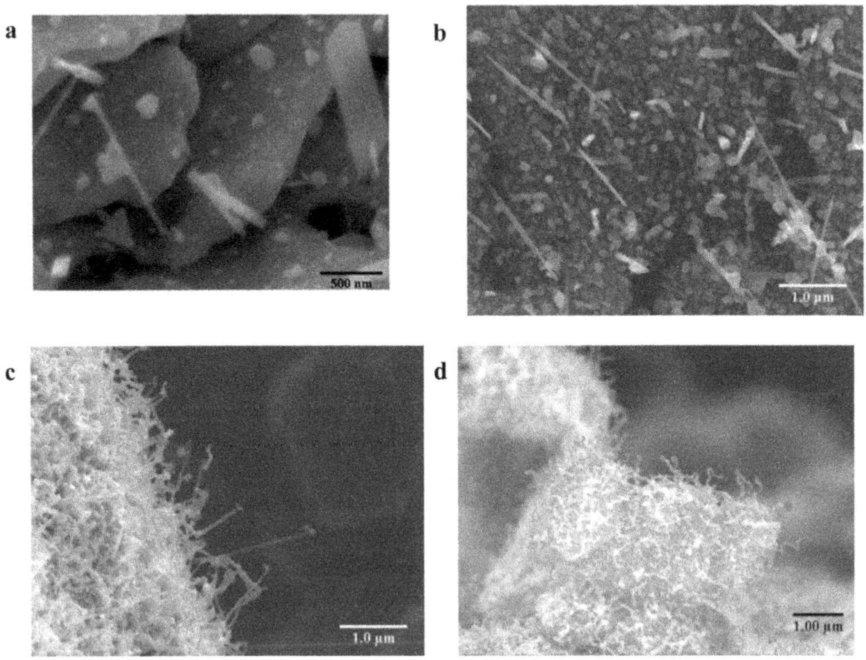

Fig. 10 Electromigration on surface of sintered Ag (**a**–**d**) Images from different areas of the sample

with D_0 as the pre-exponential factor and E_a as the activation energy. Equation (3) shows that temperature gradients are critical for the formation of voids and hillocks. In simple terms, given that the diffusion coefficient increases at higher temperatures, atoms diffusing from a high-temperature region to a lower-temperature region will pile up, forming hillocks. In the opposite case where atoms diffuse from a low temperature region to a higher one, atoms will be depleted from the region faster than incoming atoms can arrive, and voids will nucleate. In sintered Ag, the complex nature of the microstructure leads to temperature gradients forming at the intersections between grains where local current concentration occurs. This leads to the possibility of whisker formation over the entire region of the sintered Ag and not just at the anode. Figure 10 shows some examples of whisker formation after ~200 h at 2×10^8 A/m^2. The whisker size is self-limiting as once they contact other whiskers, growth stops. The interior of the sintered Ag is also affected by sintered Ag, and grain refinement occurs whereby the average grain size is lowered and becomes orientated in the direction of local current flow [29]. Eventually void formation causes electrical failure [30].

4 Conclusions

In this chapter the role of atomic diffusion in affecting microstructural change in sintered Ag has been explored. It has been shown that the presence of the free surfaces on the pores can lead to rapid mass transfer and microstructural evolution which may affect long-term high-temperature reliability above 200 °C. However, modification of surfaces can arrest microstructural evolution or the sintering process modified to render the system thermodynamically stable so that the advantages of porosity, notable lower stress transmitted to the die during thermal cycling, need not be discarded in the pursuit of stability. Finally the effects of electromigration on sintered Ag were examined, and it was found that electromigration causes whiskers and nodules to grow on the surface of sintered Ag.

References

1. S.A. Paknejad, S.H. Mannan, Review of silver nanoparticle based die attach materials for high power/temperature applications. Microelectron. Reliab. **70**, 1–11 (2017)
2. K.S. Siow, Are sintered silver joints ready for use as interconnect material in microelectronic packaging? J. Electron. Mater. **43**, 947 (2014)
3. K. Suganuma, S. Nagao, T. Sugahara, E. Yokoi, H. Zhang, J. Jiu, Silver sinter joining and stress migration bonding for WBG die-attach, in *International Symposium on 3D Power Electronics Integration and Manufacturing (3D-PEIM)*, 2016
4. K.S. Siow, Mechanical properties of nano-silver joints as die attach materials. J. Alloys Compd. **514**, 6–19 (2012)
5. K.S. Siow, Y.T. Lin, Identifying the development state of sintered silver (Ag) as a bonding material in the microelectronic packaging via a patent landscape study. J. Electron. Packag. **138** (2), 020804 (2016)
6. H.S. Chin, K.Y. Cheong, A.B. Ismail, A review on die attach materials for SiC-based high-temperature power devices. Metallogr. Mater. Trans. B. **41**, 824 (2010)
7. A.A. Bajwa, Y. Qin, R. Reiner, R. Quay, J. Wilde, Assembly and packaging technologies for higherature and high-power GaN devices. IEEE Trans. Compon. Packag. Manuf. Technol. **5** (10), 7243341, 1402–1416 (2015)
8. R. Khazaka, L. Mendizabal, D. Henry, Review on joint shear strength of nano-silver paste and its long-term high temperature reliability. J. Electron. Mater. **43**(7), 2459–2466 (2014)
9. V.R. Manikam, K.A. Razak, K.Y. Cheong, Sintering of silver-aluminum nanopaste with varying aluminum weight percent for use as a high-temperature die-attach material. IEEE Trans. Compon. Packag. Manuf. Technol. **2**(12), 6334434, 1940–1948 (2012)
10. S.A. Paknejad, G. Dumas, G. West, G. Lewis, S.H. Mannan, Microstructure evolution during 300 °C storage of sintered Ag, nanoparticles on Ag and Au substrates. J. Alloys Compd. **617**, 994–1001 (2014)
11. Y. Li, H. Jing, Y. Han, L. Xu, G. Lu, Microstructure and joint properties of nano-silver paste by ultrasonic-assisted pressureless sintering. J. Electron. Mater. **45**(6), 3003–3012 (2016)
12. S.T. Chua, K.S. Siow, A. Jalar, Effect of sintering atmosphere on the shear properties of pressureless sintered silver joint, in *Proceedings of the IEEE/CPMT International Electronics Manufacturing Technology (IEMT) Symposium*, 2015-June, 7123119
13. J. Carr, X. Milhet, P. Gadaud, S.A.E. Boyer, G.E. Thompson, P. Lee, Quantitative characterization of porosity and determination of elastic modulus for sintered micro-silver joints. J. Mater. Process. Technol. **225**, 19–23 (2015)

14. W. Rmili, N. Vivet, S. Chupin, T. Le Bihan, G. Le Quilliec, C. Richard, Quantitative analysis of porosity and transport properties by FIB-SEM 3D imaging of a solder based sintered silver for a new microelectronic component. J. Electron. Mater. **45**(4), 2242–2251 (2016)
15. A. Gillman, M.J.G.H. Roelofs, K. Matouš, V.G. Kouznetsova, O. van der Sluis, M.P.F.H.L. van Maris, Microstructure statistics–property relations of silver particle-based interconnects. Mater. Des. **118**, 304–313 (2017)
16. G. Antczak, G. Ehrlich, *Surface Diffusion: Metals, Metal Atoms, and Clusters* (Cambridge University Press, Cambridge/New York/Melbourne, 2010), p. 347
17. F. Jaumot, A. Sawatzky, Diffusion of gold in single crystals of silver. J. Appl. Phys. **27**(10), 1186–1188 (1956)
18. S. Wang, M. Li, H. Ji, C. Wang, Rapid pressureless low-temperature sintering of Ag nanoparticles for high-power density electronic packaging. Scr. Mater. **69**(11–12), 789–792 (2013)
19. G. Chen, Y.-Z. Wang, Y. Mei, L. Yu, X. Li, X. Chen, Influence of temperature and microstructure on the mechanical properties of sintered nanosilver joints. Mater. Sci. Eng. A. **626**, 390–399 (2015)
20. J. Tominaga, The application of silver oxide thin films to plasmon photonic devices. J. Phys. Condens. Matter. **15**(25), R1101–R1122 (2003)
21. T. Morita, Y. Yasuda, E. Ide, Y. Akada, A. Hirose, Bonding technique using microscaled silver-oxide particles for in-situ formation of silver nanoparticles. Mater. Trans. **49**, 2875–2880 (2008)
22. S.A. Paknejad, A. Mansourian, J. Greenberg, K. Khtatba, L. Van Parijs, S.H. Mannan, Microstructural evolution of sintered silver at elevated temperatures. Microelectron. Reliab. **63**, 125–133 (2016)
23. E.L. Cussler, *Diffusion: Mass Transfer in Fluid Systems*, 2nd edn. (Cambridge University Press, New York, 1997). ISBN 0-521-45078-0
24. S.T. Chua, K.S. Siow, Microstructural studies and bonding strength of pressureless sintered nano-silver joints on silver, direct bond copper (DBC) and copper substrates aged at 300 °C. J. Alloys Compd. **687**, 486–498 (2016)
25. F. Yu, R.W. Johnson, M.C. Hamilton, Pressureless sintering of microscale silver paste for 300 °C applications. IEEE Trans. Compon. Packag. Manuf. Technol. **5**(9), 7180312, 1258–1264 (2015)
26. S.A. Paknejad, A. Mansourian, Y. Noh, K. Khtatba, S.H. Mannan, Thermally stable high temperature die attach solution. Mater. Des. **89**, 1310–1314 (2016)
27. Cher Ming Tan, Arijit Roy, Electromigration in ULSI interconnects. Materials Science and Engineering: R: Reports **58**(1–2), 1–75 (2007)
28. X. Zhu, H. Kotadia, S. Xu, H. Lu, S.H. Mannan, C. Bailey, Y.C. Chan, Electromigration in Sn-Ag solder thin films under high current density. Thin Solid Films. **565**, 193–201 (2014)
29. Ali Mansourian, Seyed Amir Paknejad, Qiannan Wen, Gema Vizcay-Barrena, Roland A. Fleck, Anatoly V. Zayats, Samjid H. Mannan, Tunable Ultra-high Aspect Ratio Nanorod Architectures grown on Porous Substrate via Electromigration. Scientific Reports **6**, (22272) (2016)
30. J.N. Calata, G.Q. Lu, K. Ngo, L. Nguyen, Electromigration in sintered nanoscale silver films at elevated temperature. J. Electron. Mater. **43**, 109–116 (2014)

Doctrine of Equivalents and Sintered Silver (Ag) Paste as Bonding Materials

K. S. Siow

1 Introduction

Infringement occurs when a product/device/composition or process read on the claims of a granted patent in a literal or a nonliteral sense. In a literal infringement, every element of a claim of the patent is present in the alleged infringing product, and the absence of a single element is enough to render the product to be non-infringing. In the case of nonliteral infringement, the doctrine of equivalents (DOE[1]) allows the patentee to extend the original claims beyond their literal meanings to include other equivalent elements. Hence, DOE poses a clarity issue to the boundary of patent claims that is central to clear claim drafting. However, DOE has been continuously implemented and debated because of its equitable reasoning to encourage innovation via patenting, address the limitations of claim languages, and avoid "imitation" through trivial or insubstantial changes.

DOE differs from the doctrine of claim differentiation; the latter prevents two patents to have the same claim, i.e., the similar elements need to be construed to have a different meaning. Unlike DOE, the doctrine of claim differentiation cannot be used to "broaden" the claim limitation. This chapter reviews and summarizes the historical contexts, implications, and implementation of DOE based on key cases in Sect. 2. Subsequent sections select two pairs of patents related to material

[1]DOE is used as an abbreviation for doctrine of equivalents in this chapter. In science and engineering-based research, DOE is often associated with the design of experiment, which is a methodology pioneered by Box, Hunter, and Hunter, to study the relationship between factors and outputs of an industry process.

K. S. Siow (✉)
Faculty of Law, University of Turin, Torino, TO, Italy

Institute of Microengineering and Nanoelectronics, Universiti Kebangsaan Malaysia, Bangi, Selangor, Malaysia
e-mail: kimsiow@ukm.edu.my

K. S. Siow (ed.), *Die-Attach Materials for High Temperature Applications in Microelectronics Packaging*, https://doi.org/10.1007/978-3-319-99256-3_7

compositions of Ag paste from three Ag paste suppliers to determine whether there are nonliteral infringements among the technologies based on their filing dates.

2 Key Cases and Principles in Doctrine of Equivalents

The earliest case law on the application and also the "birth" of DOE was attributed to Winans vs. Denmead [1]. The *Winans* case concluded that the "frustum of a cone" includes "octagon or a pyramid" shape used by the accused infringer because of the similar principle in distributing the coal weight on the coal car body. This case was also decided when the patent claim was primarily based on central claiming, i.e., the core inventor's contribution to the concerned technology.

Although the Patent Act was amended in 1870 to include "peripheral claiming system" to highlight the invention boundaries through clear claiming [2], the principles of DOE was again reaffirmed in the *Graver* vs. *Linde Air* case. The *Graver* case established equivalence when the replacement performs "substantially the same function in substantially the same way to obtain the same result" [3]. In this case, *Graver's* product reads on Linde Air patent claims despite replacing alkaline metal silicate with manganese silicate (a non-alkaline earth metal silicate) as flux because this replacement still fulfills the three-part function/way/result test or better known as "tripartite test" or "triple identity test."

Subsequently, the Warner-Jekinson vs. Hilton Davis case established that the equivalency in DOE is evaluated at the time of infringement, not at the granting of the patent [4]. The *Warner-Jekinson* case also proposed the "all-element rule," i.e., the defendant's product must read on all elements of at least one of the patentee's claims. The case further acknowledged that "in determining equivalents, things equal to the same thing may not be equal to each other and, by the same token, things for most purposes different may sometimes be equivalents." However, the *Warner-Jekinson* case also refused to adopt a linguistic framework to perform the DOE test, stating that it was fact-dependent but instead acknowledging that the tripartite test provided an inadequate framework for other products or processes than mechanical devices.

In the context of mechanical devices, *Ethicon* case suggested that "components of an accused device may be viewed in combination to serve as an equivalent of one element of a claim invention, so long as no claim limitation is thereby wholly vitiated" [5]. This approach is extended to the method or process claim in the case of EMI Group North America Inc. vs. Intel Corp. [6], but such principles were not reported for the materials or chemical test.

Doctrine of equivalents also cannot be extended to cover claims in the prior art [7]. The *Slimford* case further clarified that if a hypothetical claim inclusive of the infringing product can be filed over the prior art, then infringement under DOE can be established [7].

The influence of prosecution history estoppel (PHE) in DOE was widely debated but finally clarified by the Festo Corp vs. Shoketsu case [8]. The case established that narrowing of claims to satisfy patentability, in addition to prior art, gives rise to estoppel and reinstated the "flexible bar" approach that narrowing amendments surrendered the equivalents, except if the patentee is able to demonstrate the lack of foreseeability at the time of filing, amendment tangentially related to equivalents, and reasonable expectation that the patentee is not able to describe insubstantial substitute in question.

Related to prosecution history is the "admission estoppel" which may be considered as claim interpretations, including all statements made by patentee about the patent during prosecution but not considered by the examiner [9]. "Public dedication rule" is also considered an admission estoppel, embodiments which are mentioned in the specification but not claimed by the patentee as dedicated to the public [10]. Finally, a pioneering patent may potentially entitle to a broader interpretation of the claims than an incremental patent, but such interpretation is often due to the "sparseness of prior arts in nascent fields of technology" [11]. Such wider interpretation is not provided by jurisprudence, but merely coincidental due to its nascent nature. The next section reviews the motivation for selecting sintered Ag as our case study and related nano-material issues before analyzing the selected patents on sintered silver (Ag) for equivalence in chemical compositions.

3 Background of Sintered Ag Technologies

The importance of sintered Ag technology is demonstrated by the large number of patents filed and granted in the last 5 years [12]. The key driver for this technology is the need to find lead (Pb)-free bonding materials for die-bonding application to conform to the EU Directives on Restriction of Hazardous Substances (ROHS) 2011/65/EU by July 2021. Industry consortiums, such as Die-Attach 5 (DA5) and International Electronics Manufacturing Initiative (iNEMI),[2] identified sintered Ag as one of the most promising technologies for this application [13]. Another driver for this interest in sintered Ag is the advent of wide bandgap semiconductors (i.e., silicon carbide and gallium nitride) that operate at the temperature above 200 °C and that are not currently served by the high lead solder materials [12]. Sintered Ag technology fulfills the requirements of elevated operating temperature as well as being Pb-free.

Generally, sintered Ag technology is applied as a paste, film/laminate, or thin film deposition [14]. This chapter only concentrates on the Ag paste technology. The

[2]DA5 consortium consists of Bosch, Infineon Technologies, NXP Semiconductor, STMicroelectronics, and Nexperia, formed to look for alternative Pb-free technologies to comply with EU Directive ROHS by 2021. iNEMI is a not-for-profit consortium of more than 90 electronics companies, associations, government agencies, and universities, formed with a mission to "forecast and accelerate improvements in the electronics manufacturing industry for a sustainable future" http://community.inemi.org/hightemp_pbfree_dieattach.

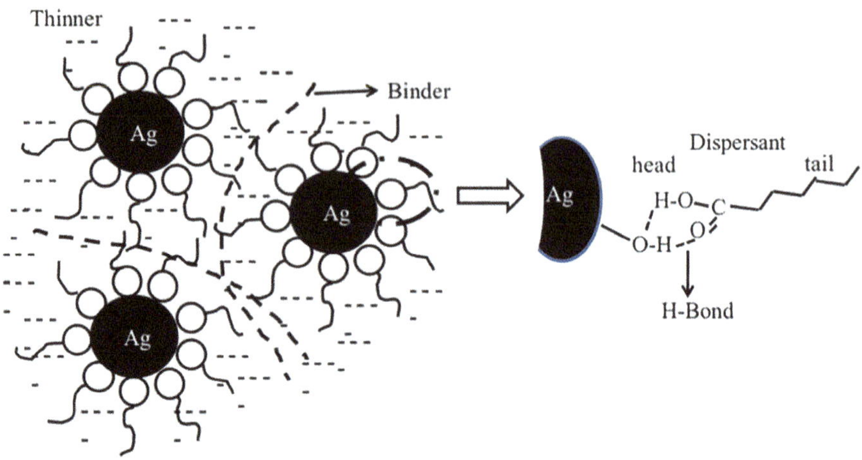

Fig. 1 Formulation of nano-Ag paste containing the Ag fillers (particles), dispersant (capping agent), binder, and thinner (solvent) [15]

main compositions of sintered Ag paste consists of fillers,[3] binders (sintering aids), and thinners (solvents). The Ag filler sizes are broadly divided into:

1. Micron-sized (i.e., diameter more than 1 μm)
2. Nano-sized particles (i.e., diameter less than 100 nm)

(Note: submicron is commonly defined as between 100 nm and 1 μm (1000 nm).)

As shown in Fig. 1, nano-Ag surfaces are typically covered by dispersant (capping agent), while the thinner (solvent) controls the rheology and viscosity of the nano-Ag paste and binder prevents cracking of the nano-Ag paste during preheating. Preheating removes the solvent from the nano-Ag paste applied on the bonding region, before further heating decomposes the capping agent to allow the nano-Ag to sinter and bond into a Ag joint. The nano-Ag particle coalesces with their adjacent nano-Ag particle because of its high reactivity associated with a large surface area to volume ratio for a nanoparticle with a diameter less than 100 nm (preferably less than 20 nm).

In terms of sintering technologies, the Ag paste is broadly classified to nano-Ag, micron-Ag, and hybrid Ag; the latter generally consists of predominantly micron-Ag with nano-Ag particles used to fill up the gaps within the micron-Ag particles [12]. Table 1 summarizes the main dimensional parameters, sintering mechanism, and composition of these Ag pastes. The sintering mechanisms will be further explained in Sects. 6 and 7.

[3]Fillers refer to the initial solid metal content which can come in the form of particles or flakes. A particle is approximately spherical with nearly identical lengths in all three dimensions, while flakes are layer-like structures with one dimension substantially smaller than the other two dimensions [14]

Table 1 Classification of sintered Ag paste used as die-bonding materials

	Types of pastes		
Parameters	Nano-Ag	Hybrid-Ag[a]	Micron-Ag
Particle size length/x (nm)[b]	<100 nm	Mixed volume ratio of nano- and micron-Ag	>1000 nm (1 μm)
Main compositions	Capping agent[c], binder, solvent	Capping agent (nano-Ag)[c], binder, solvent	Binder (optional), solvent
Sintering mechanism	Surface reactivity	Surface reactivity and in situ nano-Ag (from organo-Ag binder)	Surface diffusion

[a]Hybrid Ag paste is normally classified as micron-Ag paste because the volume ratio of micron-Ag to nano-Ag paste is typically 3:1 [16] or 7:3 [17]
[b]Particle size definition is based on common scientific literature, not any particular patents discussed in this chapter. Definition of nanoparticles (of less than 100 nm) is provided by USPTO [18]
[c]Additional external dispersant (non-bound to nanoparticles) may be added to the paste to stabilize the Ag paste for long-term storage

4 Patent Infringement Analysis of Sintered Ag Pastes

The patents analysis of sintered Ag showed that most of the material suppliers for sintered Ag paste for bonding application filed their invention as material composition approach, though one patent by Dowa used the "product (Ag paste) by process" approach to file the material composition [19]. Another type of claim found in the sintered Ag patent is the "use claim" which specified the Ag paste composition used for the specific purpose (i.e., bonding dies to the substrate [20, 21] or heat transfer purpose [22]) in their claim. Such claims and patents are considered to be weaker than a "pure" material composition patent. These Ag pastes are also not amenable to the "means-plus-function" format but, instead, drafted with a preamble as "bonding material" or some generic words, such as metallic, inorganic, conductive materials with transition words, such as "consisting" or "comprising" to describe the main constituents of the paste.

A Ag paste formulation is a "chemistry-based" patent, and chemistry is often considered an "unpredictable field" because "minor changes in a product or process may yield substantially different results" [23]. This "unpredictability" is demonstrated in sintered Ag paste when the inclusion of an ether-based solvent in the Ag paste produces better shear strength (mechanical properties) and electrical conductivity than the hydroxyl-based solvent [24]. Hence, the necessity and requirement for disclosure are typically higher for a chemistry-based patent than a mechanical-based patent.

Our analysis also showed that nano-Ag fillers in the sintered Ag paste are defined in structural terms like dimensions, instead of functional features required to form the sintered Ag joints. Thus, claims covering micron-Ag paste will also include nano-Ag particles as well as the original micron-sized Ag particles used in the Ag

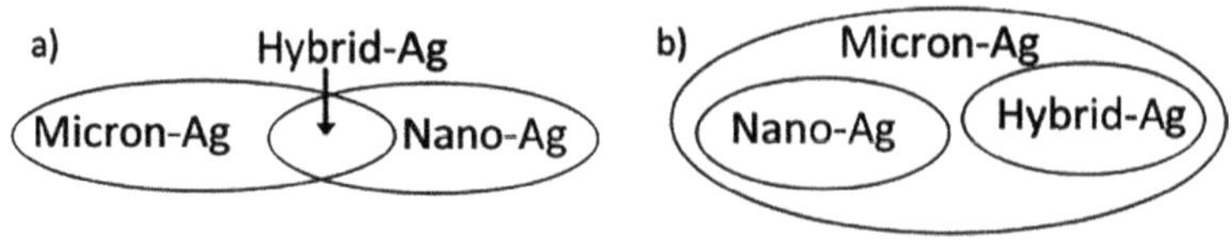

Fig. 2 (**a**) Scientific classification (left), (**b**) patent claim language classification of Ag paste and particles based on published nano-material-related patents (right)

paste because of the claim language for their physical dimension of using "less than" with no lower limit [25]. Fig. 2 illustrates the differences in classification of nano-materials (i.e., Ag pastes) based on the perspective of science and patent claim language. This uncertainty could be resolved by the reverse doctrine of equivalents though it was seldom adopted [26].[4] Reverse doctrine of equivalents, which originate from the *Graver* case, means that the infringed devices read on the patentee claims literally but operate so differently that warrants a ruling of non-infringement [3]. The requirements for clear claiming will also eliminate such possibility because claims need to be supported by the preferred embodiment in their patents.

Historically, the main components of micron-Ag pastes consisting micron-Ag flakes and solvent (e.g., hexanol) have been reported since the late 1980s [28, 29]. The synthesis of nano-Ag particle has a long history,[5] but they were not used as fillers for Ag paste for bonding application till the early 2000s [31, 32]. In patents on nano-Ag paste, the use of nano-Ag as fillers in the Ag paste faces another familiar issue related to other nano-material patents that depend on their sizes to achieve the desired properties or results. The measurement methods, reporting parameters (geometric or hydrodynamic), material polydispersity, and statistical treatments, can produce different values, resulting in different patentability or infringement analysis [33–36]. Hence, the characterization technique and protocol should form part of the specification. In the case of sintered Ag paste, the material composition needs to be diluted to a suitable concentration, before their diameters are analyzed with transmission electron microscopy, X-ray diffraction, or dynamic light-scattering techniques. Each technique provides different diameter or length because of their differences in operating principles, preparation techniques, and related assumptions.

[4]It is difficult to prove non-infringement when the accused product is considered to have "changed in principle" though the elements matched between the patentee and alleged infringer's product [27].

[5]Nano-Ag particle is known as Ag colloidal particle in the older scientific literature, and the latter is sometimes mentioned as such in the patent application. The synthesis of Ag colloidal particle was first reported in 1889 and repeated in 1969 based on a method known as Carey Lea's approach [30].

5 Infringement Tests and Methodology

Generally, infringement test follows a two-step test of determining literal infringement before applying the nonliteral infringement, also known as DOE. The intrinsic evidence of this infringement test are the claim language, specifications, and prosecution history estoppel (PHE), which will be considered first before any external evidence, such as expert evidence or dictionary, are sought [37]. Unless given otherwise in the specification or PHE, the words in the claim will be given the ordinary meaning familiar to a person having ordinary skill in the art (PHOSITA). While claim interpretation and claim construction are sometimes used interchangeably, others specify interpretation as discovering the linguistic meaning of a text [38], while construction gives the legal effect to the text [39]. Here, this chapter interpreted and constructed the necessary claims of the patents on Ag pastes to apply the DOE based on the patent claims, specifications, prosecution history estoppel, public dedication, prior art, and all-element rules. Section 2 suggested the importance of the all-element rule in determining equivalence in the DOE, but one element/component in infringer can perform the function of several elements/components in the patentee claims. In addition, this chapter also adopted the pragmatic textualist approach to apply the DOE that emphasized on the technological context of the patents with support from extrinsic evidence like the dictionary and related non-patent literature [40].

This chapter created a "legal fiction" by assuming that the infringing product possesses the properties described in the patent that reads on the patent published earlier.[6] These patents are valid patents that have achieved the patentability requirements set by USPTO. It should be mentioned here that sintered Ag paste formulation has another major application area as metallization or conductive lines for the printed electronics industry. Although the Ag paste formulation for printed electronics applications overlapped with those from bonding applications, this chapter did not consider this issue in this chapter [41]. These patents have also not been litigated based on analysis provided by the MaxVal Litigation Checker.[7]

Clearly, the selected patents in Table 2 did not show any literal infringement; the limitations are not found in the accused product. These are analyzed pair-wise based on the dimensional classification of Ag particles mentioned in Table 1, such as nano-Ag vs. micron-Ag paste (Sect. 6) and nano-Ag vs. nano-Ag pastes (Sect. 7).

The only tool available for testing equivalency in DOE, particularly in the chemistry-based patent, is the "tripartite test" which fails miserably in the

---

[6]In practice, analysis of silver paste is a difficult task because the characteristic claims (chemical and physical features of the active ingredients) of the patents need to be deciphered from the complex mixture of Ag paste and quantified with analytical tools that possess inherent technical limitations. For example, the characteristic claim could just be the capping agent on the nano-Ag particles that mixed and decomposed together with the surrounding dispersion medium (e.g., solvent) when heated during chemical analysis.

[7]The web link for checking litigation status of patents (http://maxinsight.maxval.com/LitigationChecker.aspx).

Table 2 Two classes of sintered Ag paste to be analyzed under DOE

Dimension	Priority date	Grant date	Assignee	Patent no.
Micron-Ag	31 May 2010	6 Nov 2012	Hitachi	US8303854B2
Nano-Ag	23 Oct 2009	14 Oct 2014	Dowa	US8858700B2
Nano-Ag	5 Aug 2013	10 Oct 2017	Bando	US9783708B2

Table 3 Claim comparison between Hitachi patent (micron-Ag) and Dowa (nano-Ag) pastes. Only main independent claim is compared in this table

Hitachi US8303854B2	Dowa US8858700B2
A sintering silver paste material comprising:	A bonding material comprising:
1. *Silver* particles having an average particle size of *0.1–2.0 μm* 2. An organic silver complex solution kneaded together into a slurry state, wherein: (a) The organic silver complex solution is a solution containing a *liquid fatty acid* silver salt compound (b) A *solid fatty acid silver salt* dissolved in *an organic solvent having a boiling point of 150 °C or higher* and comprising C, H, and O (c) Silver content in the organic silver complex solution is 30 wt % or more	1. *Metal nanoparticles* 2. A polar solvent in an amount of 5–20% by mass with respect to the metal, wherein: (a) The metal nanoparticles have an *average particle diameter of 100 nm or less* (b) At least one of the metal nanoparticles is surface coated with an *unsaturated fatty acid having 6–8 carbon atoms* (c) At least one of the metal nanoparticles is surface coated with a *saturated fatty acid* having 6–8 carbon atoms

"substantially same way" [7]. Neither the US Supreme Court nor other jurisprudence offers any linguistic framework for DOE beyond the notion of substantial difference [4]. However, this substantial difference and tripartite tests demand the identification of PHOSITA which is defined as a person skilled in the material formulation with a PhD in chemistry and metallurgy, instead of an engineer using the Ag paste in the manufacturing environment. Based on this framework, our subsequent infringement analysis was based on the claim charts containing:

1. The particle size distribution and shape (if necessary)
2. Related chemical formulations (i.e., capping agent, binders (organo-Ag) and solvents)

6 Case 1: Nano-Ag Vs. Micron-Ag Pastes

Besides analysis provided in Sect. 4.0, it is necessary to determine whether formulation for nano-Ag paste reads on the micron-Ag paste. Here, this chapter selected Hitachi (micron-Ag) as the original patentee, while Dowa nano-Ag paste was deemed to be infringing on the Hitachi claims. As shown in Table 3, their main independent claims do not show any literal infringement in terms of elemental identification, though Dowa paste only mentions Ag in their preferred embodiment.

In addition, the PHE of Hitachi patent application showed that Hitachi dropped the term "approximately spherical" for their nanoparticle sizes, while the diameter[8] in the Dowa patent would be interpreted, by a PHOSITA, as spherical particles. These claims on shape highlighted the difference between the Dowa product and Hitachi claims.

Their claims also did not show an overlap in their average particle size; 100 nm serves as a demarcation between Hitachi claims and Dowa product. Dowa claims further emphasized that their preferred embodiment has an average diameter of 10 nm and 30 nm, while Hitachi maintained the similar average of particle sizes between 100 nm and 2000 nm for claims and specification. In reality, it is possible for the two particle size distributions to overlap because the claims only mention the average sizes while the actual particle size has a distribution of sizes [43, 44]. Furthermore, dimensional miniaturization has been raised earlier as insufficient for inventiveness for granting a patent, though DOE analysis does not include the issue of inventiveness [45].

Hence, this chapter used the tripartite test to analyze the sintering mechanism of nano-Ag (Dowa) and micron-Ag (Hitachi) joints to establish their equivalences; nano-Ag paste did not fulfill the criteria of "substantially similar way" compared to the micron-Ag paste. As shown in Table 2, nano-Ag particles of the nano-Ag paste coalesce and sinter to form a Ag joint during the bonding process because of the inherent reactivity of its nano-sized particles when their capping agent decomposed during the heating cycle.

On the other hand, micron-sized Ag particles are not reactive because of their large sizes. Any organic compounds found on the micron-Ag particles used to make the micron-Ag paste are not capping agents, but lubricants left from the particle-manufacturing (milling) process, though the chemical analysis may show some similarities in terms of functional groups between micron- and nano-Ag particles [14, 46]. Instead, micron-Ag paste relies on their binder, such as organosilver or endothermically decomposable silver (EDS) compounds (e.g., silver oxide), to decompose to nano-Ag particles to act as bonding phases between the micron-sized Ag particles during the sintering-bonding step [47].[9] In our analysis, the fatty acid Ag salts acted as the binder for the Hitachi micron-Ag paste, while similar (un) saturated fatty acids acted as the capping agent for the Dowa Ag pastes based on the all-element rules.

---

[8]*Diameter* – "value represented by the diameter of a sphere having the same volume as measured by conventional methods known to those of skill in the art, including sieving and laser diffraction particle size analysis" [42]. In spite of similar length scale, characterization of nano-Ag particles, typically, used other measurement method, i.e., transmission electron microscopy or scanning electron microscopy.

[9]Some micron-Ag paste does not have any decomposable silver compounds, but instead moderate pressure, up to 10 MPa, is often applied to form the sintered Ag joint during the bonding process. Such applied pressure was also used by nano-Ag and hybrid Ag paste to densify the Ag joint to meet the required mechanical and thermal properties.

7 Case 2: Nano-Ag Vs. Nano-Ag Pastes

The earliest patent applications and most cited document for sintered nano-Ag paste formulation for bonding application were disclosed by Ebara in 2002 [31, 32]. The main claim for the Ag paste formulation of this Ebara patent application is:

1. Dispersion in an organic solvent of composite metallic nanoparticles
2. Said composite metallic nanoparticles each having such a structure
3. That a metal core of a metal particle having an average particle diameter of not more than 100 nm
4. Combined and coated with an organic material
5. Said dispersion being in a liquid form

However, this patent application was not granted and abandoned to prioritize the bonding method during the patent prosecution [48]. Based on this background, this chapter compared the two nano-Ag patents from Dowa and Bando, for the possibility of nonliteral infringement, i.e., whether Bando Ag paste reads on Dowa claims, as shown in Table 4.

Table 4 Claim comparison between Dowa patent (nano-Ag) and Bando (nano-Ag) pastes. Only main independent claim is compared in this table

Dowa US8858700B2	Bando US9783708B2
A bonding material comprising	A conductive paste, comprising
1. *Metal nanoparticles*, and 2. A polar solvent in an amount of 5–20% by mass with respect to the metal, wherein: (a) The metal nanoparticles have an *average particle diameter of 100 nm or less*, (b) At least one of the metal nanoparticles is surface coated with an *unsaturated fatty acid having 6–8 carbon atoms*, and (c) At least one of the metal nanoparticles is surface coated with a *saturated fatty acid* having 6–8 carbon atoms	1. *Inorganic particles* having alkylamine with 6 or less of carbon number on at least a portion of a surface 2. *A polymer dispersant* that has a pigment affinity group in a main chain and/or: (a) A plurality of side chains, and, that comprises (b) A polymer with a *comb structure* having a plurality of side chains constituting a solvation portion, (c) A polymer having *a plurality of pigment affinity* portions made from a pigment affinity group in the main chain or (d) A *straight-chain polymer* having a pigment affinity portion made from a pigment affinity group in one terminal of the main chain; and 3. A *dispersion medium*, wherein: (a) A weight reduction percentage at the time of heating solid content of the conductive paste *from room temperature to 500 °C* by thermal analysis is *5% by weight or less*, and (b) The weight reduction percentage at the time of heating solid content of the conductive paste from 200 °C. to 500 °C is *5% by weight or less*, and (c) The content of *the alkylamine with 6 or less of carbon number in the conductive paste is 0.1–15% by weight*

Table 5 Name, structural formula, and boiling point of glycol ethers used by Dowa and Bando nano-Ag pastes

Dowa (US8858700B2)	Bando (US9783708B2)
Diethylene glycol butyl ether (DEG butyl ether)	Diethylene glycol **di**butyl ether (DEG dibutyl ether)
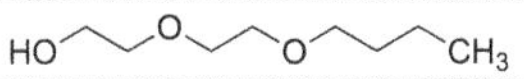	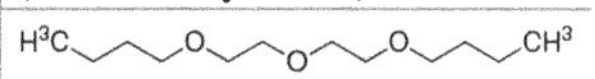
Boiling point: 226–234 °C	Boiling point: 256 °C

In spite of different terminology, the terms for "bonding material" and "metal nanoparticles" in Dowa patents and Bando terms for "conductive paste" and "inorganic particles" refer to the nano-Ag particles used in their nano-Ag paste, as clarified in their respective embodiments.[10] Their range of nanoparticle diameters overlapped; Bando embodiment preferred their particle diameter to be from 1 to 200 nm, while Dowa claim clearly mentioned that their diameter is less than 100 nm.

In terms of "polymer dispersant," Bando claimed the characteristic features of their polymer dispersant added into their nano-Ag paste. Bando's polymer dispersant is also claimed by Dowa as a dependent claim no. 2 that mentioned "the bonding material according to claim 1, further comprising a dispersant added thereto." Furthermore, Dowa embodiment provided the same family of polymer dispersant (i.e., DisperBYK) as stated in the Bando specification.

In terms of "dispersion medium," Dowa claimed the mass percentage of 5–20% of the polar solvent in their Ag paste. This percentage of polar solvent for Dowa patent is replicated in Bando-dependent claims (Claim 7 and 8) on "dispersion medium" that used a mixture of polar solvents and hydrocarbon ranging from 0.5 wt% to 10 wt%. On the same issue of "dispersion medium," the weight percentage loss for Bando Ag paste during the heating cycle (5 wt% or less from room temperature to 500 °C, and from 200 °C to 500 °C) was likely to overlap with Dowa claims on the weight percentage of their polar solvent of 5–20 wt%. The reason is attributed to the same glycol ether family mentioned in their dependent claims (Bando) and specifications (Dowa). Further information on these dispersion mediums are compared in Table 5. Their difference in boiling point of 20 °C was insignificant compared to their sintering temperature of less than 350 °C and 300 °C for Dowa and Bando patent, respectively. These sintering temperatures were sufficiently high to decompose the dispersant to initiate the sintering and bonding processes. The preceding discussion on polymer dispersant and dispersion medium suggested that Bando Ag pastes read on Dowa claims, except for the elements of capping agents in the Dowa claims.

[10]Although the "inorganic particles" was not included in the Bando claims, their specifications suggested gold, silver, platinum, palladium, and copper as examples, but more than half of the disclosed embodiments referred to silver. In a "real" infringement case, these different metallic elements can be easily distinguished with ordinary laboratory equipment like energy-dispersive X-ray spectroscope attached to scanning electron microscope.

As mentioned in Sect. 3, the capping agents prevent the nano-Ag particles from self-sintering at room temperature till the peak sintering temperature is reached. Self-sintering of the nano-Ag particles will result in micron-sized Ag particles in the Ag paste. These micron-sized Ag particles in the original nano-Ag paste will not form the Ag joint because of its low driving force to coalescence, compared to the large surface area associated with nano-Ag particles. Hence, it is important to select suitable capping agents to decompose just before the peak sintering temperature.

Here, the question on equivalent element rest on whether the capping agent with the "(un) saturated fatty acid" (Dowa patent claims) can be replaced by "alkylamine" (for Bando Ag paste) for their range of sintering temperature. The functional group of the fatty acid mentioned in Dowa patent is the carboxylate group, which is only mentioned as such in the best mode section. During patent prosecution, Dowa narrowed their claims from fatty acid to saturated and unsaturated acid, but this information was not affecting the subsequent infringement analysis. Base on their claims in Table 4, Dowa suggested (un) saturated fatty acids with six to eight carbon atoms, which have boiling points from 200–250 °C to fit to their range of specified sintering temperature mentioned earlier.

On the other hand, Bando patents suggested alkylamine as their functional group, having carbon atoms of six or less for their capping agent. These compounds have the boiling point of 130 °C or less, while their sintering temperature is specified as less than 300 °C without any lower limit. This chapter speculated that the sintering temperature for Bando Ag paste is likely to be more than 250 °C because DEG dibutyl ether (Table 5) has the boiling point of more than 250 °C. The boiling points of typical organic compounds with carboxyl and alkylamine capping agents are plotted in Fig. 3.

Prior art literature suggested that the boiling point of the capping agent did not address the full mechanism of decomposition and subsequent Ag sintering process. The decomposition of any capping agent on the nano-Ag particles to initiate the sintering steps depends on two stages [44]:

1. Detachment of capping agent that depends on the *binding energies* of functional groups (i.e., amine or carboxyl groups) facing the nano-Ag particles (E_b)

Fig. 3 Boiling point (°C) of alkylamine (primary) and fatty acid (carboxylate group) with different length of carbon atoms based on data provided by www.sigmaaldrich.com

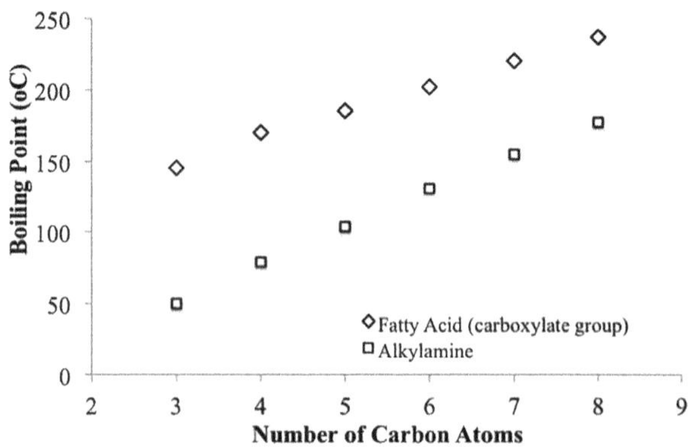

2. *Vaporization energies* of the detached capping agent that depend on the *boiling point of the molecules*, containing the functional group in the capping agent (E_v)

The total energy to remove the capping agents depends on the multiplicative sum of E_b and E_v and the density of the capping agents (D) on the nano-Ag particles ($D^*(E_b + E_v)$) [44]. Here, this chapter could not compare the decomposition process of the capping agents with amine and carboxyl groups in the absence of information on the density of capping agents on the nano-Ag particles and their vaporization energies (which are related to their boiling points).

Hitherto, this chapter do know that the capping agents, regardless of amine or carboxylate group (fatty acids), have the same function which is to disperse the adjacent nanoparticles from agglomerate based on steric hindrance of the carbon atom length. Dowa selected capping agent with carbon atoms of six to eigth while Bando chose capping agent with carbon atom length of six or less. The decomposition steps for Bando and Dowa Ag pastes were based on the multiplicative sum of density (D) and ($E_b + E_v$). Others have also replaced fatty acid (carboxylate) with amine group as the functional group for their capping agent on the nano-Ag particles [49]. Hence, there is a possibility that these two capping agents are considered equivalent based on the substantially similar function/way/ results analysis under the doctrine of equivalents.

8 Conclusions and Recommendations

Based on the case studies and preceding discussion, this chapter conclude the following points:

1. Nano-Ag paste does not read on the "older" micron-Ag paste because of the different "way" in which the nano-Ag paste form the sintered Ag joint based on the tripartite test.
2. In the comparison of two nano-Ag pastes, there is a possibility of equivalence for the use of functional group like "fatty acid, i.e., carboxyl group," and "amine" as capping agent on the nano-Ag particles based on the tripartite test because of their similar function, way, and results.

This infringement analysis shows that the tripartite test, while unsatisfactory, remains the primary tool of analysis for the application of DOE in sintered Ag pastes. Our future research will compare claims from two micron-Ag pastes to complete the review of equivalence in sintered Ag paste formulation.

Acknowledgment The author greatly appreciate WIPO Scholarship for providing the financial support to attend his Master of Laws in Intellectual Property at the University of Turin-WIPO programme, and Universiti Kebangsaan Malaysia Research Grant (GUP-2017-055 "Production of Metallic Conducting Nanowires for Industrial Applications") for this research chapter. He also acknowledged Prof. Craig Nard (Case Western University), Prof. Alessandro Cogo (University of Turin), and Dr. Maximiliano Marzetti (University of Turin) for their advices and guidance in writing this research work.

References

1. Winans v. Denmead, vol. 56 US, ed: US Supreme Court, 1854, p. 330
2. J.S. Cianfrani, An economic analysis of the doctrine of equivalents. Virginia J Law Tech **Spring**, 1–26 (1997)
3. Graver Tank & Mfg. Co. v. Linde Air Products Co., vol. 339 US, ed: US Supreme Court, 1950, p. 605
4. Warner-Jenkinson Co. v. Hilton Davis Chem. Co., vol. 520 US, ed: US Supreme Court, 1997, p. 17
5. Ethicon Endo-Surgery Inc. v. United States Surgical Corp, vol. 149 F.3d 1309, ed: US Supreme Court of Appeals, Federal Circuit, 1998, p. 1309
6. EMI Group North America Inc v. Intel Corp vol. 157 F.3d ed: US Court of Appeals, Federal Circuit, 1998, p. 887
7. Slimfold Manufacturing Co. v. Kinkead Industries, Inc, vol. 810 F.2d, ed: United States Court of Appeals, Federal Circuit, 1987, p. 1113
8. Festo Corp. v. Shoketsu Kinzoku Kogyo Kabushiki Co. Ltd., vol. 535 US, ed: US Supreme Court, 2002, p. 722
9. G.P. Belvis. *The Doctrine of Equivalents and 112 Equivalents*, (Brinks Gilson & Lione, 2017), www.brinksgilson.com/files/98.pdf. Accessed on 28 Oct 2017
10. Johnson & Johnston Associates Inc. v. R.E. Service Co, vol. 285 F.3d, ed: US Court of Appeals, Federal Circuit, 2002, p. 1046
11. Texas Instruments Inc. v. United States International Trade Commission, vol. 846 F.2d, ed: US Court of Appeals, Federal Circuit, 1988, p. 1369
12. K.S. Siow, Y.T. Lin, Identifying the development state of sintered ag as a bonding material in the microelectronic packaging via a patent landscape study. J. Electron. Packag. **138**, 020804-1–020804-13 (2016)
13. S.P. Lim, B. Pan, H. Zhang, W. Ng, B. Wu, K.S. Siow, S. Sabne, M. Tsuriya, High-temperature Pb-free die attach material project phase 1: survey result, in International Conference on Electronics Packaging, (Yamagata, Japan, 2017), pp. 51–56
14. K.S. Siow, Are sintered silver joints ready for use as interconnect material in microelectronic packaging? J. Electron. Mater. **43**, 947–961 (2014)
15. G. Bai, Low-temperature sintering of nanoscale silver paste for semiconductor device interconnection, PhD (Materials Science and Engineering), Department of Materials Science and Engineering, Virginia Polytechnic Institute and State University Blacksburg, Virginia, 2005
16. K. Kiełbasiński, J. Szałapak, M. Jakubowska, A. Młozniak, E. Zwierkowska, J. Krzemiński, M. Teodorczyk, Influence of nanoparticles content in silver paste on mechanical and electrical properties of LTJT joints. Adv. Powder Technol. **26**, 907–913 (2015)
17. T. Kim, Y. Joo, S. Choi, *Power Module Using Sintering Die Attach and Manufacturing Method Thereof* (Samsung Electro-Mechanics Co., US8630097B2, 2014)
18. United States Patent Classification System, Class 977, nanotechnology § 1. (USPTO, 2010), http://www.uspto.gov/go/classification/uspc977/defs977.pdf. Accessed on 2 Nov 2017
19. Y. Saito, S. Sasaki, *Silver particle-containing composition, dispersion solution, and paste and method for manufacturing the same* (Dowa Electron. Mater. Co., US9255205B2, 2013)
20. M. Boureghda, N. Desai, A. Lifton, O. Khaselev, M. Marczi, B. Singh, *Methods of Attaching a Die to a Substrate* (Alpha Metal Inc., US8555491B2, 2008)

21. T. Ogashiwa, M. Miyairi, *Method of Bonding* (Tanaka Kikinzoku Kogyo K.K., US7789287B2, 2010)
22. I.J. Rasiah, *Electrically Conductive Thermal Interface* (Honeywell Int. Inc., US7083850B2, 2006)
23. In re Soni, vol. 54 F.3d, ed: US Court of Appeals, Federal Circuit, 1995, p. 746
24. H. Zhang, W. Li, Y. Gao, H. Zhang, J. Jiu, K. Suganuma, Enhancing low-temperature and pressureless sintering of micron silver paste based on an ether-type solvent. J. Electron. Mater. **46**, 5201–5208 (2017)
25. R. Voigt, E. Michelson, Nanotechnology-related inventions: infringement issues. Nanotechnol. Law Bus. **2**(1), 45–53 (2005)
26. A. Wasson, Protecting the next small thing: nanotechnology and the reverse doctrine of equivalents. Duke Law Technol. Rev. **10** (2004)
27. A.L. Durham, *Patent Law Essentials: A Concise Guide* (Praeger, Westport, 1999)
28. H. Schwarzbauer, *Method of Securing Electronic Components to a Substrate* (Siemens AG., US4810672B2, 1987)
29. W. Baumgartner, J. Fellinger, *Method of Fastening Electronic Components to a Substrate Using a Film* (Siemens AG., US4856185, 1989)
30. G. Frens, J.T.G. Overbeek, Carey Lea's colloidal silver. Colloid Poly. Sci. **233**, 922–929 (1969)
31. H. Nagasawa, K. Kagoshima, N. Ogure, M. Hirose, Y. Chikamori, *Bonding Material and Bonding Method* (US20040245648A1, 2004)
32. K.S. Siow, M. Eugénie, Patent landscape and market segments of sintered silver as die attach materials in microelectronic packaging, in 37th International Electronics Manufacturing Technology (IEMT) & 18 Electronics Materials and Packaging (EMAP) Conference, (2016), pp. 1–6
33. K.S. Siow, A.A.O. Tay, P. Oruganti, Mechanical properties of nanocrystalline copper and nickel. Mater. Sci. Technol. **20**, 285–294 (2004)
34. A. Zattoni, D.C. Rambaldi, P. Reschiglian, M. Melucci, S. Krol, A.M.C. Garcia, A. Sanz-Medel, D. Roessner, C. Johann, Asymmetrical flow field-flow fractionation with multi-angle light scattering detection for the analysis of structured nanoparticles. J. Chromatogr. A **1216**, 9106–9112 (2009)
35. O.T. Mefford, M.R.J. Carroll, M.L. Vadala, J.D. Goff, R. Mejia-Ariza, M. Saunders, R.C. Woodward, T.G. St. Pierre, R.M. Davis, J.S. Riffle, Size analysis of PDMS−magnetite nanoparticle complexes: Experiment and theory. Chem. Mater. **20**, 2184–2191 (2008)
36. S.B. Rice, C. Chan, S.C. Brown, P. Eschbach, L. Han, D.S. Ensor, A.B. Stefaniak, J. Bonevich, A.E. Vladar, A.R.H. Walker, J. Zheng, C. Starnes, A. Stromberg, J. Ye, E.A. Grulke, Particle size distributions by transmission electron microscopy: An interlaboratory comparison case study. Metrologia **50**, 663–678 (2013)
37. Markman v. Westview Instruments, Inc, vol. 517 US ed: US Supreme Court, 1996, p. 370
38. W.A. DeVries, Meaning and interpretation in history. Hist. Theory **22**, 253–263 (1983)
39. L.B. Solum, The interpretation-construction distinction, Georgetown Public Law and Legal Theory Res. Paper No. 11-95, (2010), p. 95
40. C.A. Nard, A theory of claim interpretation. Harv. J. Law Technol. **14**, 1–82 (2000)
41. K.S. Siow, Mechanical properties of Nano-Ag as die attach materials. J. Alloys Compd. **514**, 6–14 (2012)
42. Apotex v. Cephalon, vol. Civil Action No. 2:06-cv, ed: US District Court, 2010, p. 2768
43. M. Matsui, T. Tomura, T. Watanabe, K. Shimoyama, *Conductive Paste* (Bando Chem. Ind., US9783708B2, 2013)
44. M. Tobita, Y. Yasuda, E. Ide, J. Ushio, T. Morita, Optimal design of coating material for nanoparticles and its application for low-temperature interconnection. J. Nanopart. Res. **12**, 2135–2144 (2010)
45. F.A. Fiedler, G.H. Reynolds, Legal problems of nanotechnology:an overview. South. Calif. Interdisc. Law. J. **3**, 593–629 (1994)
46. D. Lu, Q.K. Tong, C.P. Wong, A study of lubricants on silver flakes for microelectronics conductive adhesives. IEEE Trans. Compon. Packag. Technol. **22**, 365–371 (1999)

47. H. Zhang, Y. Gao, J. Jiu, K. Suganuma, In situ bridging effect of Ag O on pressureless and low-temperature sintering of micron-scale silver paste. J. Alloys Compd. **696**, 123–129 (2017)
48. US 20040245648 11 Members in Patent Family, 17, (Global Dossier USPTO, 2017), https://globaldossier.uspto.gov/#/result/publication/US/20040245648/1142. Accessed on 30 Oct 2017
49. S. Ghosal, R. Pandher, O. Khaselev, R. Bhatkal, R. Raut, B. Singh, M. Ribas, S. Sarkar, S. Mukherjee, S. Kumar, R. Chandran, P. Vishwanath, A. Pachamuthu, M. Boureghda, N. Desai, A. Lifton, N.K. Chaki, *Sintering powder* (US20150353804, 2013)

Sintered Copper (Cu): Chemistry, Process, and Reliability

Y. Yamada

1 Short Introduction to Joints for Power Semiconductor Devices

In addition to gold (Au) solder, which is a well-established technology, there are ongoing research studies into other bonding technologies for wide bandgap semiconductor devices that can operate at temperatures exceeding 200 ° C, as shown in Fig. 1. These semiconductors include silicon carbide, gallium nitride, and other materials. For connecting circuit components, researchers are examining high melting point solders containing zinc or bismuth, sintered materials incorporating silver (Ag) or copper (Cu), and alloys of tin (Sn) and Cu [1, 2]. The greatest progress has been made with Ag-based sintered materials, which are ready for industry-wide use. The base material is rather expensive, however, and is prone to electromigration. This chapter describes joining technologies based on the sintering of Cu-based materials, which offer high thermal and electrical conductivity similar to Ag. The thermal properties and reliability of Cu-based materials are also discussed.

2 Preparation of Cu Nanoparticle Joints

The Cu nanoparticles employed for joining are coated with fatty acids and amines. This study examined three types of nanoparticles. The oleic acid ($C_{17}H_{33}COOH$) + oleic amine ($C_{18}H_{35}NH_2$) combination was called C18 and had a mean particle diameter of 64.8 ± 19.9 nm, the dodecane acid ($C_{11}H_{23}COOH$) + dodecane amine ($C_{12}H_{25}NH_2$) combination was called C12 and had a mean particle diameter of

Y. Yamada (✉)
Daido University, Nagoya, Japan
e-mail: yamada-y@daido-it.ac.jp

K. S. Siow (ed.), *Die-Attach Materials for High Temperature Applications in Microelectronics Packaging*, https://doi.org/10.1007/978-3-319-99256-3_8

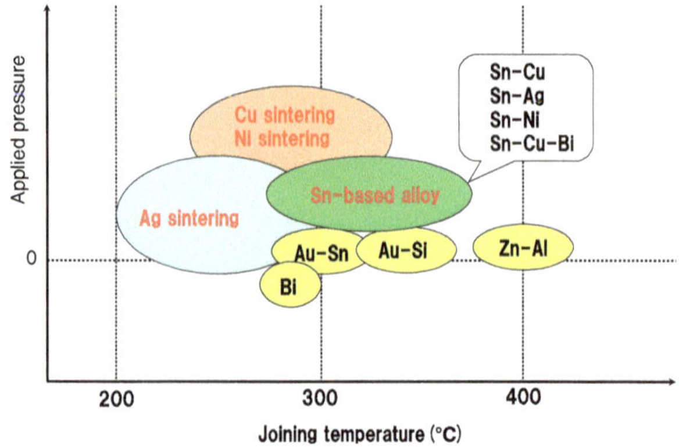

Fig. 1 Overview of joining technologies for power semiconductor devices at high temperature operation

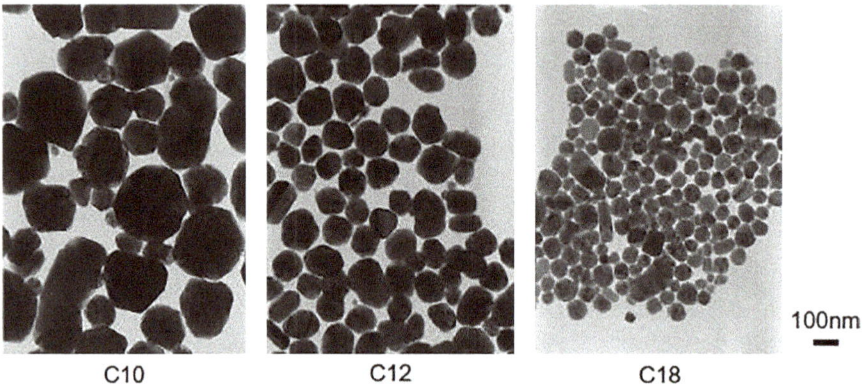

Fig. 2 TEM images of Cu nano-particles

135 ± 29.0 nm, and the decane acid ($C_9H_{19}COOH$) + decylamine ($C_{10}H_{21}NH_2$) combination was called C10 and had a mean particle diameter of 231 ± 77.9 nm. Figure 2 shows the TEM images of these Cu nanoparticles. Joining was performed under a pressure of 1 MPa in a H_2 atmosphere at a temperature of 250–350 °C.

3 Thermal Properties

3.1 Measurement of Thermal Resistance and Estimation of Thermal Conductivity

The reliability assessment of a Cu joint is based first of all on the assumption that the joint has been precisely formed, with almost no voids or delaminations. Our examination of the fabricated specimens began with an assessment of their thermal properties [3]. As shown in Fig. 3, the specimens were fabricated with simple

Fig. 3 Structure of specimens

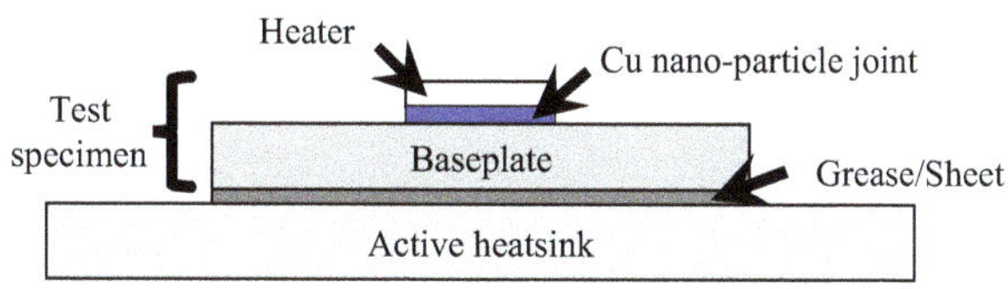

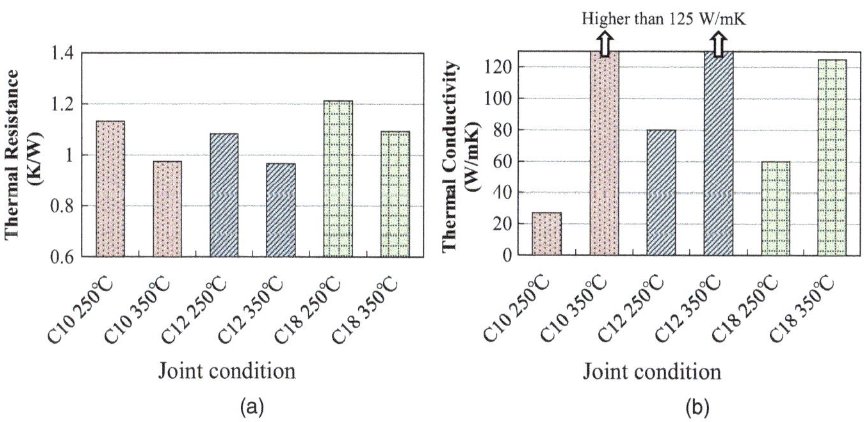

Fig. 4 Thermal properties. (**a**) Thermal resistance. (**b**) Estimated thermal conductivity

structures while taking strict measures to eliminate other elements, to limit the observations to the characteristics of the joints. The specimens had a layered structure. Based on the specimen structure from top to bottom, they consisted of a heater chip, a joint consisting of Cu nanoparticles, and a baseplate.

The structure was fixed to a water-circulated cooler using thermal grease. Metal gages were inserted between the two surfaces on each side to measure the thickness of the grease without blocking the path of heat flow. The heater chip was made from Al_2O_3, and its electrical resistance was several tens of ohms, which is rather high, so it was easy to use with low currents of about 1–2 A. Cooling water was circulated through the cooler at a constant temperature and flow rate. Constant power was applied to the heater chip. The temperature at the center of the upper surface of the chip was measured using a thin thermocouple to obtain thermal resistance.

As shown in Fig. 4a, the thermal resistance varied with the joining temperature and with the type of nanoparticles, but generally, the joints created at high temperatures had a lower thermal resistance. The reason for this is that when materials are joined at high temperatures, sintering takes place, resulting in a fine structure. Also, a finite element model was constructed in order to estimate the thermal conductivity of the Cu nanoparticle joints in a steady-state thermal analysis; the joint thickness was accounted for, and the thermal conductivity of the joint was the unknown variable. The results are shown in Fig. 4b. The specimens with the lowest thermal resistance were estimated to have thermal conductivities of at least 125 W/mK. The C18 joint had somewhat inferior thermal properties, so only C10 and C12 were examined in the subsequent reliability tests.

3.2 Further Simulation

Conventional lead or tin-based solder joining materials have a lower thermal conductivity than substrates or baseplates, at about 30–70 W/mK, so these conventional joints behave as thermal resistors, based on these thickness. In contrast, the base metals of Cu or Ag solder joining materials have thermal conductivities of over 400 W/mK, which is higher than that for any other materials used in these electronics, so it can be expected that heat spreads out horizontally in those joints. Therefore, we performed an estimate of the thermal properties of these soldered joints in a thermal finite element (FE) analysis [4].

The specimen structure was as described above, and the thermal properties were evaluated. As shown in Fig. 5, the parameters were the thickness and thermal conductivity of the joint and the thermal conductivity of the baseplate.

First, the thermal conductivity of the baseplate was assumed to be 207 W/mK, and two joint thicknesses were examined while varying the thermal conductivity of the joints to determine the thermal resistance. Figure 6 presents the behavior of the thermal resistance, in inverse relation to conductivity.

Figure 7 shows how the thermal resistance varied with the thickness of the joint, holding the thermal conductivity of the joint and the baseplate constant at 400 W/mK and 207 W/mK, respectively. The thermal resistance reached a minimum for a certain thickness, i.e. 0.5 mm.

In order to observe this behavior more closely, the thermal resistance of each layer of the joint is shown in Fig. 8. The thermal resistance of the joint increased in proportion to the thickness even though the thermal resistance of the baseplate decreased. This anomaly can be attributed to an increase in the specific area of the baseplate, which conducts the heat away, in the horizontal direction.

This behavior disappeared when the thermal conductivity of the baseplate was set to 400 W/mK, the same value as of the joint, as shown in Fig. 9. This effect may only occur when the thermal conductivity of the joint is higher than that of the baseplate.

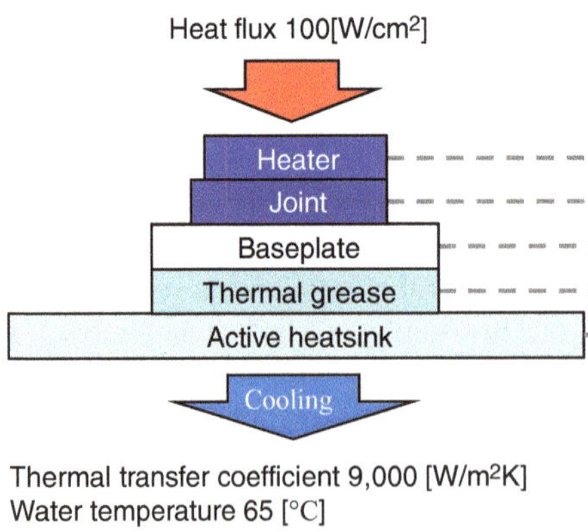

Material	Thermal conductivity [W/mK]	Thickness [mm]
Al_2O_3	20	0.5
Cu	70~400	0.1~1.0
x	100,207,400	3
Silicone	4.5	0.06
Al(A5052)	140	5

Fig. 5 Physical properties and boundary condition in an analytical model

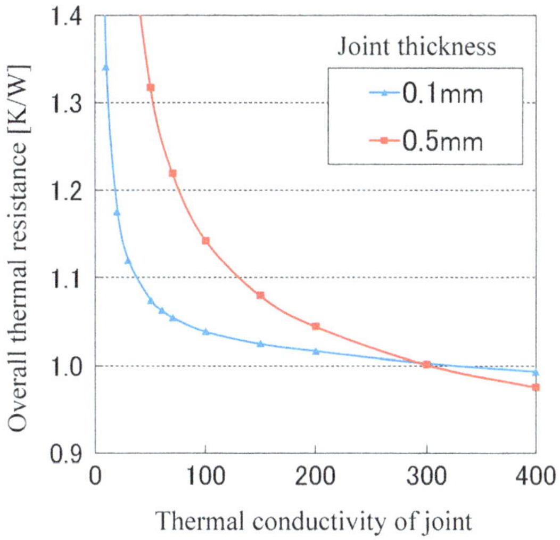

Fig. 6 Thermal conductivity and thermal resistance of the joint. Thermal conductivity of the baseplate: 207 W/mK

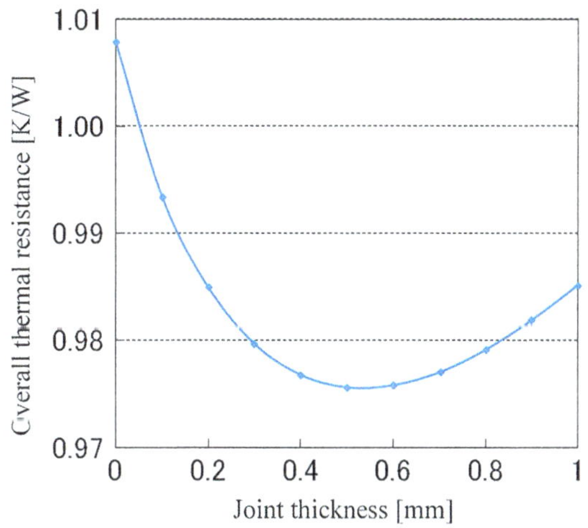

Fig. 7 Joint thickness and thermal resistance. Thermal conductivity of the joint: 400 W/mK. Thermal conductivity of the baseplate: 207 W/mK

It is possible to extend these findings to the actual circuits when the joints are made from Cu or similar materials on direct bonded copper. A substrate with a baseplate is made from a material such as CuMo or Al-SiC, which has a low coefficient of thermal expansion (CTE) and low thermal conductivity.

Generally, thermal stress concentrations form at thin joints, which degrade their reliability. Nonetheless, as the above cases suggest, the thermal resistance reaches a minimum for a constant joint thickness; this is also desirable from the point of view of reliability. Thus, it is possible that highly conductive joints made from Cu-based materials will have advantages over the conventional materials.

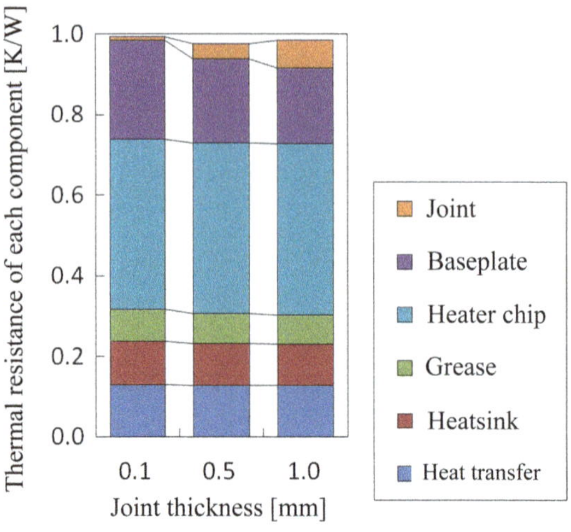

Fig. 8 Breakdown of thermal resistance. Thermal conductivity of joint: 400 W/mK. Thermal conductivity of baseplate: 207 W/mK

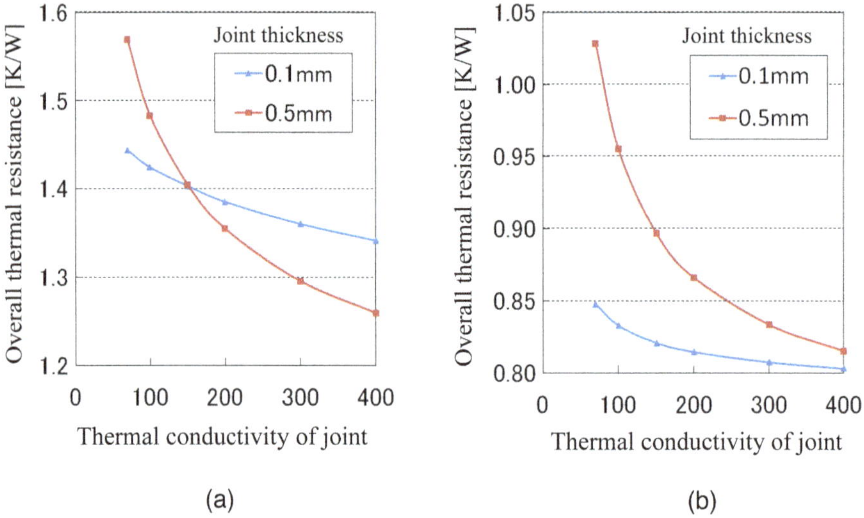

Fig. 9 Thermal conductivity and thermal resistance of joint. (**a**) Baseplate: 100 W/mK. (**b**) Baseplate: 400 W/mK

4 Reliability

4.1 Power Cycle Test

When a high current flows through the electronic circuit of the power semiconductor module, the temperature of the module begins to rise within a short time due to the self-generated heat. A power cycle test is an accelerated test mimicking such situations.

A power cycle test was conducted with the layered structure described earlier [5]. The heater chip was Al_2O_3 and the baseplate was Cu-65Mo, so the difference in the CTE was fairly low, at about 1.5 ppm/K. Reliability was tested at high temperatures of more than 200 °C. The evaluation was carried out as follows. As for the evaluation of the thermal properties, voltage pulses were provided from a DC source using a computer, while the temperature, voltage, and current were recorded continuously on a data logger. Temperature changes were observed during 60-s periods. In the 65 °C/200 °C test, the voltage was on for 10 s and off for 50 s. In the 65 °C/250 °C test, it was on for 20 s and off for 40 s. The 65 °C/200 °C test was continued for 3000 cycles; once there were no further changes, the 65 °C/250 °C test was initiated. Figure 10 shows the temperature profiles during these tests.

Figure 11 shows the results of these tests. No changes in temperature were found in either test when using the C12 particles. When using the C10 particles, no changes were seen in the 65 °C/200 °C test, while in the 65 ° C/250 °C test, there was a sudden temperature increase, fracturing the heater chip. For comparison, a test was also performed using the Sn-0.7Cu solder (melting point 228 °C), but a rapid temperature change was observed at low cycles of the 65 °C/200 °C test, so this test was stopped.

Cross sections of the test specimens were observed after the tests, and these are shown in Fig. 12. Vertical cracks were seen in the joint for the C12 specimen. The cause of these cracks is unclear, but since they were not oriented in directions that impede the flow of heat, they probably did not cause any temperature changes. In the C10 specimen, prominent delamination was found at the interfaces. The soldered joint showed lateral cracking with considerable growth. The above results indicate that joints containing Cu nanoparticles provided power cycling reliability that was superior to that for tin-based solder systems at high temperatures exceeding 200 °C. Also, on comparing C12 and C10, since C12 has smaller particles, it undergoes more sintering at any given joining temperature and exhibits greater strength at the joint interfaces.

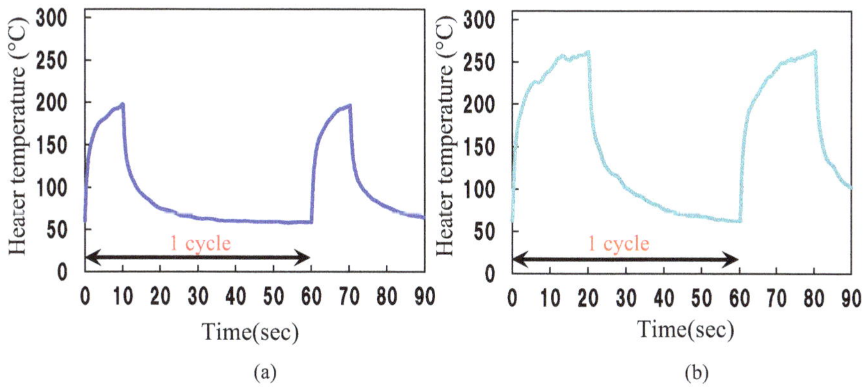

Fig. 10 Condition of the power cycle test. (**a**) 65 °C/200 °C test. (**b**) 65 °C/250 °C test

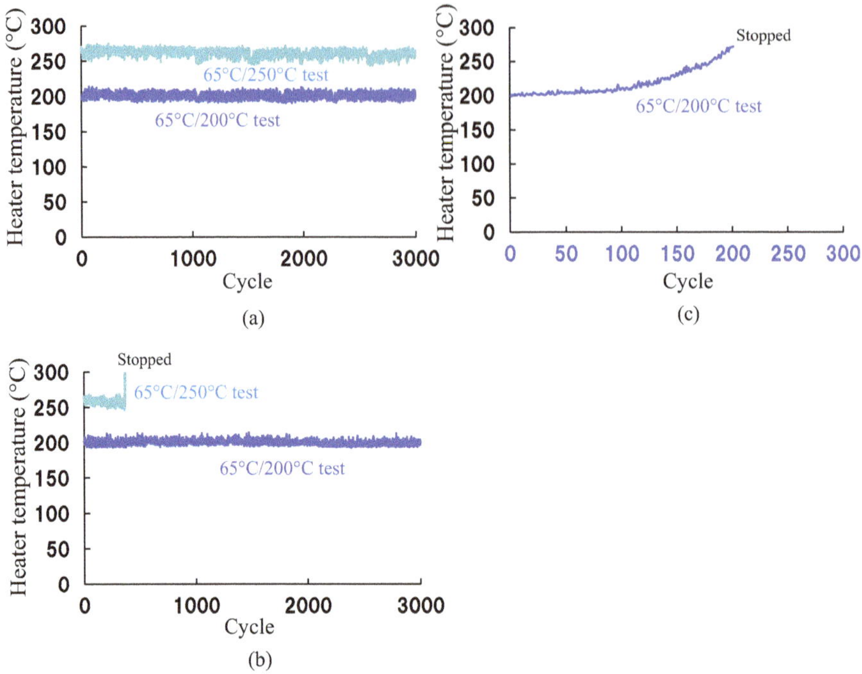

Fig. 11 Results of the power cycle test. (**a**) C12, joining at 300 °C. (**b**) C10, joining at 300 °C. (**c**) Joining with the Sn-0.7Cu solder

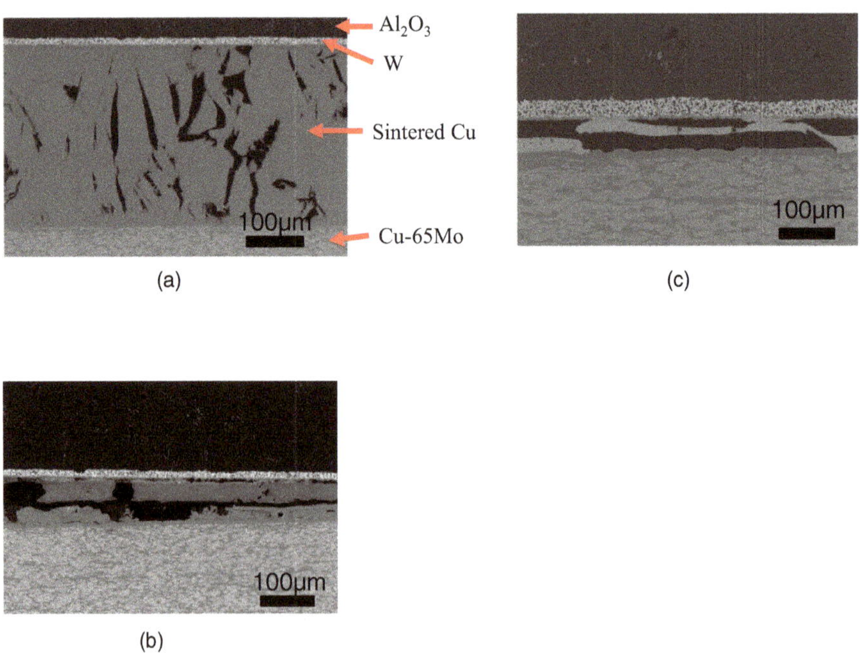

Fig. 12 Cross section of the joint after the power cycle test. (**a**) C12, joining at 300 °C. (**b**) C10, joining at 300 °C. (**c**) Joining with the Sn-0.7Cu solder

Next, we carried out the same test, using different joining temperatures, on C12, which had shown superior power cycling reliability. Three joining temperatures of 250, 300, and 350 °C were used. Figure 13 shows the results. No changes in temperature were seen during the power cycle test for any of the specimens, and the same results were obtained. In the cross sections shown in Fig. 14, some vertical cracking is again observed. In order to examine the possibility of even greater reliability, a 65 °C/300 °C power cycle test was performed. The temperature difference was large here, so the cycle was lengthened to 120 s, corresponding to 40 s on and 80 s off. As shown in Fig. 15, no changes in temperature were found. This test confirmed the potential of this material for use under these stringent conditions.

So far, we have examined power cycling reliability assuming a small difference between the CTE for the chip and that for the baseplate [6]. In the next stage, we varied the CTE for the chip and baseplate to obtain four levels of difference and repeated the power cycle tests in order to examine the reliability limits for the joint. Table 1 shows the materials used for the baseplate; these were Cu-65Mo (8 ppm/K), Cu-40Mo (11 ppm/K), Al-SiC (14 ppm/K), and just Cu (17 ppm/K). All were subjected to 3000 cycles of the 65 °C/200 °C test and then the same number of cycles of the 65 °C/250 °C test. In this experiment, the electrical resistance of the heater chip was used as an index of the degeneration. Three specimens were selected that required joining temperatures of 250 °C, 300 °C, and 350 °C, and the test was

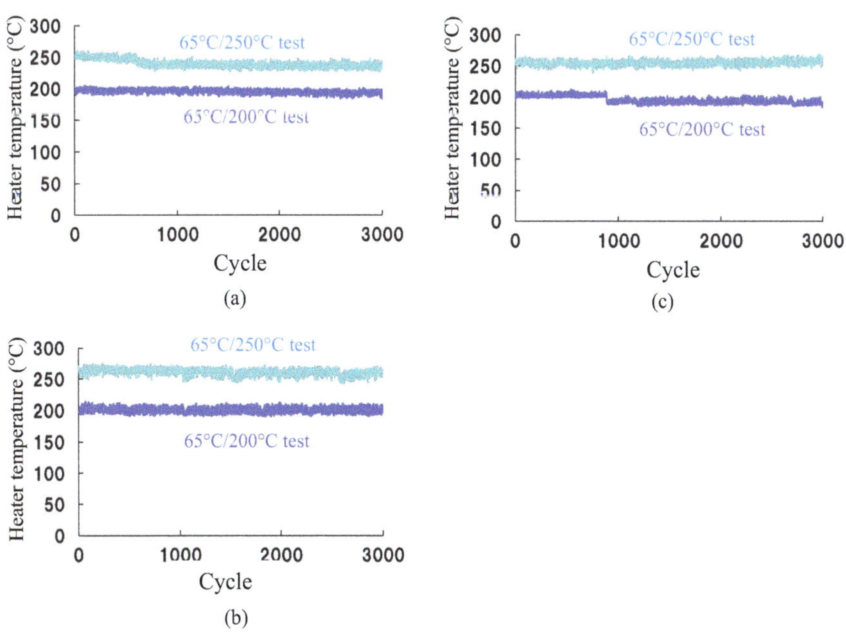

Fig. 13 Results of the power cycle test. Dependency of joining temperature. (**a**) C12, joining at 250 °C. (**b**) C12, joining at 300 °C. (**c**) C12, joining at 350 °C

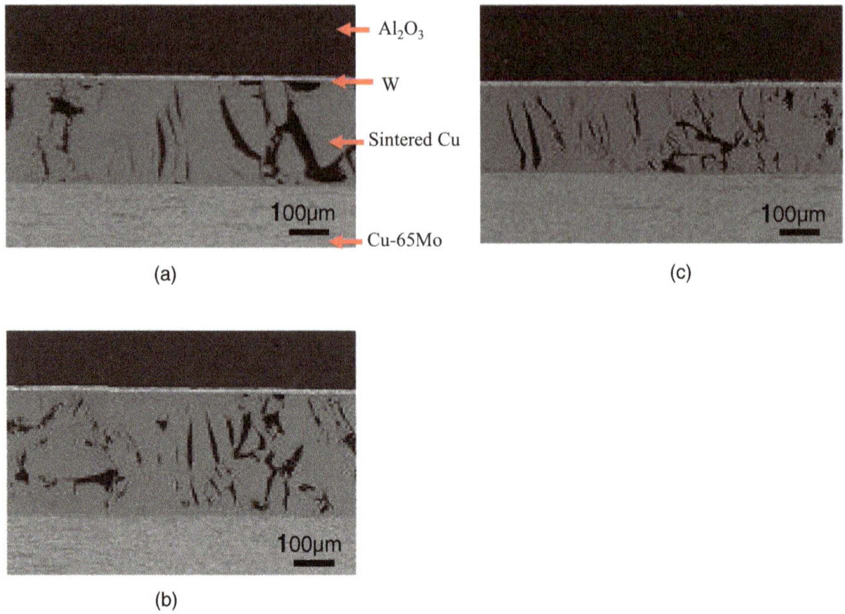

Fig. 14 Cross section of joint after the power cycle test. (**a**) C12, joining at 250 °C. (**b**) C12, joining at 300 °C. (**c**) C12, joining at 350 °C

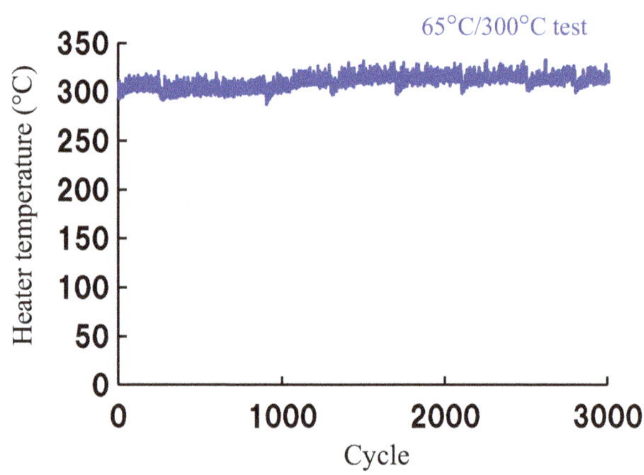

Fig. 15 Results of 65 °C/300 °C power cycle test on C12, jointed at 300 °C

stopped when any material showed a prominent change in resistance during the test. Since the thermal conductivities differed because of the baseplate, the applied voltages were adjusted to ensure that all had identical maximum temperatures at the outset of the tests. The results for these tests are given in Figs. 16, 17, and 18 and in Table 2.

No dramatic changes were seen for Cu-65Mo, Cu-40Mo, or Al-SiC. The specimens joined by Cu at 250 °C and 350 °C showed an increase in resistance,

Table 1 Specimens with different ΔCTE

Component	Size (mm)	Specimen (ΔCTE of chip [ppm/K])
Heater chip	6 × 6 × ^{t}0.5	Al_2O_3
Joint	6 × 6 × (^{t}0.075~0.14)	Cu nano-particle
Baseplate	20 × 40 × ^{t}3	Cu-65Mo (1.5)
		Cu-40Mo (4.9)
		Al-SiC (7.3)
		Cu (10.3)

Joint condition: temperature 250, 300, 350 °C, pressure 1 MPa

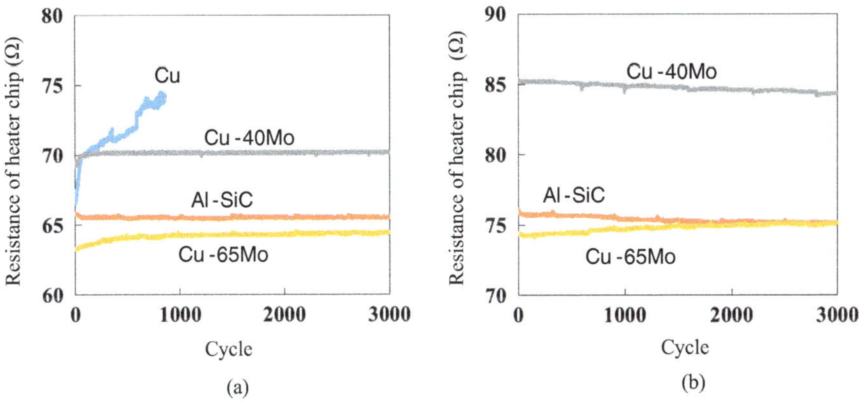

Fig. 16 Results of the power cycle test with different ΔCTE (joining temperature: 250 °C). (**a**) 65 °C/200 °C test. (**b**) 65 °C/250 °C test

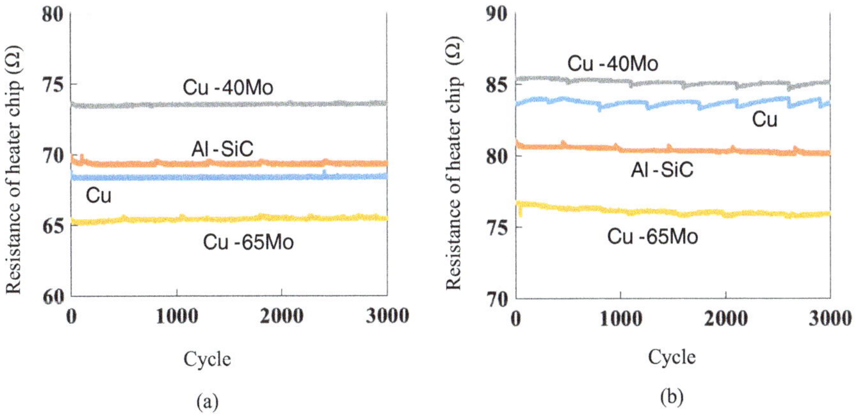

Fig. 17 Results of the power cycle test with different ΔCTE (joining temperature: 300 °C). (**a**) 65 °C/200 °C test. (**b**) 65 °C/250 °C test

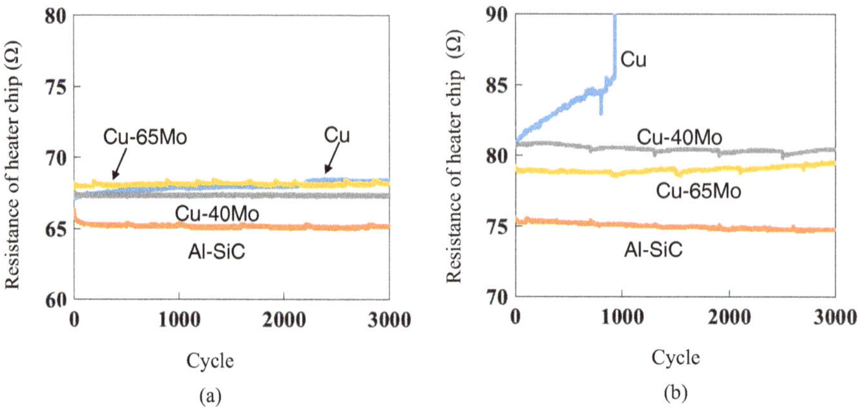

Fig. 18 Results of the power cycle test with different ΔCTE (joining temperature: 350 °C). (**a**) 65 °C/200 °C test. (**b**) 65 °C/250 °C test

Table 2 Results of power cycle tests for specimens with different ΔCTE

※Cycle number when heater chip failed

Baseplate (ΔCTE ppm/K)	250℃ join		300℃ join		350℃ join	
	65℃/200℃ test	65℃/250℃ test	65℃/200℃ test	65℃/250℃ test	65℃/200℃ test	65℃/250℃ test
Cu-65Mo (1.5)	≧3000	≧3000	≧3000	≧3000	≧3000	≧3000
Cu-40Mo (4.9)	≧3000	≧3000	≧3000	≧3000	≧3000	≧3000
AlSiC (7.3)	≧3000	≧3000	≧3000	≧3000	≧3000	≧3000
Cu (10.3)	857※		≧3000	≧3000	≧3000	931※

Joint deteriorated Chip cracked

suggesting deterioration; however, the modes of deterioration differed between those joints. Observations of the cross sections of the joints after the test revealed crack growth in the specimens joined at 250 °C, as shown in Fig. 19. These cracks are attributed to the high thermal stresses in the weak joints. In contrast, there was little damage in the joints formed at 350 °C, although several cracks were observed in the Al_2O_3 chips. This was attributed to the strengthening of the joint by its formation at the higher temperature; as a result, thermal stresses arose that exceeded the failure

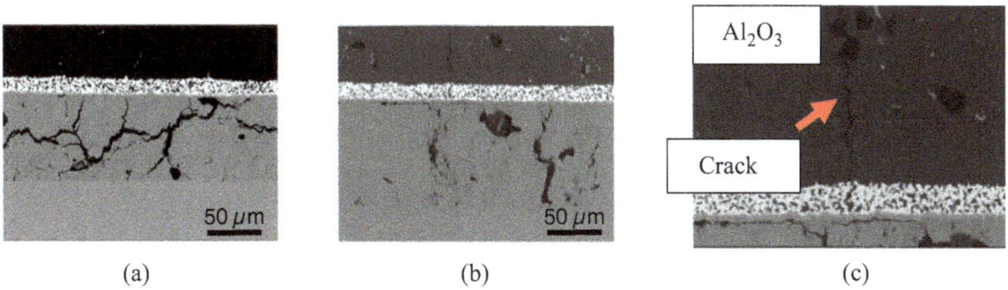

Fig. 19 Cross section of specimens after the power cycle test. (**a**) Joining at 250 °C. (**b**) Joining at 350 °C. (**c**) Periphery of the heater chip joining at 350 °C

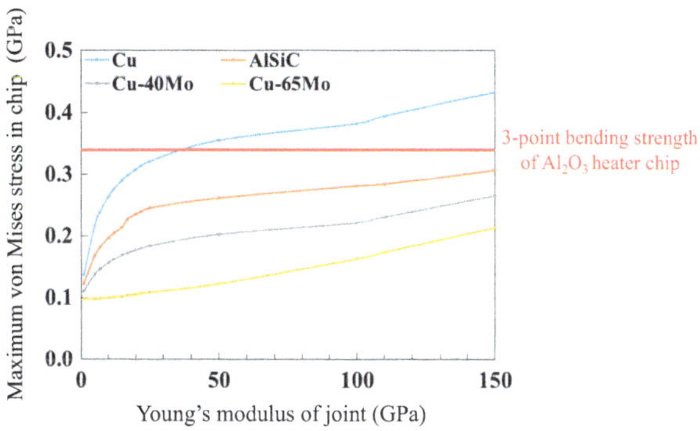

Fig. 20 Young's modulus of joint and von Mises stress

strength of the heater chip, causing cracking in the chip. An FE analysis of the structure predicted that high stresses would occur in the strip. As shown in Fig. 20, of the four types of baseplates, the thermal stresses were clearly the highest in the Cu baseplate. When the joint was assumed to have a high Young's modulus, the stresses in the chip were prominently higher. The analysis indicated little sintering when the joining was performed at 250 °C, and the resulting Young's modulus was low. In the 350 °C model, sintering progressed well and the Young's modulus was high; this would explain why the thermal stresses in the chip reached high values.

4.2 *Thermal Cycle Test*

Thermal cycle reliability was next examined [7]. The structure of the samples was similar to earlier, as shown in Fig. 2. An off-the-shelf thermal shock tester was employed, and each specimen was removed from the tester after a specified number of cycles so that the joint cross section could be evaluated. The temperatures were −40 °C/150 °C and −40 °C/200 °C. The specimens were exposed to each temperature for 20 min, for a total of 40 min per cycle. Figure 21 shows the temperature profile employed in the experiment. The specimens were evaluated using scanning

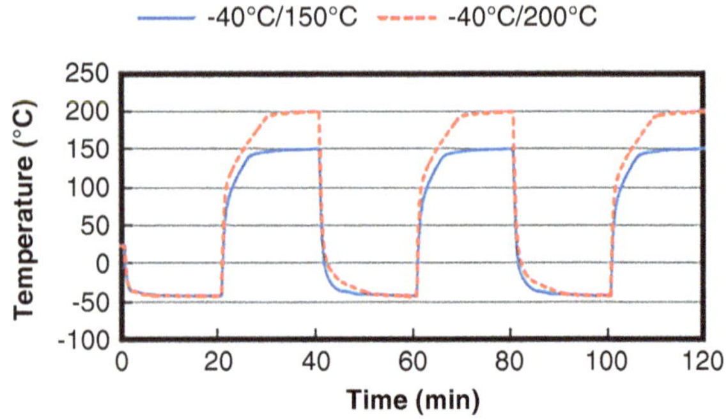

Fig. 21 Thermal cycle test conditions

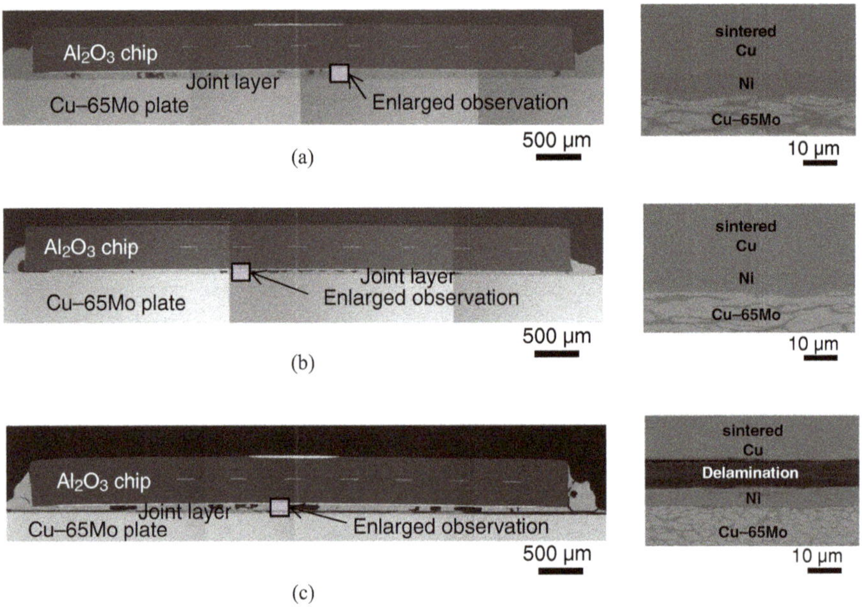

Fig. 22 Sintered at 300 °C. (**a**) After initial processing. (**b**) After 1000 cycles of −40 °C/150 °C test. (**c**) After 1000 cycles of −40 °C/200 °C test

electron microscopy (SEM) after mechanical polishing of the cross section and were graded by the presence or absence of cracking or delamination. C12 was used for the joints since these particles had shown superior reliability in the power cycle test.

Figure 22 shows SEM micrographs of joints fabricated at 300 °C. They were first subjected to 1000 cycles of the −40 °C/150 °C test and then 1000 cycles of the −40 °C/200 °C test. No deterioration was noted after the −40 °C/150 °C test, but delamination from the baseplate was found after the −40 °C/200 °C test. Figure 23 shows the results of the same examination of specimens formed at 350 °C. These exhibited high reliability, with no deterioration of the joint after the −40 °C/200 °C test. For

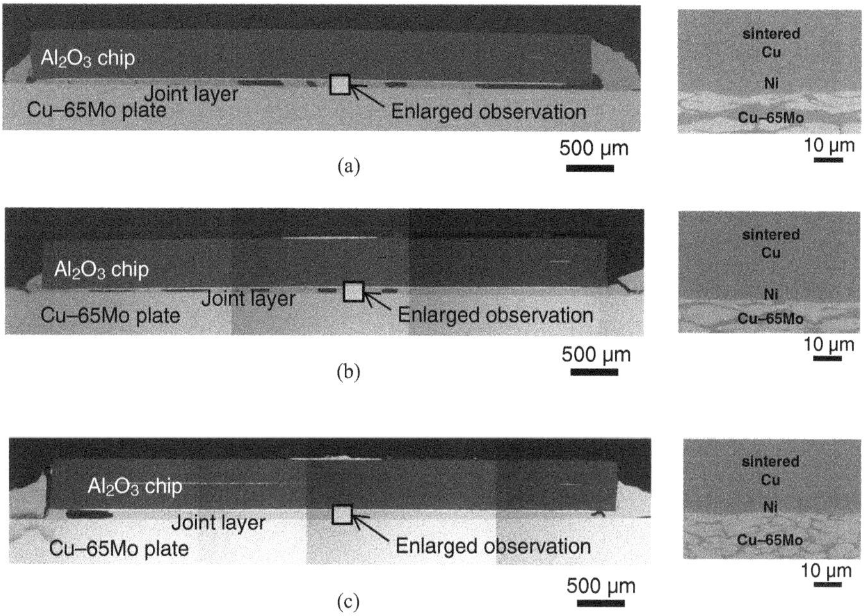

Fig. 23 Sintered at 350 °C. (**a**) After initial processing. (**b**) After 1000 cycles of the −40 °C/150 °C test. (**c**) After 1000 cycles of the −40 °C/200 °C test

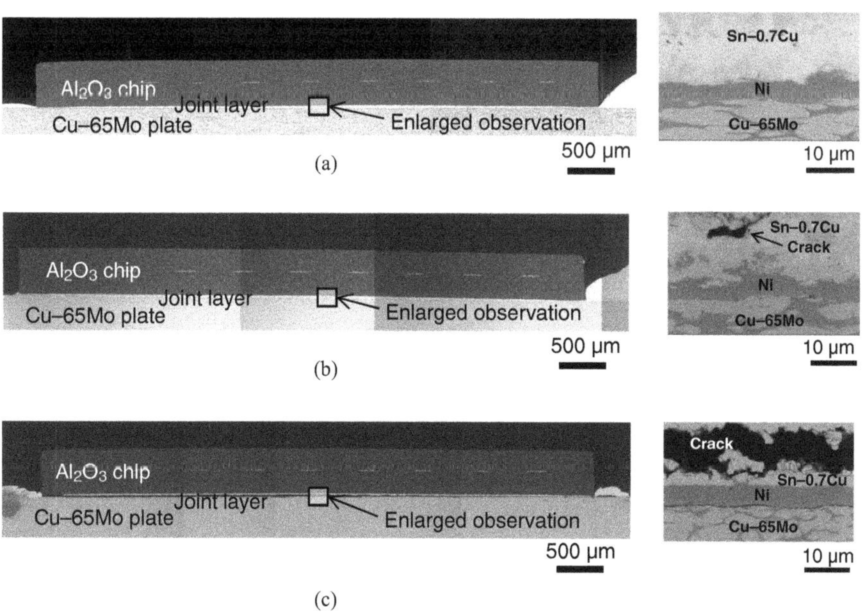

Fig. 24 Soldered using the Sn-0.7Cu solder. (**a**) After initial processing. (**b**) After 1000 cycles of the −40 °C/150 °C test. (**c**) After 1000 cycles of the −40 °C/200 °C test

comparison, the same test was carried out on joints of the Sn-0.7Cu solder, as described earlier for the power cycle test. However, as shown in Fig. 24, some cracking was found after the −40 °C/150 °C test, and large cracks were found in the solder layer after the −40 °C/200 °C test. These results indicate that joints formed using Cu nanoparticles are more reliable than solder joints during thermal cycling, similar to the results from the power cycle test.

5 Summary

The thermal properties and reliability of environmentally-friendly and low-cost Cu nanoparticle joints were investigated for their potential use in high-temperature power semiconductor devices operating at temperatures above 200 °C. They were found to have thermal properties superior to those for tin-based solders and good reliability under the power or thermal cycling test.

Acknowledgment This study was carried out at Toyota Central R&D Labs, Inc. and Daido University. The former partner prepared the nanoparticles and joint specimens, and the latter partner carried out the thermal and reliability evaluations.

References

1. Y. Yamada, Y. Takaku, Y. Yagi, I. Nakagawa, T. Atsumi, M. Shirai, I. Ohnuma, K. Ishida, Reliability of wire-bonding and solder joint for high temperature operation of power semiconductor device. Microelectron. Reliab. **47**, 2147–2151 (2007)
2. Y. Yamada, Y. Takaku, Y. Yagi, I. Nakagawa, T. Atsumi, M. Shirai, I. Ohnuma, K. Ishida, Pb-free high temperature solder joints for power semiconductor devices. Trans. JIEP **2**(1), 79–84 (2009)
3. T. Ishizaki, T. Satoh, A. Kuno, A. Tane, M. Yanase, F. Osawa, Y. Yamada, Thermal characterization of Cu nanoparticle joints for power semiconductor device. Microelectron. Reliab. **53**, 1543–1547 (2013)
4. T. Ishizaki, M. Yanase, A. Kuno, T. Satoh, M. Usui, F. Osawa, Y. Yamada, Thermal simulation of joints with high thermal conductivities for power electronic device. Microelectron. Reliab. **55**, 1060–1066 (2015)
5. T. Ishizaki, A. Kuno, A. Tane, M. Yanase, F. Osawa, T. Satoh, Y. Yamada, Reliability of Cu nanoparticle joint for high temperature power electronics. Microelectron. Reliab. **54**, 1867–1871 (2014)
6. T. Ishizaki, D. Miura, A. Kuno, R. Nagao, S. Aoki, Y. Ohshima, T. Kino, M. Usui, Y. Yamada, Power cycle reliability of Cu nanoparticle joints with mismatched coefficients of thermal expansion. Microelectron. Reliab. **64**, 287–293 (2016)
7. T. Ishizaki, M. Usui, Y. Yamada, Thermal cycle reliability of Cu-nano-particle joint. Microelectron. Reliab. **55**, 1861–1866 (2015)

Transient Liquid Phase Bonding

J. R. Holaday and C. A. Handwerker

Abbreviations

CTE	Coefficient of thermal expansion
HTP	How melting temperature phase
IGBT	Insulated-gate bipolar transistor
IMC	Intermetallic compound
LPDP	Liquid phase diffusion bonding
LTP	Low-melting temperature phase
MSL	Moisture sensitivity level
NIST	National Institute of science and technology
RoHS	Reduction of hazardous substances act
SAC	Sn-Ag-Cu
SLID	Solid-liquid interdiffusion
TLPB	Transient liquid phase bonding
TLPS	Transient liquid phase sintering
WEEE	Waste electrical and electronic equipment

Key Phrases

- Transient liquid phase bonding (TLPB)
- Transient liquid phase sintering (TLPS)
- Solid-liquid interdiffusion (SLID)
- Low-temperature diffusion bonding

The original version of this chapter was revised. The correction to this chapter is available at https://doi.org/10.1007/978-3-319-99256-3_11

J. R. Holaday (✉) · C. A. Handwerker
Purdue University, West Lafayette, IN, USA
e-mail: handwerker@purdue.edu

K. S. Siow (ed.), *Die-Attach Materials for High Temperature Applications in Microelectronics Packaging*, https://doi.org/10.1007/978-3-319-99256-3_9

- Thermodynamic equilibrium requirements for TLPB
- Kinetic requirements for TLPB
- Power electronics
- Alternatives to high-Pb solder
- Phase diagrams

1 Introduction: Challenges in Replacing High-Pb Solders and Creating Even Higher Temperature Bonding Technologies

Since the beginning of integrated circuit manufacturing and circuit board assembly, Sn-Pb soldering has been the workhorse system for providing reliable interconnections, time and time again, as the underlying technologies, geometries, and number of interconnects per circuit board have changed. The global electronics industry has transitioned to lead-free components in consumer electronic devices due to the known risks of lead to human health as reflected in the European Union's Reduction of Hazardous Substances Act (RoHS) [68] and the Waste Electrical and Electronic Equipment (WEEE) [69] Directive and similar bans in other regions. Compliance with EU regulations has driven the transition toward a more restricted set of materials in consumer products regardless of where they are sold, particularly for solder alloys used in circuit board assembly and packaging [9, 68, 69, 71]. However, the transition to Pb-free electronics is far from complete: the RoHS exemptions for high-Pb, high-temperature alloys used in wafer bumping, substrate bumping, and die attach will likely continue until a practical replacement becomes available [24, 79].

At the same time, new devices and applications are demanding ever higher performance solder alloys for assembly than currently exist, leading to continuing research in new Pb-free alloys and interconnect solutions that meet these enhanced performance challenges. More compact auto engine compartments require placing control systems closer to the engines and actuators, which increases exposure of vulnerable integrated circuitry to higher temperatures, vibration, and thus, conditions for fatigue and failure. Semiconductors based on GaN, III-Vs, SiC, and diamond operate at significantly higher temperatures than Si, whose maximum operating temperature is approximately 200 °C. This creates the need for new, RoHS-compliant bonding solutions for die attach, insulated-gate bipolar transistor (IGBT) attach, and power electronics to replace high-Pb solders, Pb-free solders, and thermal interface materials used today [13, 15, 26–28, 40, 49, 53].

The challenges in developing a high-temperature Pb-free solder or alternative interconnect solution for die attach, wafer and substrate bumping, and other joining applications can be further illustrated by examining the use of the Sn-Pb system in terms of its processing temperatures, thermal and mechanical behavior, and the microstructures created under certain processing conditions. Tin-lead alloys have many advantageous properties for electronics, including forming a simple binary eutectic with a eutectic temperature of 183 °C, excellent and reproducible mechanical properties, damage tolerance, and thermomechanical fatigue and impact/drop shock resistance, good electrical conductivity, excellent wetting characteristics, and a classic eutectic solidification morphology with two primary solid solution phases. As shown in the Pb-Sn phase diagram in Fig. 1, the large solid solubility of Sn in Pb (18.9 wt. % at 183 °C) leads to a range of useful, tunable solidus temperatures from 183 °C to close to the melting point of Pb (327 °C) and, therefore, with a tunable "mushy" (liquid + solid) zone width based on alloy composition [55]. Soldering is first performed at the chip or die attach level with high-Pb, Pb-Sn solders with a high solidus temperature and narrow mushy zone on melting. A commonly used high-temperature solder composition of Pb-Sn is 95Pb-5Sn (wt. %), with a solidus temperature of 320 °C. Once solidified, packaged components are soldered onto circuit boards without melting the high-Pb solder by using the Sn-Pb eutectic composition of 62.13Sn-37.87Pb (wt. %), with its eutectic temperature of 183 °C or using Pb-free solders, such as in the Sn-Ag-Cu system with a eutectic composition of 95.6Sn-3.5Ag-0.9Cu wt. % and a eutectic temperature (T_e) of 217 °C [54]. Numerous other board level assembly solders are available commercially and widely used, including the Sn-Ag eutectic (T_e = 221 °C), near eutectic Sn-Cu alloys (T_e = 227 °C) with various alloying additions, and for even lower temperature assembly, Sn-Bi eutectic alloys (T_e = 139 °C).

While there are many alloy compositions with liquidus temperatures in the 260–310 °C range, there are no drop-in, Pb-free solders which have *solidus* temperatures in the same range as high-Pb solders. For example, Sn-Cu alloys with 2–3 wt. % Cu have liquidus temperatures in that range but a solidus temperature of 227 °C. Based on an examination of the binary phase diagrams with one low-melting point element (T_m < 400 °C), or with a low-temperature eutectic, and the community's experience with designing Pb-free alloys to replace the Sn-Pb eutectic, it is unlikely

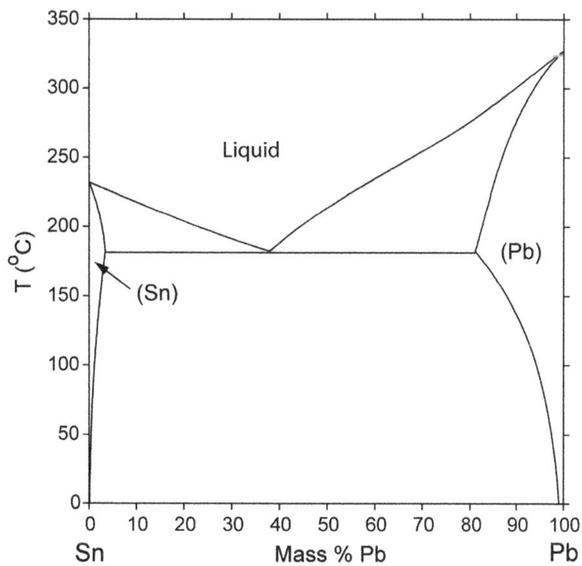

Fig. 1 Pb-Sn phase diagram [55]

such a solder is possible other than the ones reviewed by Chen et al., Zeng et al., and Suganuma et al. ([12, 77, 82]).

Non-solder alternatives have been developed resulting in a range of commercial products being currently offered, including sintering of Ag and Cu nanoparticles, core-shell nanoparticles, mixtures of coarser Cu with Ag nanoparticles, Ag-filled adhesives, Bi-Ag-X alloys, transient liquid phase sintering, and foil-based transient liquid phase bonding. For any to be acceptable substitutes for high-Pb solders in a given application or for a higher temperature application, the interconnects/bonds must:

1. Perform within an acceptable range of the electrical, thermal, and mechanical requirements for the application.
2. Form at temperatures less than or equal to commercial high-Pb solders it is replacing (<300 °C) or at any temperature limits determined by the systems to be joined.
3. Perform reliably for typical operation and use conditions, including during thermal cycling to low and high temperatures depending on the operating ranges, which can reach 400 °C for some applications.
4. Exhibit acceptable mechanical, thermal, and electrical performance after long-term aging and use.
5. Be manufacturable in high-volume electronics with an extremely low level of manufacturing defects commensurate with performance requirements.

In this chapter, we focus specifically on approaches for creating transient liquid phase systems that meet these criteria and defining the ranges of compositions and processing conditions appropriate for their use in specific applications. Of particular importance is our current understanding of (1) what it will take to achieve the manufacturability required for a Pb-free technology that could replace high-Pb solders in current applications, (2) how these can be extended to higher temperature applications, and (3) how transient liquid phase bonding compares with other alternative technologies.

2 Transient Liquid Phase Bonding: Key Thermodynamic Concepts

The defining characteristic of transient liquid phase bonding is the formation of a solid bond by isothermal solidification reactions as opposed to melting, wetting, and solidification of a liquid alloy during cooling, i.e., soldering. In conventional soldering, a relatively low-melting temperature component, such as a Sn alloy, is heated above its liquidus temperature. The liquid metal wets the surfaces to be joined, with some intermetallic formation possible, and cooled with the majority of liquid remaining. During cooling, a solid joint is formed by solidification. There can be significant undercooling of liquid tin, leading in most cases to formation of Sn dendrites even for a eutectic alloy. TLPB technologies produce solid bonds between a low-melting temperature phase (LTP) and the two high-melting temperature substrate phases (HTP) by several possible routes: (1) by intermetallic compound (IMC) formation that consumes at least one component of the LTP, changing the composition of the liquid and continuing to react until no liquid remains; (2) by interdiffusion into the HTP to form a solid solution; and (3) by precipitation of another phase in ternaries and higher-order phase diagrams as the phase equilibrium reactions at the liquid-solid interface change from a two-phase (liquid + IMC) to a three-phase equilibrium (liquid + IMC + precipitate).

The history of using these reactions for nonelectronics applications from ancient times to the 1980s is described in a review paper by MacDonald and Eagar [48]. In the comprehensive TLPB study of Ag-Sn-Ag, Li, Agyakwa, and Johnson reviewed the literature for electronics applications in detail, starting with the pioneering work of Bernstein and Bartholomew from the 1960s in bonding Ag, Au, and Cu with indium as the initial liquid for "using low-temperature processing conditions to produce high-temperature bonds for a multitude of semiconductor and non-semiconductor brazing and bonding applications" [41]. They reviewed other research and its relevance to electronics, including the Roman and Eager studies of commercial solder alloys as the low-temperature phases to bond Ag, Au, and Cu substrates as well as bonding of Ni, Pd, Pt, and Zr substrates and the use of ternaries and quaternaries in the Cu-In-Sn-Bi system [73]. Transient liquid phase bonding processes in Sn-based systems have been referred to by various names and acronyms by different research groups. The terms "transient liquid phase sintering" (TLPS), "liquid phase diffusion bonding" (LPDB), "transient liquid phase bonding" (TLPB), and "solid-liquid interdiffusion" (SLID) have all been used to describe bonding of two substrates via isothermal solidification reactions between a low-melting temperature phase and the high-melting temperature substrates [16, 50, 53, 72]. In this paper, the term TLPB will be used, although all can be used interchangeably.

Two binary and two ternary phase diagrams are used here to illustrate these three TLPB routes for isothermal solidification. The Cu-Sn binary system is the classic TLPB system, given the ubiquity of Cu and Sn in electronic interconnects. The Cu-Ni binary system, a simple isomorphous system, is used to illustrate how interdiffusion alone can lead to isothermal solidification. A ternary system is required for precipitation to occur when the other two processes deplete the liquid from one of the components leading to the formation of a new phase. The first ternary presented here is the Sn-Ag-Cu system in which IMC formation of the low-melting liquid phase with the substrate eventually leads to precipitation of an IMC between Sn and the non-reacting component. After complete isothermal solidification, the melting temperature of the bond is not as high as the melting points of the individual phases because of additional reactions that can occur with ternary alloys (and those with more elemental components.) The second ternary is the Sn-Bi-Cu system in which Bi precipitates as the Sn-Cu IMCs grow. This ternary system is shown because the melting temperature of solidified bonds can be substantially lower than the melting temperatures of the individual phases. The possible reactions, the temperatures over which they can occur, and the temperatures of stability of the final bonds can be understood by showing how the compositions shift as reaction, diffusion, and precipitation proceed.

2.1 *Cu-Sn Binary System: Reaction, Intermetallic Formation, and Interdiffusion*

The binary Cu-Sn phase diagram is shown in Fig. 2, with the five possible temperatures ranges and reactions for Cu bonding with a transient Sn-Cu liquid phase [56]. The Cu solubility of the liquid (Sn) changes with temperature, as does the solubility of Sn in (Cu). (Note that the use of parentheses denotes a liquid solution or solid solution phase, such as (Sn), which is a liquid that contains both Sn and Cu, or (Cu) which is solid Cu into which Sn is dissolved.) Although only the two lower temperature regimes are applicable to electronics manufacturing, all ranges are included for completeness, particularly to illustrate how to interpret binary phase diagrams in the context of TLPB. The important features to note in each regime are what the solidus and liquidus temperatures are across the phase diagram, what phases are stable and at what composition, and how they change from the initial compositions as intermetallic forms isothermally.

Above the eutectic temperature of Sn-Cu (227 °C) up to approximately 415 °C (*A* in Fig. 2), Sn liquid reacts with Cu substrates to form two intermetallics, Cu_6Sn_5 and Cu_3Sn, as shown schematically in Fig. 3. For a limited amount of Sn relative to the Cu substrates, the liquid (Sn) can be completely consumed by the formation of these two intermetallics. At the point when the liquid is consumed, the two intermetallics and Cu will remain, which means that the joint can be heated up to 415 °C, the lowest melting temperature of the three phases remaining, i.e., of Cu_6Sn_5, without liquid formation. As a condition of thermodynamic equilibrium in binary systems, only two phases can be in equilibrium except at invariant points. An invariant point in a binary is at a specific temperature and composition at which three phases can coexist, such as the eutectic composition and temperature. An additional type of invariant point in this binary is the peritectic at 415 °C when Cu_6Sn_5 decomposes into liquid (Sn) and Cu_3Sn. In the case of *A*, continued annealing after isothermal solidification leads to the disappearance of one of the two phases; in this case to the disappearance of Cu_6Sn_5 when the initial amount of Cu is significantly greater, the amount of Sn and the final equilibrium is between Cu_3Sn and (Cu), a Cu-Zn solid solution. This TLPB reaction was studied in great detail by Li et al. [40]. Starting at 348 °C, an additional phase between Cu and Cu_3Sn is stable with the composition $Cu_{41}Sn_{11}$. Continued annealing of isothermally solidified joints at temperatures between 348 and 415 °C can lead not only to disappearance of Cu_6Sn_5 but also to the disappearance of Cu_3Sn, leaving behind $Cu_{41}Sn_{11}$ and (Cu).

Between 415 and 638 °C (*B* in Fig. 2), reaction of liquid (Sn) with Cu leads to formation of Cu_3Sn in direct contact with the Sn-Cu liquid, with additional solid phases present at equilibrium between Cu_3Sn and Cu. The melting temperature of the isothermally solidified bond is then 677 °C due to the stability of Cu_3Sn above 638 °C. Between 638 and 757 °C (*C* in Fig. 2) reaction between liquid Sn and Cu leads to the formation of solid solution γ (gamma phase), the stable phase in contact with (Sn). The melting temperature of the solidified joint is determined by the specific composition of solid solution γ that is formed. Continued annealing in this range can lead to the disappearance of γ and Cu_3Sn, leaving behind β and (Cu). In *D*, β is the stable intermetallic phase with a wide range of solid solubility in contact with the liquid and (Cu). Ranges *A–D* are where intermetallic formation leads to isothermal solidification. In *E* in Fig. 2, there are no stable intermetallic phases between the (Sn) liquid phase and (Cu). This means that, in this temperature range, isothermal solidification only occurs by diffusion of Sn into the Cu substrate to form (Cu). This is the same concept shown for the Cu-Ni system described below. Note that diffusion of Sn into Cu to form the solid solution (Cu) may also contribute to isothermal

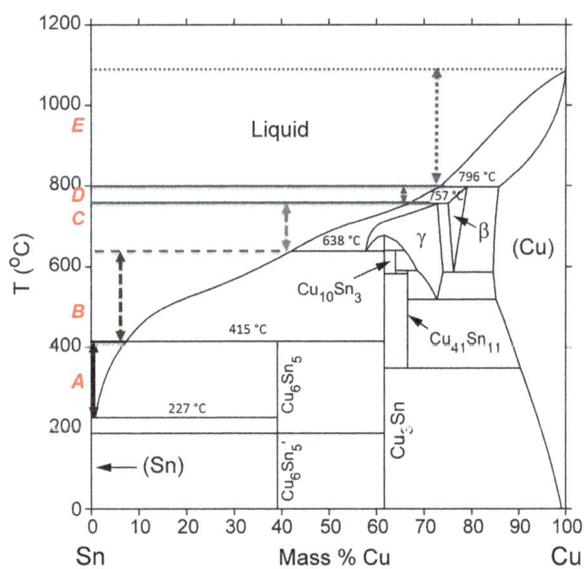

Fig. 2 Cu-Sn phase diagram from NIST annotated with the five ranges over which transient liquid phase bonding can be performed. (Modified from NIST [56])

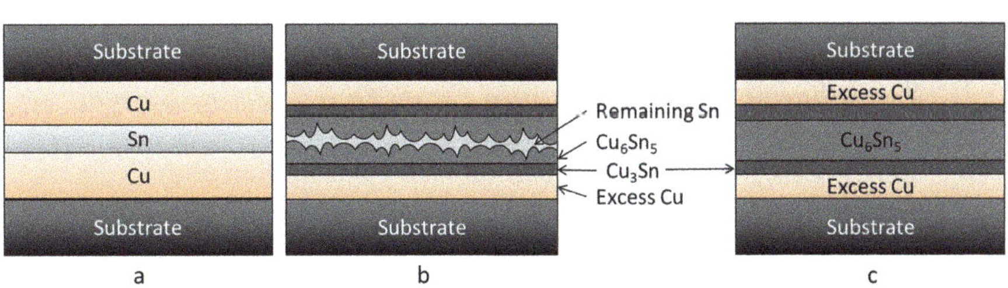

Fig. 3 (**a**) Schematic of planar geometry for transient liquid phase bonding in the Cu-Sn system (**b**) after partial reaction in region *A* in Fig. 2 (**c**) after complete reaction

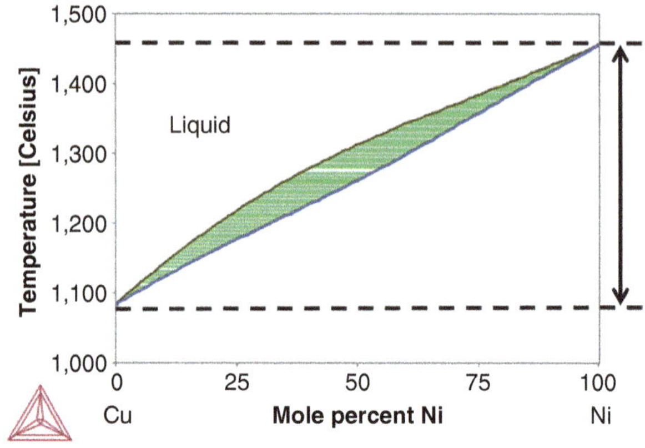

Fig. 4 Cu-Ni phase diagram calculated using ThermoCalc with the two-phase region indicated where isothermal solidification can occur by interdiffusion

solidification at lower temperatures even when there are intermetallic compounds present. The contribution of Sn diffusion into (Cu) is dependent on the relative rates of intermetallic formation and interdiffusion of Sn in Cu, i.e., on temperature and the IMC and Cu microstructures.

2.2 *Cu-Ni Binary System: Diffusion Alone*

The Cu-Ni system is an isomorphous system, with only a liquid phase and a face-centered cubic (FCC) solid solution phase, as shown in the phase diagram in Fig. 4. When pure liquid Cu is brought into contact with pure Ni in this temperature range, dissolution of the Ni into the Cu liquid will occur until the liquid becomes saturated with Ni. Simultaneously, Cu diffuses into Ni to form the solid solution. Isothermal solidification of a Ni-Cu-Ni structure can occur if the relative volume of the initial Cu liquid is small compared to the Ni such that the final composition of the bond is completely within the FCC-solid solution phase. A thermodynamic and kinetic model for isothermal solidification by diffusion in a single solder phase developed by Illingworth et al. has relevance for electronics if temperatures are high enough for solid-state diffusion in the HTP to be significant [30].

2.3 *Precipitation of an Additional Phase in a Ternary System*

2.3.1 Sn-Ag-Cu System: IMC Formation with Precipitation of a Second IMC

In this analysis of the Sn-Ag-Cu system, we will first discuss reactions at temperatures where the Cu_6Sn_5 and Ag_3Sn are the stable phase that form between a Cu or Ag substrates and (Sn) liquid. Four isothermal sections in the Sn-Ag-Cu phase diagram

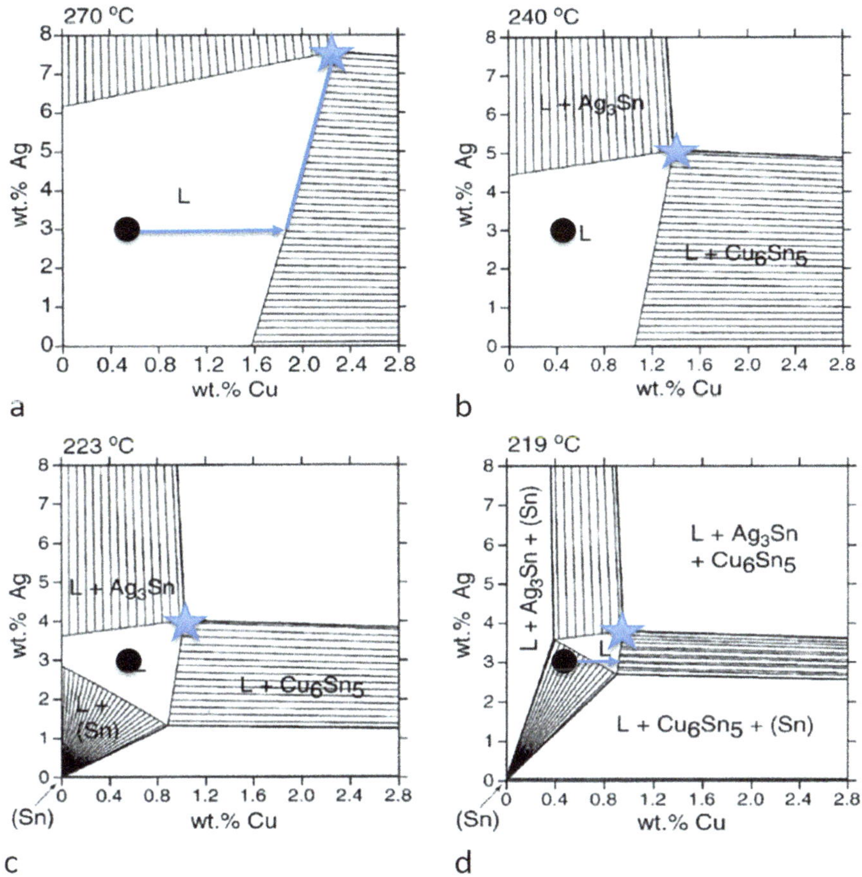

Fig. 5 Isothermal sections of the Sn-Ag-Cu ternary (**a**) 270 °C, (**b**) 240 °C, (**c**) 223 °C, and (**d**) 219 °C. (Modified from Moon et al. [54])

[54] are used here to show the reaction pathways for isothermal solidification of a Sn-3.0Ag-0.5Cu (wt. %) alloy (SAC305) in contact with Cu at 270, 240, 223, and 219 °C (Fig. 5a–d). In Fig. 5a–c, the liquid dissolves additional Cu until it becomes saturated with Cu, while the Ag and Sn contents remain constant before IMC formation begins. For Fig. 5d, the initial alloy contains some solid Sn but melts to form a liquid in contact with Cu_6Sn_5. The composition of the saturated liquid solution in contact with Cu_6Sn_5 is a function of temperature, as represented by the tie lines. As the Cu_6Sn_5 and the Cu_3Sn layers between the Cu_6Sn_5 and Cu grow, the liquid becomes enriched in Ag as shown by the arrows in Fig. 5 until the terminal liquid composition is reached for each temperature, as indicated by the starred compositions. As the Cu-Sn IMCs continue to grow, Ag_3Sn precipitates in the liquid to maintain the terminal liquid concentration.

If SAC305 is used to bond two Ag substrates at 270 °C, the path is reversed as seen in Fig. 6a. The Ag_3Sn intermetallic forms at the interface between the liquid and

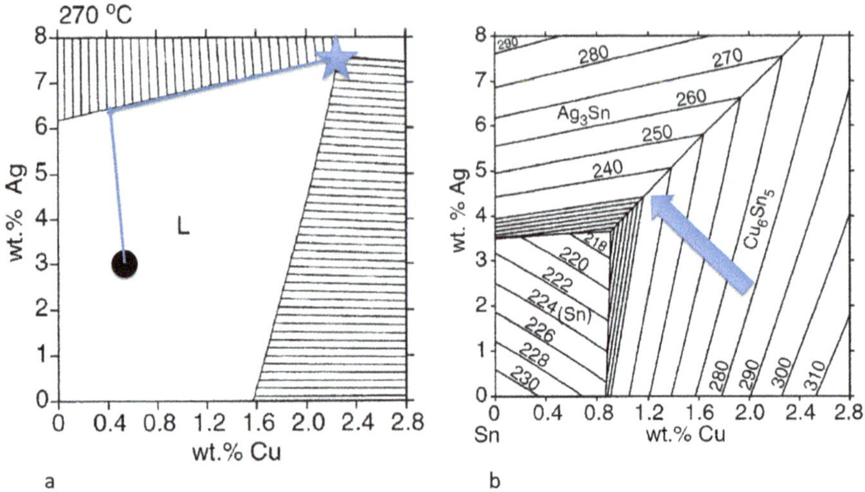

Fig. 6 (**a**) Isothermal section of the Sn-Ag-Cu ternary phase diagram showing the reaction path for SAC305 with a Ag substrate at 270 °C and (**b**) is the liquidus projection of the Sn-Ag-Cu ternary diagram at 355 °C with the terminal liquid compositions occurring on the join between the Cu_6Sn_5 and Ag_3Sn primary phase fields, indicated by the arrow, as modified from Moon et al. [54]

the Ag substrates. The Cu concentration in the liquid increases, reaching the same terminal liquid in equilibrium with Ag_3Sn and Cu_6Sn_5. With further Ag_3Sn growth, Cu_6Sn_5 precipitates in the liquid until the liquid is consumed. The temperature dependence of the terminal liquid solution can be easily seen by following the Cu_6Sn_5-Ag_3Sn two-phase trough in the liquidus projection indicated by the arrow in Fig. 6b. The final structures contain precipitated intermetallic particles along the bond's centerline. With further annealing, there may be additional solid-state transitions that change the tie lines between solid phases but are not considered here.

However, the melting temperature of the resulting joint is not at 414 °C due to additional transitions that can occur at lower temperatures in this ternary [57]. The reaction $L + Cu_3Sn \leftrightarrow Cu_6Sn_5 + Ag_3Sn$ occurs at approximately 356 °C, as shown in the isothermal ternary sections in Fig. 7b.

2.3.2 Sn-Bi-Cu: IMC Formation with Bi Precipitation

A second example of a reaction to precipitate an addition phase is also presented here for the Sn-Cu-Bi system. This system was chosen because of the type 2 reaction that occurs in this system above approximately 200 °C, in which Bi + Cu_6Sn_5 transforms into Cu_3Sn and liquid. What this means is that any reaction path below 200 °C in which the final microstructure contains Cu_6Sn_5 and Bi will form a liquid at higher temperatures and not at the melting point of Cu_6Sn_5 in the binary Cu-Sn diagram [58]. These reactions are shown in two isothermal sections for the Sn-Bi-Cu phase diagram shown in Fig. 8. At temperatures below approximately 200 °C (Fig. 8a),

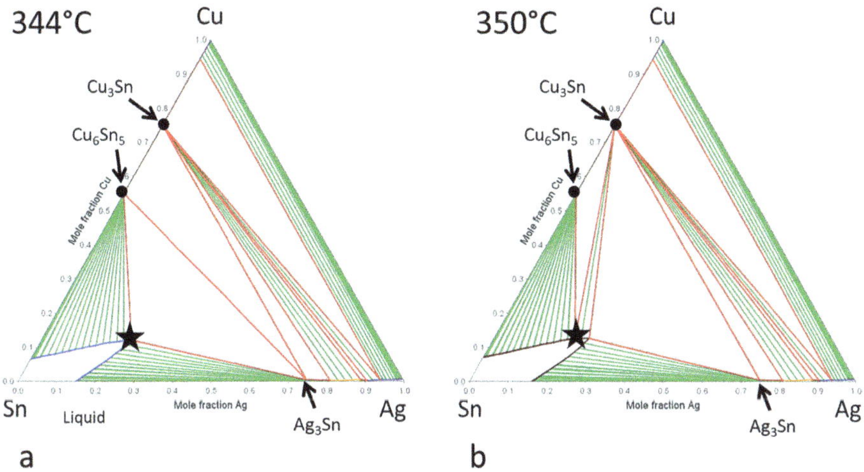

Fig. 7 (**a**) Calculated isothermal sections of the Sn-Ag-Cu ternary diagram showing the Class II reaction occurring at (**a**) 344 °C and (**b**) 350 °C. Showing the terminal liquids in equilibrium with Cu_6Sn_5 (stars). In (**a**) this composition forms a three-phase triangle with Cu_6Sn_5 and Ag_3Sn, whereas in (**b**) the three-phase triangle is between Cu_6Sn_5 and Cu_3Sn

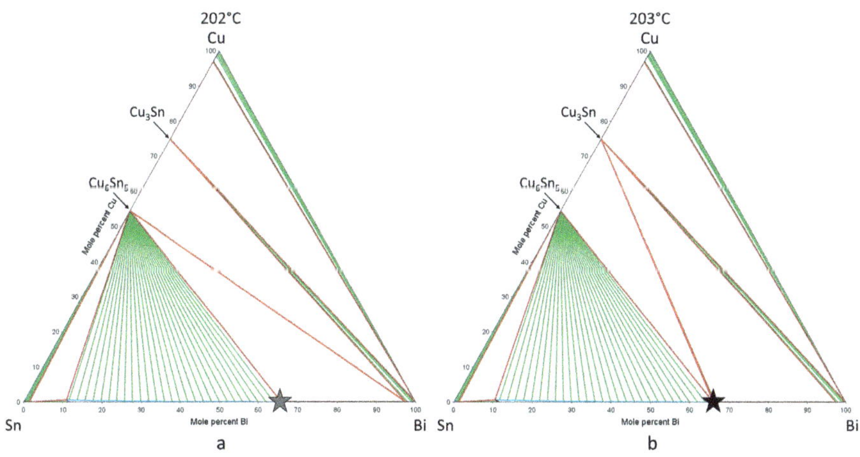

Fig. 8 Calculated isothermal sections of the Cu-Sn-Bi ternary phase diagram showing Class II reaction between (**a**) 202 °C and (**b**) 203 °C

Sn-Bi liquids in contact with Cu form Cu_6Sn_5 and Cu_3Sn on Cu, with the liquid decreasing its Sn concentration until the terminal liquid composition (indicated by the gray star) is reached. Further reaction to form the intermetallics leads to the precipitation of Bi to maintain the constant terminal liquid composition. Isothermal solidification can go to completion in this way. If the as-solidified bond is heated to slightly higher temperatures (Fig. 8b), the Class II reaction will lead to liquid formation, with no solid Bi remaining. This is because, as seen by comparing Fig.8a, b, Cu_6Sn_5 is not in equilibrium with solid Bi at this temperature. Cu_6Sn_5 is only in equilibrium with Cu_3Sn and the terminal liquid (black star). In this case, further growth of Cu_3Sn at the expense of Cu_6Sn_5 may occur until it is completely consumed. At this point the liquid is in equilibrium with Cu_3Sn and Bi, and Bi can precipitate. Isothermal solidification will then occur by growth of Cu_3Sn and Bi precipitation until all the liquid is consumed. The final as-solidified bond is in the three-phase stability triangle of Cu, Bi, and Cu_3Sn.

These examples illustrate the critical relationship between the thermodynamic properties of systems, the processing temperatures which can be used, and their resulting application and use temperatures. The binary phase diagrams provide all relevant information on what phases might occur and at what temperature. For ternaries and higher-order systems, "Class 2" reactions may occur that can dramatically decrease the temperatures where the liquid forms. In order to understand what may occur in ternary and higher-order systems, one cannot, therefore, estimate their behavior based on the behavior of the individual binaries. In the sections below, more examples will be explained in detail and how they determine the possible processing and application/use temperatures for specific compositions.

3 Transient Liquid Phase Bonding: Key Kinetic Concepts

For TLPB technologies to be adopted, the rate of reaction to form an isothermally solidified bond must be rapid enough to be consistent with commercial manufacturing, i.e., it must be quick enough to be practical. Four parameters come together in a simple geometric and kinetic model for how quickly such a reaction can occur. These parameters are (1) IMC growth rate as a function of temperature, time, and composition, (2) the total volume of low-melting phase that must be consumed for isothermal solidification to occur, (3) the interfacial area where the reaction is taking place, and (4) the uniformity of the microstructure.

3.1 IMC Growth Rates

Extensive observations of IMC growth kinetics have been performed by the Pb-free solder research community. These studies are typically focused on IMC formation at the interface between copper contacts, any metallization, and the bulk solder alloy. Interdiffusion studies related to Kirkendall void formation also provide a valuable source of data for TLPB design. Study of IMC formation specifically for TLPB has also been performed for the key IMC phases typically utilized. Here we introduce some of the relevant literature results for Ag_3Sn, Cu_3Sn, Cu_6Sn_5, and Ni_3Sn_4.

There have been many studies reporting the kinetics of Cu_6Sn_5 and Cu_3Sn growth in the Cu-Sn system. Li et al. measured IMC growth in Cu/Sn/Cu arrays using a combination of differential scanning calorimetry and measurement of IMC growth from cross sections [40]. Measurement of IMC thickness was then used to determine the residual Sn content from DSC heat flow measurements. This technique was used to gather a comprehensive dataset for IMC growth in Cu-Sn and to develop a model that includes consumption of Cu_6Sn_5 via Cu_3Sn formation. Based on this dataset, the change in total IMC thickness with time was found to be proportional to $Kt^{1/2}$ – the typical equation assumed to correspond to lattice diffusion control, with K that includes the temperature dependence. With this dataset and the derived equation, it is then possible to predict the necessary processing time for complete isothermal solidification of a known layer thickness in the Cu-Sn system and for complete conversion to Cu_3Sn. The Arrhenius method was used with the resulting kinetics constants for 260, 300, and 340 °C to arrive at an activation energy of 84.6 ⊥ 25.8 kJ/mol for Cu_3Sn and 19.7 ⊥ 13.1 kJ/mol for Cu_6Sn_5. A summary by Laurila et al. [38] provides further background on reactions between Sn and Cu and the effects of ternary additions [38].

Li et al. [41] also studied the growth kinetics of Ag_3Sn in Ag/Sn/Ag couples at 260, 300, and 400 °C using a combination of DSC experiments coupled with microstructural characterization using an SEM [41]. Ag_3Sn growth was shown to conform to a time dependence of $t^{1/3}$ where the exponent of 1/3 is assumed to correspond to a grain boundary diffusion-type controlled growth. The IMC growth rate and morphology were shown to be consistent with grain boundary diffusion/molten channel-controlled growth, i.e., the growth morphology showed irregular, facetted growth front, with a reported activation energy of 37.2 ± 18.9 kJ/mol [41].

The kinetics of Ni-Sn formation are generally slower than for the Cu-Sn or Ag-Sn intermetallics, with Ni_3Sn_4 IMC growth reported to dominate at lower temperatures and to be parabolic. Tomlinson and Rhodes investigated interfacial IMC formation between Sn and Ni, Ni-P, and Ni-B surfaces from 180 to 220 °C. An activation energy of 128 kJ/mol was observed [78].

As seen in Fig. 2a for the simple case of Cu-Sn, a Sn film of thickness L is placed between two planar Cu substrates of area A. Intermetallic formation to form Cu_6Sn_5 and Cu_3Sn only occurs at the interfaces between the Cu and Sn, so the consumption of the Sn to isothermally solidify the structure and form the bond is limited by the area of contact, which in this case is $2A$, with the effective thickness of Sn to be consumed by intermetallic formation of $L/2$, with a total volume of liquid Sn to be consumed of $2A*L/2 = A*L$. Reaction occurs by interdiffusion as described by He et al., Kumar et al., and Park et al., with the reaction rate decreasing as the layer thickness increases and increasing as the temperature increases [29, 37, 66]. The simplest form of the relationship between IMC thickness and time at a given temperature is that the thickness of the IMC layer formed after time t is proportional to the product of a kinetic coefficient (K) and $t^{1/2}$. Therefore, when this equation applies, the time required for isothermal solidification goes up as the square of the layer thickness, as shown in Fig. 9.

The $Kt^{1/2}$ equation is frequently interpreted as a measurement of lattice diffusion for forming the combined growth of Cu_6Sn_5 and Cu_3Sn, but this does not reflect the actual diffusion pathways that lead to IMC growth. Even in the growth of a single

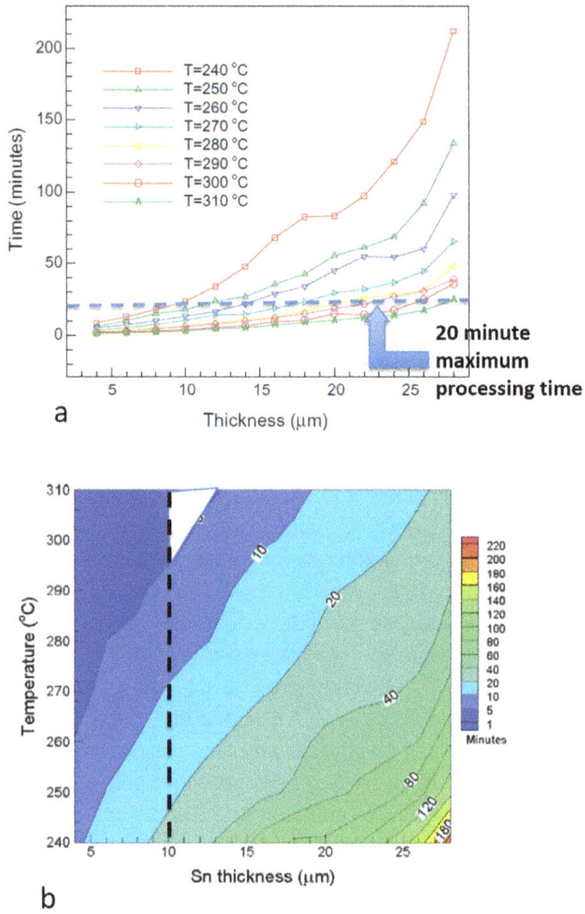

Fig. 9 (**a**) Sn elimination time with respect to temperature and initial Sn layer thickness, with the maximum processing time line drawn at 20 min to illustrate a possible practical limit. (**b**) Contour map representing Sn elimination time with respect to temperatures (240–310 °C) and Sn layer thicknesses 4–28 μm. The white-shaded area shows the range of temperatures and thicknesses near 10 μm possible for a maximum 5-min processing time. (Modified from Park et al. [66])

IMC, the rate of reaction is determined by the specific contributions of grain boundary, lattice, and interfacial diffusion, as well as other microstructural and physical characteristics of the bonds that affect their evolution with time and temperature. When two IMCs are evolving, the rate of growth of both IMCs is coupled.

Park et al.'s phase field calculations take these contributions into account and successfully captures the scalloped morphology of the Cu_6Sn_5 [67]. Phase field modelling allows the identification of relative diffusion rates for the various diffusion pathways rather than relying on a single diffusion term. Accurate representation of observed morphology indicates realistic diffusion terms for grain boundary diffusion, interfacial diffusivity between phases, bulk diffusivity, etc. These models were in turn used to create processing maps for Cu-Sn TLPB (Fig. 9) in terms of processing temperature, Sn bond line thickness, and time for complete conversion. The maximum acceptable processing time determines the maximum bond line thicknesses and temperatures that can be used for a given system with its kinetics and geometry [66]. Using the results from Park et al. in Fig. 9a, only the temperatures and bond line thicknesses below the horizontal dotted line can produce isothermal solidification in less than 20 min. As another example, assume that a practical lower limit of bond line thickness is 10 μm and an upper limit of processing time is 5 min. Only the conditions represented by the small shaded triangle in the top left corner of the Park et al. processing map in Fig. 9b produce isothermal solidification for temperatures up to 310 °C.

The final critical point to be made here is that the growth rate of an intermetallic can be dramatically affected by alloy composition and the presence and stability of grain boundaries/interfaces in the IMCs. This can occur even when comparing two systems in which the IMCs being formed are the same. Although the $t^{1/2}$ and $t^{1/3}$ dependence is assumed to reflect growth only by lattice diffusion and grain boundary/interface diffusion, respectively, growth of IMC layers occurs by a combination of grain boundary, latticc, and surfacc or intcrfacial diffusion, with grain boundarics serving as fast diffusion paths between the two bounding phases. Having a higher density of grain boundaries, i.e., a smaller grain size, in the IMC layer will lead to faster growth. The grain size at any given stage in processing will also be determined by the rate of coarsening of the IMC by surface/interfacial diffusion and grain boundary motion. One example is TLPB in the Sn-Cu-Ni system. Using Cu-Ni solid solution substrates of different compositions in contact with initially pure Sn, several papers have reported that the intermetallic growth rate of Cu_6Sn_5 can be higher by orders of magnitude when it forms on Cu-Ni alloys rather than on Cu [1, 2, 16, 70, 72, 80]. Furthermore, the Cu_3Sn IMC appears to become unstable for Cu-xNi compositions for ($x \geq 10$). Using a particle-based TLPB approach, Choquette and Anderson determined that, depending on the volume fraction of liquid, $(Cu,Ni)_6Sn_5$ growth could be fast enough to completely solidify bonds/solder joints after 90 s above 250 °C [16]. Another example is the Cu-Sn-Bi system, in which the effective diffusion coefficient for Cu_3Sn appears to be two orders of magnitude faster than in the Cu-Sn system at the same temperature. In both cases, the IMC growth morphologies were different from their Cu-Sn counterparts, with a significantly smaller, yet stable IMC grain size for the former as shown in Fig. 10 [1, 3, 70], and facetted,

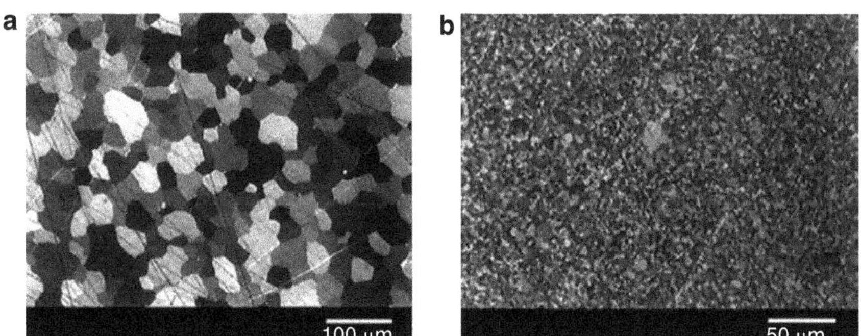

Fig. 10 Micrograph showing the grains of the Cu_6Sn_5 phase in diffusion couple annealed at 215 °C for 400 h for (**a**) binary Cu/Sn and (**b**) ternary Cu(5Ni)/Sn [70]

irregular IMC growth for the latter. The practical implication for these observations is that growth rate and microstructural evolution experiments are needed for calibration of any new materials system. The results from these two systems are described in greater detail below.

3.2 *Effect of Reaction Geometry on Isothermal Solidification Rate: Changing the Volume of Low-Temperature Phase and Interfacial Area for Reactions*

The time for isothermal solidification can be reduced by increasing the interfacial area over which the reaction is occurring while keeping a fixed volume of low-melting point phase. For example, if the reaction configuration is changed by adding a Cu foil of the same size as the substrates being joined while keeping the total Sn volume constant, as illustrated in Fig. 11a–c, the interfacial area for

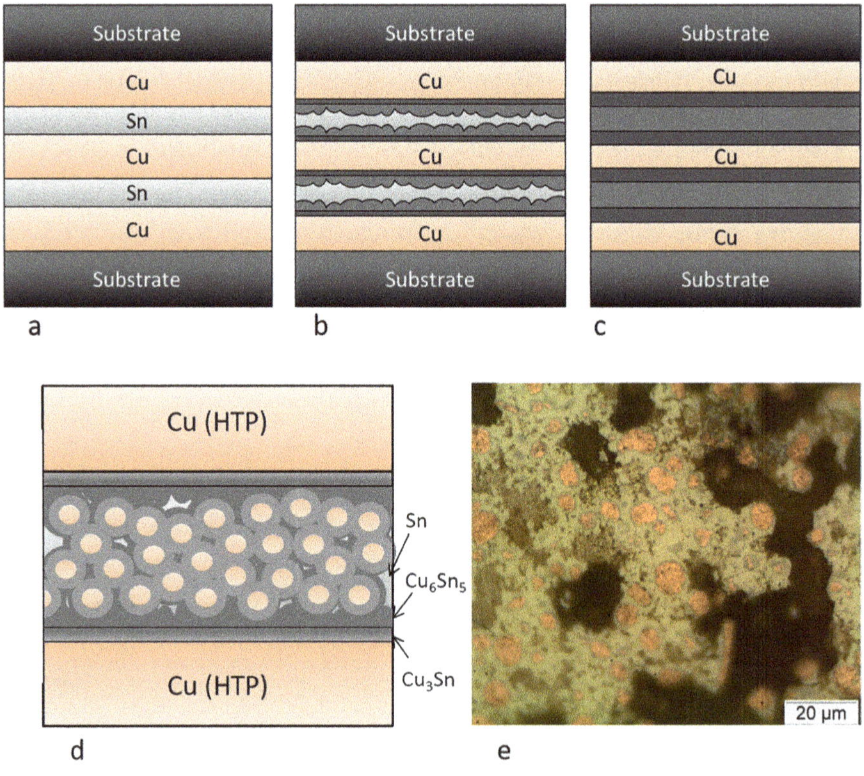

Fig. 11 (**a**) Schematic for multilayer Cu-Sn TLPB, (**b**) partially reacted, (**c**) completed reacted, (**d**) schematic of Cu particle-based Cu-Sn transient liquid phase sintered bond, (**e**) cross section from Bi-Cu-Sn TLPS bond formed from Cu HTP particles and Sn-Bi TLP after 20 min. Annealing at 210 °C

reaction doubles, and the Sn layer thickness that must be consumed by IMC formation is thus reduced by a factor of 2. According to the simple equation above, this would reduce the reaction time by a factor of 4. Bajwa and Wilde demonstrated that creating a multilayer geometry can be an effective strategy for reducing the time required for isothermal solidification, achieving mechanical, thermal, and electrical properties of multilayer Ag-Sn-Ag bonds that were equivalent to sintered nanosilver [4].

Another geometry that reduces the reaction time is by combining particles of the high-temperature phase with low-temperature phase to increase the interfacial area available for reaction. This geometry, referred to here as transient liquid phase sintering (TLPS), is shown schematically for the Sn-Cu system in Fig. 11d. For a constant total volume of copper and of tin, reducing the Cu particle size increases the total area over which reaction can occur, thus decreasing the effective Sn layer thickness with the time for complete conversion being proportional to the Sn layer thickness squared. A typical microstructure of a TLPS bond using Cu HTP particles and a Sn-Bi matrix is shown in Fig. 11e. Note that there is significant porosity remaining in addition to a dense network of Cu particles, IMC, and residual Sn.

4 Manufacturing and Design Constraints

In addition to the reaction itself, manufacturing processes for a specific application impose additional constraints that must be considered in choosing TLPB formulations and selecting the materials to be bonded. Processing takes place above the liquidus of a relatively low-melting temperature phase, commonly Sn or a Sn-based alloy, which melts and reacts with a high-temperature phase to isothermally solidify the bond. As with soldering, there are many other additional processing steps that affect the microstructure, properties, performance, and reliability of the resulting bonds:

1. Fluxing or preprocessing of the substrates.
2. Application of the foil, preform, or paste onto one substrate and then capping with the additional substrate. Some pressure is required to set the initial configuration. Additional pressure may be needed in subsequent steps.
3. Heat-up of the assembly to be bonded, with or without pressure.
4. De-gassing of the flux during heat-up, with rearrangement and densification in the solid state occurring in particle–/paste-based systems.
5. Melting of the LTP.
6. Wetting of the HTP (substrates and particles) by the liquid phase and capillary flow to redistribute the liquid.
7. Dissolution of HTP into the LTP until LTP is saturated.
8. Reaction at the maximum temperature to form the desired intermetallic and precipitate phases until no liquid remains and without formation of porosity.
9. Cooling of the assembly to room temperature without cracking or debonding.

Table 1 LTP to HTP ratios and fractions for binary intermetallic compounds

Phase	Maximum volume ratio of LTP to HTP	Maximum volume fraction of LTP	Maximum mass ratio of LTP to HTP	Maximum mass fraction of LTP
Ag_3Sn	0.52:1	0.34	0.37:1	0.27
Cu_3Sn	0.76:1	0.43	0.62:1	0.38
Ni_3Sn	0.67:1	0.45	0.81:1	0.45
Cu_6Sn_5	1.89:1	0.66	1.56:1	0.61
Ni_3Sn_4	3.22:1	0.78	2.70:1	0.73

All steps must be successful for useful bonds to form.

Since TLPB is a reaction-based process, maintaining appropriate stoichiometry is necessary to ensure complete conversion of the LTP to desired IMC and precipitate phases. For example, if we consider the complete conversion of Sn to form Cu_6Sn_5, the molar ratio of Cu:Sn must be greater than 6:5. If we take into account the density and atomic volumes of Cu and Sn, the minimum molar ratio can be converted into volume- and mass-based ratios and fractions that are useful for selecting HTP and LTP particle sizes and estimating minimum reaction time for isothermal solidification from IMC growth rate data. Ratios and fractions for conversion into the most common Ag-, Cu-, and Ni-Sn-based IMCs formed during TLPB are presented in Table 1. In some cases, the fraction of LTP required to produce bonds exhibiting reasonable adhesion and density exceeds the maximum LTP/HTP ratio for complete conversion. Increasing the ratio of LTP to HTP typically improves wetting of the surfaces to be bonded. For TLPS systems, it also promotes particle rearrangement and densification. For a constant interfacial area of HTP, increasing LTP volume increases the thickness of IMC per unit area that must be formed and therefore increases the minimum process time.

Various techniques for assembling TLPB structures have been used in research studies, and a few have been incorporated into commercial products. Screen-printable TLPS pastes containing a polymer added for increased stability have been developed in the Sn-Cu-Bi system [63] as a replacement for lead-free solders, Cu-via fill, TIMs, and high-Pb solder paste. A TLPB foil, an LTP preform, or a TLPS preform could be applied using die attach equipment currently used for die attach adhesive films. Plating and other deposition techniques of the LTP have also been used instead, or in addition to foil, to promote substrate wetting. Additionally, novel or hybrid approaches may be employed, such as the multilayer described above. All design choices result in trade-offs in process requirements, bond quality, application space, and market appeal.

Foil or LTP preform configurations simplify the design process by minimizing flux and densification issues, but process time increases with bond line thickness since the two interfaces being bonded function as the only source of HTP. This becomes of particular concern when metallization layers are thin. For example, electroless Ni layer thicknesses are limited to 3–6 μm according to IPC Standard 4552 [31]. For example, for complete conversion of two 3 μm Ni layers to Ni_3Sn (one on each surface to be bonded) without complete consumption of the Ni, the total bond line thickness of Sn must be less than 5 μm. This is a challenge for initial bonding due to issues of substrate planarity, deposition control, and substrate wetting. If the reaction layer is only Ni_3Sn_4, the maximum allowable Sn thickness would be approximately 20 μm. Acceptable reflow times and temperatures, ability to provide pressure, and post-assembly annealing capabilities also constrain the composition and maximum TLP thickness.

The key advantage of a LTP-HTP paste is that the maximum bond line for complete conversion is independent of bond line thickness due to the presence of the HTP in the bulk of the bond. This is in contrast to a foil-based format where the bonding interface is the only source of HTP, as described above. Since reaction rates are a function of HTP surface area, LTP volume, and interdiffusion rates, paste-based technologies (TLPS) provide an opportunity for controlling the reaction rate by independently changing HTP volume, HTP interfacial area, and LTP volume. In addition, paste-based formulations introduce the variables of particle size, particle size distribution, and morphology, flux and binder characteristics and amounts, LTP to HTP ratios, and overall bond line thickness. There can be issues in reducing the particle size of the HTP to increase the interfacial area and thus decrease the processing time; for example, poor wetting and oxidation may be more pronounced at smaller particle sizes. Additional issues are presented in the next section.

5 Wetting and Microstructural Inhomogeneities

In the previous section, changing the LTP to HTP ratio and the LTP layer thickness lead to straightforward conclusions about the kinetics. For a constant interfacial area of HTP, increasing LTP volume increases the thickness of IMC per unit area that must be formed and therefore increases the minimum process time. In addition, increasing the ratio of LTP to HTP typically improves wetting of the surfaces to be bonded. For systems containing HTP particles, it also promotes rearrangement and densification. However, uniform, predictable wetting of the HTP is required for these relationships to be true and ultimately for producing a commercially viable TLPB technology.

5.1 Incomplete Wetting

If the low-temperature phase does not completely wet the high-temperature phase, the actual thickness of the low-temperature phase to be consumed by isothermal solidification will be larger than the optimal for that system, meaning that the solidification time required goes up by $(thickness)^2$ or $(thickness)^3$. Furthermore, the bond will contain significant voids and stress-concentrating defects. This partial wetting situation is shown in Fig. 12a for a fluxed 10 μm Sn foil placed between two singulated, lead frame Cu sheets (5.6 mm × 5.6 mm and acid etched). The trilayer

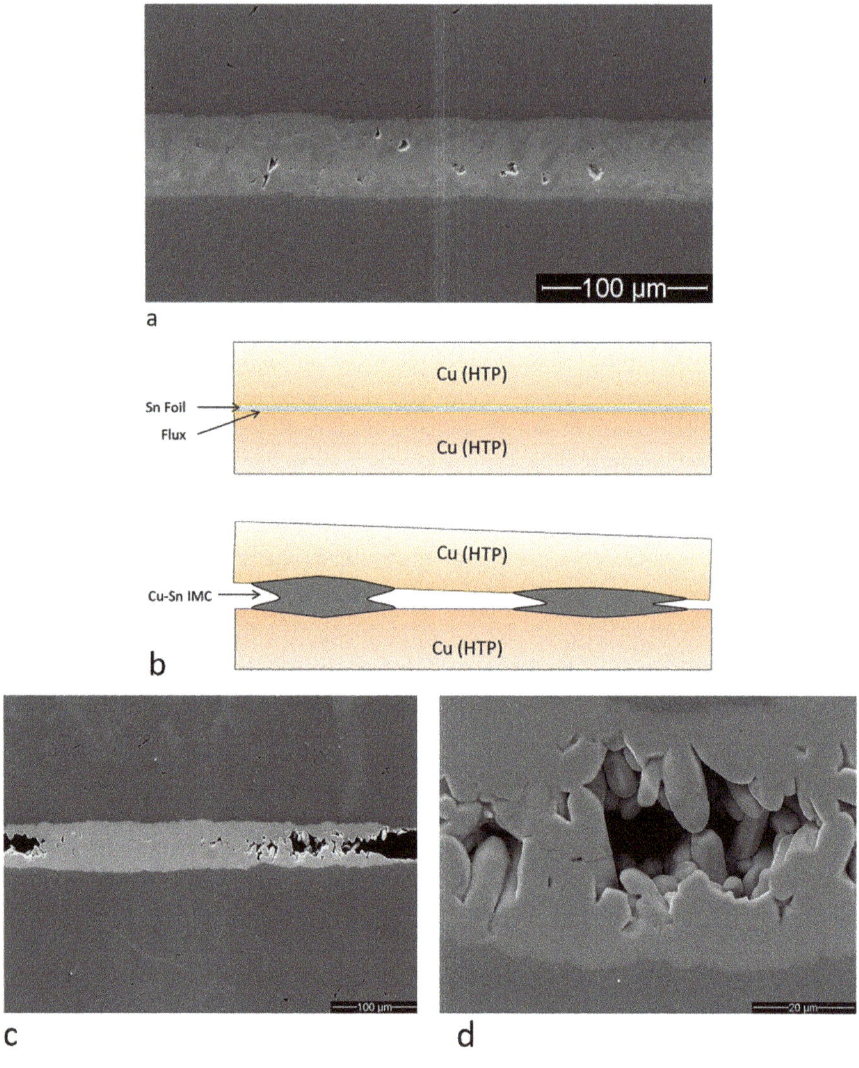

Fig. 12 Bond in foil-based TLPB Sn-Cu using 10 μm pure Sn foil, with no-clean RMA flux and low pressure (28 kPa) joining Cu lead frame substrates at 260 °C for 40 m showing (**a**) a region of good wetting, (**b**) schematic TLPB cross section illustrating ideal wetting condition (top) and final partial wetting condition (bottom) with exaggerated defects (non-planarity, irregular growth fronts, and hot tearing), (**c**) edge of wetted region showing hot tearing, and (**d**) exposed Cu-Sn IMCs

assembly was heated using a hot plate to 260 °C and held for 40 min. As a result of nonuniform melting and wetting, the sheets contained wetted and non-wetted regions as shown in Fig. 12b–d, with the wetted regions being considerably thicker than the initial 10 μm foil thickness. Along the edges of the wetted regions, the low-temperature phase retracts as the IMC forms, leading to bare IMC where the liquid was previously in contact with the substrate and an incipient "crack" along the interface as shown in Fig. 12d. This solidification phenomenon is known as "hot tearing." Wetting in TLPB with foil systems can be improved by creating a more uniform temperature field during heating, applying pressure to redistribute the liquid between the two substrates, using a reducing atmosphere, improving planarity of the substrates, and ensuring there is enough liquid to completely wet the surfaces during heating before IMC growth begins to form bridges between the two substrates.

Complete wetting in the Cu-Bi-Sn system with the target bond line thickness was achieved by preheating two fluxed substrates on a hot plate, dropping two 500 μm diameter solder balls on one substrate where the solder balls then wet and spread across the substrate, then putting the second substrate in contact with the molten solder, and pressing the two together with slight pressure. The total time above liquidus was 5–10 s. The joined substrates were cooled and then subsequently annealed to isothermally solidify the bonds. Several other studies have shown that fully wetted structures are possible in foil-based systems if the pressure is high enough, if the surfaces being joined are sufficiently planar for the LTP volume used, and if the other effects are controlled.

5.2 Liquid Redistribution and Porosity Formation in TLPS Systems

Porosity of the same scale as the low-melting phase particle size can also occur in a particle-based TLPS systems due to the melting and redistribution of the low-melting point phases into agglomerated HTP particle clusters. Furthermore, both pockets of liquid and voids can be created due to the simple geometrical arrangement of particles, as shown in Figs. 11e and 13. This represents a significant deviation from an ideal uniform layer thickness of low-temperature melting phase. Flanagan et al. [23] and Greve et al. [26, 27] noted that such porosity can remain in the isothermally solidified bond if the Cu and low-temperature phase particles are mixed together to form a paste [23]. When the low-temperature phase particles melt, the liquid wets the Cu particles, pulling the Cu particles together due to capillary forces, but not necessarily with much Cu particle rearrangement. When the IMC forms between adjacent Cu particles, the structure becomes rigid. This leaves behind voids where the low-temperature phase particles had been and which cannot be removed due to the rigid interconnected structure of the IMC. In addition, the role of paste formulation is important as well, including particle size differences, LTP composition differences, and fluxes. Pan and Yeo [65] performed a process optimization study with three different pastes containing Cu, Sn alloy particle sand Sn-Bi alloy particles, flux, and polymer additive. They demonstrated that reflow parameters could be controlled to significantly decrease formation of large voids and improve the resulting mechanical and thermal properties of bonded structures, as shown in Fig. 14.

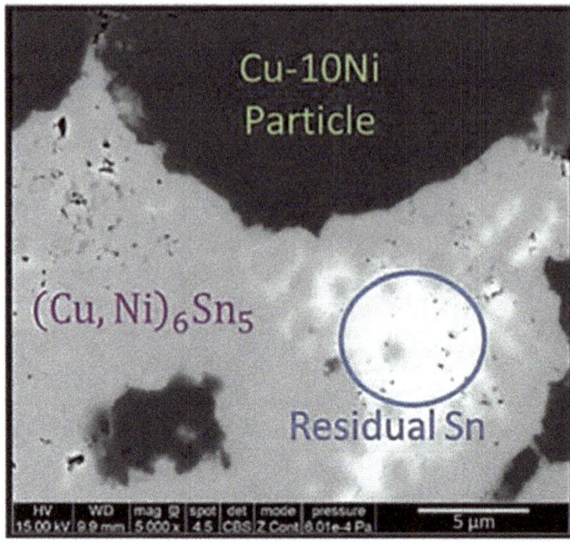

Fig. 13 Cross section of as-fabricated Cu-Ni-Sn TLPS bond using Cu-10wt.%Ni HTP particles and Sn-Ni LTP, showing formation of (Cu,Ni)-Sn IMC and unreacted Sn at particle junctions. Pockets of residual tin between HTP particles in a TLPS Cu-Ni-Sn bond [16]

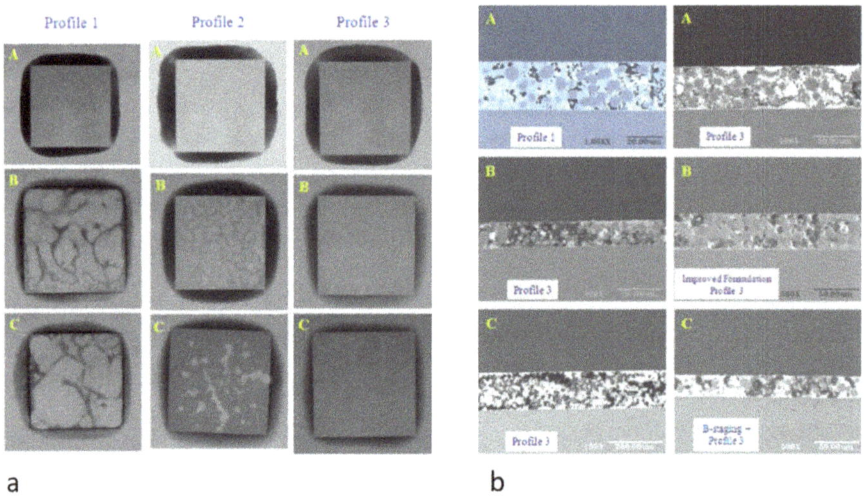

Fig. 14 Process optimization showing the results of three paste formulations using three different thermal reflow profiles (**a**) X-ray images of TLPS samples after processing (**b**) SEM pictures of different TLPS samples under different profiles demonstrates significant variation in voids and microstructure based on processing for the same materials [65]

In a recent study, Chen et al. eliminated these issues with TLPS by using Cu-Sn core-shell particles to form a preform at 30 MPa and applying light loads during processing/reaction [11]. From this process, high-density bonds could be formed with minimal porosity and acceptable mechanical, thermal, and electrical properties for IGBT applications. It should be noted that process optimization studies must be performed for all new formulations in the specific applications and geometries being considered.

5.3 *Porosity Formation During Isothermal Solidification and Solid-State Transformations*

Porosity formation in TLPB systems can have several other root causes that should be avoided. These include (1) Sn consumption by solid-state IMC formation during heat-up before a liquid forms, (2) volume changes during bond formation and IMC equilibration after isothermal solidification, (3) debonding due to consumption of the HTP, and (4) flux entrapment. In terms of porosity formation during solid-state reactions, Bosco and Zok demonstrated that if IMC formation in the Cu-Sn system is completed before forming any liquid, significant porosity forms along the midline of the bond [8]. The critical limits for porosity formation are determined by the LTP thickness, the target reaction temperature, and the heating rate to the reaction temperature. This is particularly an issue for electroplated LTP phases on metal substrates, where in systems such as Sn-Cu, IMC growth begins at room temperature. This may have contributed to porosity formation in thin Sn-Ag TLPB bonds formed at a heating rate of approximately 10 °C/min up to 300 [75]. Midline porosity formation was also observed by Chuang et al. for Sn-Ni bonds that were partially reacted above the LTP liquidus temperature and then subsequently annealed below the solidus temperature to consume the Sn [18]. As with Bosco and Zok, they attributed porosity formation to the impingement of a few large IMC particles along the midline that limited volume change accommodation as IMC growth continued in adjacent areas. They discovered that adding 2.4 wt. % Ag to Sn eliminated void formation, likely by rapid coarsening of the Ag_3Sn to fill in the pores.

A second cause of void formation is due to volume changes as the phase fractions of the IMCs change during equilibration after isothermal solidification. Budhiman et al. observed pronounced void growth along the midline in as-solidified Ni-Sn TLPB bonds composed of Ni_3Sn_4 and Ni formed at 300 °C was subsequently reacted to form Ni_3Sn_2 and Ni_3Sn at 600 °C, as shown in Fig. 15 [10]. Void formation was exacerbated by the squeezing out of Sn to the edges of the bond during initial bonding where it continued to form IMC during the high-temperature anneal. A third root cause of porosity is dewetting of the underlying substrate when a thin, plated HTP layer forms an IMC not only with the HTP but also with the underlying underlayer metallization [17]. Flux entrapment can also create a physical barrier to impingement of the two reaction fronts, leading to midline porosity formation [17].

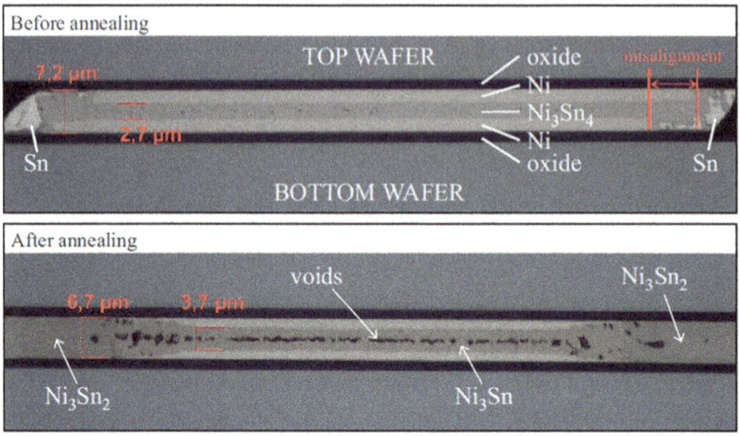

Fig. 15 Optical microscopy cross sections of Ni-Sn TLPB bond showing (top) Ni_3Sn_4/Ni bond fabricated at 300 °C for 20 min using a pressure of 0.5 bar. Note unreacted Sn forced out of bond during fabrication and small line of midline porosity in Ni_3Sn_4 (bottom). Significant void formation after annealing at 600 °C for 24 h accompanying transformation to Ni_3Sn and the formation of Ni_3Sn_2 where residual Sn had been [10]

6 Commercial Electronics Applications and Technical Demands

6.1 Application Temperatures

Transient liquid phase bonding has been proposed for a variety of electronics applications in which high-temperature compatibility is necessary. Automotive, aerospace, and deep well applications are commonly cited in literature as industries demanding operation at increased temperatures [15]. Changes in semiconductor market for power electronics also demands new interconnect solutions. These power electronics devices are more efficient but tend to operate at higher temperatures with greater thermal fluxes due to their small footprint compared to Si-based devices. Backside attach or die attach is a common application of high-Pb solders which is currently exempt from RoHS regulations which have driven most of the global consumer market toward the use of Pb-free solders.

Electrification in both the automotive and aerospace industries is expected to increase demand for interconnect technologies that can sustain high-temperature applications under harsh conditions [15]. As noted by Chin et al. from a report by DaimlerChrysler, devices installed in the engine compartment are expected to withstand operating temperatures up to 200 °C. The proliferation of sensors and electronics controls throughout the vehicle also means that interconnect technologies will need to tolerate even higher temperature extremes, up to 300 °C for wheel-mounted components, 500 °C for combustion systems, and 800 °C in exhaust systems. Thermal management for power electronics is increasingly a concern for the aerospace industry as fly-by-wire technologies become more common.

6.2 Application Requirements

The possible application temperatures span the ranges for high-Pb solders (room temperature to 200 °C) up to 800 °C for exhaust systems. Regardless of the high-temperature application, the following performance, properties, and processability requirements must be met, as summarized from reviews of Chin et al. [15] and Manikam et al. [49] with some additions and modifications [15, 49]. They are:

- For solders:
  - Appropriate solidus and liquidus temperatures for the specific application, for example, $\geq$ 260 °C for high-Pb solder replacement and even higher for other applications
- For non-solder alternatives:
  - Processable at reasonable temperatures, with no liquids forming near the use temperature
- Acceptable wettability and adhesion
- Low toxicity
- Reasonable cost
- Thermal conductivity >20–30 W/m · K
- High electrical conductivity
- Good corrosion resistance
- Ductility for stress relaxation within the bond
- Thermal shock reliability
- Thermomechanical fatigue resistance
- Ability to accommodate CTE mismatch between the die and substrate
- Long-term stability under use temperatures
- Manufacturable in high-volume manufacturing
- Reworkable

The inability to meet these requirements with an alternative in high-volume manufacturing for a broad range of applications has led to the continuation of RoHS Exemption 7(a) for "lead in high-melting temperature type solders (i.e., lead-based alloys containing 85 % by weight or more lead)."

The properties, performance, and processing requirements listed above can be examined in terms of intrinsic versus extrinsic behavior for a given technology. For example, the thermal and electrical conductivities of pure single crystal metal are intrinsic properties. In most applications, however, dopants, impurities, second phase particles, and defects such as dislocations and grain boundaries create scattering sites that lead to a decrease in conductivity. The microstructure creates most of the extrinsic behaviors of materials, having a dramatic effect on electrical and thermal conductivity, as well as on most of the other requirements listed above. For example, the presence and spatial distribution of porosity can dramatically decrease thermal and electrical conductivity, as is well known for solder interconnects. Consider the following three cases of solids with porosity: (1) 50% porosity distributed as isolated pores in a dense matrix, (2) a particle compact with 50% porosity but no particle-particle bonding, and (3) a dense solid with an interface crack with a substrate. In the first case, the thermal and electrical conductivities would be decreased by approximately 50% from the bulk values. For the second case, there are only point contacts between the particles, leading to very high contact resistance/low conductivity [34]. In the third case, if the crack spans the entire interface, the conductivity of the structure is dictated by the conductivity across the air gap. Microstructure and microstructure evolution dictate the extrinsic behavior of TLPB interconnects and must be optimized for the specific application.

A common method for taking multiple criteria into account when down-selecting technologies for specific applications is through a weighted decision matrix assessment. A weighted decision matrix for comparing die attach material candidates is illustrated in Table 2, a draft matrix created by the iNEMI high-temperature, Pb-free die attach material project. The importance of specific characteristics depends on the demands of the application and specific characteristics of the full system. In some cases, there will be pass-fail level that a technology must meet, such that failing by one characteristic eliminates it from further consideration. The numerical scores (0–5) are assigned to different performance levels set by the application, where a higher score is better. Each score is multiplied by its weigh factor (not shown here), and all scores are summed for a given technology to come up with an overall score that can be compared with alternatives. In many cases, the real value of setting up a weighted decision matrix is to clarify the requirements for an application, examine where various technologies fall short, and determine whether other parts of the design or processing might be used to overcome such limitations. In some cases, it is found that certain criteria are not absolute and can be traded off for others, i.e., the weighting factors can be modified after further analysis of the results.

The scoring system shown here values high-melting temperatures, high operating temperatures, high thermal and electrical conductivities, and low CTE, as seen by the relative scores. The scoring for reliability, corrosion resistance, electromigration, and other complex criteria will generally be based on performance in a standardized test as compared with a benchmark material or on a series of laboratory tests that evaluate properties affecting the more complex criteria. For example, for reliability these might consist of thermal cycling tests of an assembly in terms of cycles to failure under specific testing conditions for the former and of stress-strain behavior in a tensile test, fracture toughness, various elastic moduli, ductility, impact testing, and creep rate for the latter.

Another method for performing a quantitative, yet more visual comparison on a characteristic-by-characteristic basis is using a "spider" plot as illustrated in Fig. 16. For candidate systems that meet the pass-fail down-selection criteria, the characteristics are distributed around the outer circumference of a facetted circle. The score for each characteristic for a technology is marked on one of the six levels as determined by their scores (0–5). The order of the characteristics can be changed to reflect priorities, and the trade-offs in individual characteristics can become more apparent than by using a composite score.

Table 3 lists some of the key properties for three common Sn-based intermetallic compounds that can be used in comparisons of technologies.

Table 2 Scoring factors related to key characteristics for commercial applicability intended for comparing different die attach technologies

Characteristic	Units	0	1	2	3	4	5
Is it a solder material?		*X*	No	Yes			
Cost relative to high lead solder (Pb5Sn)		*X*	>100×	10–100×	5–10×	2–5×	<2×
Melting temperature of final bond	°C	*X*	<260	≥260	≥280	≥300	≥350
Service temperature	°C	*X*	<150	<175	<200	<250	>250
Stiffness (modulus)	GPa	*X*	Greater than 100	80–100	50–80	20–50	Less than 20
Feasible applications		*X*	Transistor	Diodes	Thyristors	Rectifiers, inverters	Power MOSFET, IGET
Surface metallization compatibility		*X*	Cu	Cu, Ag	Cu, Ag, NiAu	Cu, Ag, NiAu, NiPdAu	Cu, Ag, NiAu, NiPdAu, Ag thick film
Thermal conductivity	W/mK	*X*	0–10	10–25	25–50	50–100	> 100
Electrical resistivity	μΩ*m	*X*	Greater than 100	Less than 100	Less 50	Less than 1	Less than 0.2
Moisture sensitivity level (MSL)		*X*	6	3	2a	2	1
Reliability (cycles to failure)	°C	*X*	125 peak	150 peak	165 peak	175 peak	200 peak
CTE	ppm per °C	*X*	>80	40–80	20–40	10–20	< 10
Processing temperature	°C	*X*	<200	200 ~250	250 ~300	300 to 400	>400
Pressing pressure	MPa	*X*	> 50 MPa	>10 MPa	1–10 MPa	<1 MPa	None need
Processing atmosphere		*X*	Other:	Vacuum	Forming gas	N_2	Air
Pressing time		*X*	>1 h	<1 h	<30 min	<10 min	< 1 min

(continued)

Table 2 (continued)

Characteristic	Units	0	1	2	3	4	5
Maturity		*X*	R&D level	–	Small-scale mass production	–	Large-scale mass production
Material type		*X*	Paste	–	Preform	Wire	Film
If it is a paste, paste type		*X*	Dispensing	–	Printing	–	Both
Cleaning		*X*	Other	Chemical wash	Water wash	–	None need
Rework		*X*	No, component would be destroyed	Rework possible, but will require more steps	–	Mostly compatible	Yes, compatible with conventional rework techniques
Die size compatibility		*X*	<4 mm^2	4–16 mm^2	16–36 mm^2	36–100 mm^2	>100 mm^2
Electromigration potential		*X*	Likely				Unlikely

7 General Thermodynamic Framework for Design of TLPB Technologies

The processing temperatures and times, operation temperatures and times, and durability requirements in terms of phase stability of the isothermally solidified bonds can be translated to thermodynamic constraints based on composition and temperature. Extensive databases and literature exist detailing the equilibrium phases and reaction kinetics for a wide range of low-melting temperature metals in contact

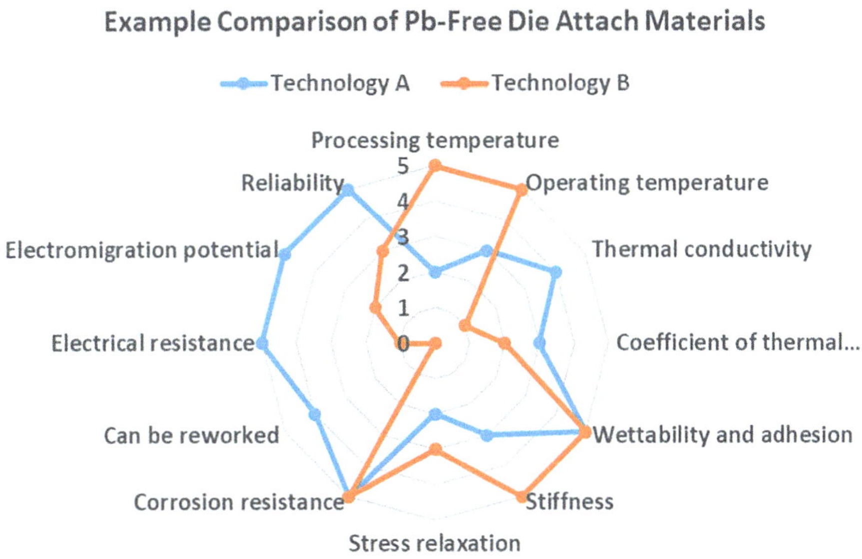

Fig. 16 Schematic spider plot comparing two die attach technologies using multiple characteristics

Table 3 Properties of intermetallic phases commonly utilized for TLPB (room temperature properties published by NIST) [22]

Phase	Cu_6Sn_5	Cu_3Sn	Ni_3Sn_4
Transition temperature (°C)	415	638	798
Vickers hardness (kg/mm^2)	378 ± 55	343 ± 47	365 ± 7
Toughness (MPa$\sqrt{m}$)	1.4 ± 0.3	1.7 ± 0.3	1.2 ± 0.1
Young's modulus (GPa)	85.56 ± 1.65	108.3 ± 4.4	133.3 ± 5.6
Poisson's ratio	0.309 ± 0.012	0.299 ± 0.018	0.330 ± 0.015
Thermal expansion (PPM/°C)	16.3 ± 0.3	19.0 ± 0.3	13.7 ± 0.3
Thermal diffusivity (cm^2/sec)	0.145 ± 0.015	0.240 ± 0.024	0.083 ± 0.008
Heat capacity (J/gram/°C)	0.286 ± 0.012	0.326 ± 0.012	0.272 ± 0.012
Resistivity (μΩ*cm)	17.5 ± 0.1	8.93 ± 0.10	28.5 ± 0.1
Density (g/cm^3)	8.28 ± 0.02	8.90 ± 0.02	8.65 ± 0.02
Thermal conductivity (W/cm*°C)	0.341 ± 0.051	0.704 ± 0.098	0.196 ± 0.019

with high-melting temperature metals. Here we present a framework for combining these data with thermodynamic assessments for TLPB design. Simple criteria are introduced for evaluating specific binary, pseudo-binary, and multicomponent systems. In addition, "processing regime maps" and "reaction diagrams" are used as two visualization techniques to extend interpretation of thermodynamic data by using diffusion couple/interdiffusion experiments to identify missing or metastable phases during reaction for designing ternary and high-order multicomponent TLPB systems.

The general thermodynamic criteria for TLPB are based on (1) the melting temperature of the LTP for the desired reaction with an HTP; (2) the formation of selected IMC(s) that incorporate appropriate component(s) of the liquid; (3) for ternaries and above, three-phase equilibria linking a terminal liquid composition, the IMC, and any other phases that precipitate during isothermal solidification; and (4) and the temperature of liquid-forming reaction for product phases for specific operating temperatures in specific applications.

Criterion (1) applies to both binary and multicomponent systems. For instance, processing in common binaries such as Cu-Sn, Ni-Sn, and Ag-Sn must occur above the melting temperature of pure Sn, 232 °C. In the case of ternaries and multicomponent systems, processing must occur above the solidus temperature of the LTP and usually above the LTP liquidus temperatures such that a sufficient liquid fraction exists to for wetting of the HTP.

Application of criterion (2) requires that the relevant liquid compositions be in equilibrium with an IMC phase incorporating the desired component of the liquid at processing temperatures. As some solubility of the HTP component(s) may exist in the liquid, it is equilibria of the HTP-saturated liquids that are relevant. When the IMC layers form by interdiffusion, local equilibrium is generally assumed to be satisfied at each interface. In a ternary, one of the components may not be incorporated into the IMCs. Such a system would invoke criterion (3) which requires that a solid phase form isothermally from the non-reacting component through its equilibrium with both the terminal liquid composition and the desired IMC phase. Criterion (4) indicates that the phase(s) formed during processing must remain stable and solid up to the intended operation temperature. Consideration of criterion (4) has implications for processing conditions for creating specific phases that are not only stable during processing but also during operation.

The three most important binaries for TLPB – Cu-Sn, Ag-Sn, and Ni-Sn – are examined below based on these criteria. In binaries, the possible processing and use ranges based on the four criteria can be read directly off the phase diagrams: the positions of three-phase invariants (eutectic and peritectic reactions) are points in the two-dimensional temperature-composition phase diagram plot. The identification of these invariants in the binary systems helps understand how more complex reactions occur when an additional element is added to form a ternary.

In ternary systems, the equilibria from the binaries are not the only reactions that occur. For example, for a three-component system, there can be points of four-phase equilibria, such as the ternary eutectic in which a liquid transforms at a specific composition and temperature into three phases. Similarly, there are peritectics, eutectoids, and peritectoids. Of particular importance to TLPB are additional four-phase invariants where two solid phases transform into a liquid plus another solid phase. The four-phase invariant reactions involving a liquid phase are classified as follows:

- Class I invariants: ternary eutectic reaction where liquid $\leftrightarrow \alpha + \beta + \gamma$
- Class II invariants: quasi-peritectic reaction where liquid + $\alpha \leftrightarrow \beta + \gamma$
- Class III invariants: ternary peritectic where liquid + $\alpha + \beta \leftrightarrow \gamma$

These are identified in specific systems by comparing isothermal sections of the ternary phase diagram, as well as through thermodynamics calculations through CALPHAD-based software. For TLPB design, four-phase reactions all in the solid state (eutectoid, quasi-peritectoid, and peritectoid) are also important since formation of a higher temperature phase by continuing interdiffusion may require the disappearance of a phase that had previously formed during isothermal solidification. A range of ternary examples will be shown to illustrate the phases that result in different composition and temperature regimes and how that limits processing conditions and composition ranges.

In this section, we show that phase diagrams serve as the beginning point in TLPB design. Phase diagrams show what phases to expect within certain composition and temperature ranges at *equilibrium*. Achieving equilibrium means that, in a two-component system, there will only be two phases present. The TLPB process is an interdiffusion process, inherently non-equilibrium, and multiple phases generally remain in the isothermally solidified bond, depending on the temperature, time, and starting compositions and geometries. Furthermore, some equilibrium phases in the phase diagram may not form during interdiffusion and isothermal solidification. In the following subsections, TLPB alloy design is discussed in terms of these invariant reactions, predicted phase sequences from phase diagrams, and the phases as a function of composition, temperature, time, and sample geometry.

7.1 Binary TLPB Systems

7.1.1 Cu-Sn

The Cu-Sn phase diagram (Fig. 2) was described above in Sect. 2.1. Consider a Cu-Sn TLPB bond starting from pure Sn in contact with Cu. The five ranges for isothermal solidification are shown as *A*, 227–415 °C for Cu_6Sn_5; *B*, 415–638 °C for Cu_3Sn; and three higher ranges for *C*, γ; *D*, β; and *E*, (Cu). As described in detail in the [56], there are 13 invariant equilibria in the Cu-Sn system with five involving a liquid phase and are therefore relevant to reacting pure Sn with pure Cu:

- Liquid $\leftrightarrow$ Cu_6Sn_5 + (Sn) at 227 °C
- Liquid + Cu_3Sn $\leftrightarrow$ Cu_6Sn_5 at 415 °C
- $\gamma \leftrightarrow$ liquid + Cu_3Sn at 638 °C
- Liquid + $\beta \leftrightarrow \gamma$ at 757 °C
- Liquid + (Cu) $\leftrightarrow \beta$ at 796 °C

Section 2.1 above provides additional details about the expected reactions during isothermal solidification as a function of temperature. One additional feature to note in the Cu-Sn system is the extended temperature range for operation if a bond reacted in *A* until it contains no Cu_6Sn_5 or *B*. In this situation, the lowest melting point IMC in the system is Cu_3Sn. If the Cu_3Sn is heated above 638 °C, the Cu_3Sn remains stable until 677 °C when it transforms congruently into γ. This provides an additional 39 °C of solid-phase stability for bonds isothermally solidified with no remaining Cu_6Sn_5.

The microstructure and phase content in a specific bond will evolve after isothermal solidification as a function of annealing temperature and time, which can be interpreted based on thermodynamics. Bosco and Zok demonstrated that annealing a Cu-Sn-Cu TLPB bond with a 20 μm thick Sn layer at 400 °C produced only the $Cu_{41}Sn_{11}$ phase (δ) in contact with (Cu) (*C*). When the as-solidified bond was heated to 550 °C, δ transformed into γ (*D*), which then transformed peritectically into δ and a dispersion of (Cu) particle, as expected from the phase diagram. After annealing the original δ bond at 600 °C, there was only (Cu), a solid solution of Sn in Cu (*E*) [7, 8].

7.1.2 Ag-Sn

The Ag-Sn binary phase diagram shown in Fig. 17 indicates three temperature ranges for isothermal solidification: *A*, 220–480 °C for Ag_3Sn; *B*, 480–724 °C for ζ Ag; and *C*, 724–962 °C for (Ag) [59]. In *A* two IMC phases Ag_3Sn and ζ Ag are shown for reaction between Sn and Ag; however, typically only Ag_3Sn is observed during interdiffusion experiments. The lack of a second IMC layer during

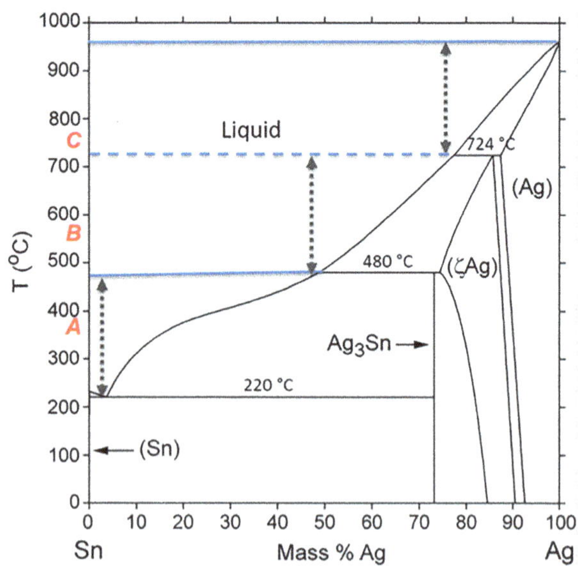

Fig. 17 Ag-Sn phase diagram annotated for three separate TLPB reaction regimes. (Modified from NIST [59])

interdiffusion simplifies analysis of the Ag-Sn binary in *A*. There are only three ternary invariants in this system:

- Liquid ↔ Ag_3Sn + (Sn) at 220 °C
- Liquid + ζ Ag ↔ Ag_3Sn at 480 °C
- Liquid + (Ag) ↔ ζ Ag at 724 °C

This makes the use of Ag-Sn straightforward and simple in TLPB systems in terms of phase evolution. For more details on the phase evolution in Sn-Ag, see [44]. As noted above, processing conditions must be controlled to avoid porosity formation that result from its rapid intermetallic growth rate compared with Cu-Sn, as well as from changes in volume with continued annealing after isothermal solidification.

7.1.3 Ni-Sn

The Ni-Sn binary system is highly relevant to TLPB due to the use of ENIG, Au/Ni, and Au/Ni-P surfaces finishes in electronics. The Ni-Sn phase diagram in Fig. 18 shows three eutectic reactions, one peritectic reaction, two eutectoid reactions, and

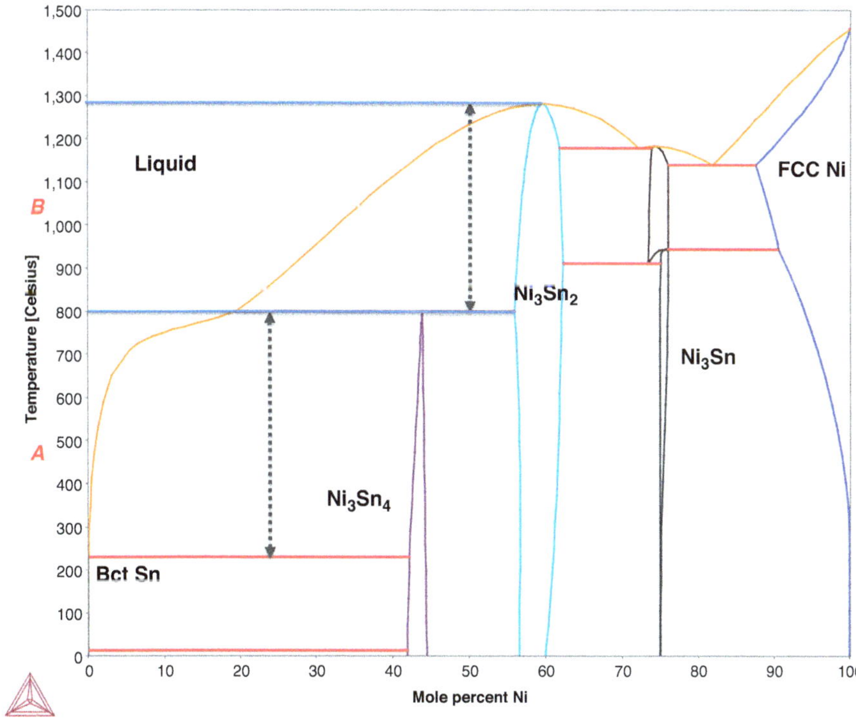

Fig. 18 Phase diagram for Ni-Sn calculated using ThermoCalc and annotated for two specific TLPB reaction regimes

three congruent points. If we only consider reactions between Sn and Ni_3Sn_2 as relevant for TLPB, the invariant reactions are:

- Liquid ↔ Ni_3Sn_4 + (Sn) at 232 °C
- Liquid + Ni_3Sn_2 ↔ Ni_3Sn_4 at 795 °C
- Liquid ↔ Ni_3Sn_2 at 1294 °C

This leads to two relevant temperature regimes for TLPB: *A*, between 232 and 795 °C, and *B*, between 795 and 1294 °C. Although the simplicity of the binary system makes it very attractive, Ni is often used in conjunction with an Au surface finish layer and co-deposited with P or W. Having P turns the reaction into a ternary and can add additional stability issues, including needle-like IMC growth, formation of Ni-P and Ni-P-Sn ternary compounds, and IMC spalling during reaction [33, 43]. In addition, the tendency for porosity formation noted above for Ni-Sn will require careful geometric design and process control for its successful application. As observed for the solid-state reaction [18], it may be possible to eliminate some porosity by the use of a Sn-Ag LTP rather than just Sn.

7.2 *Ternary Systems*

Ternary systems offer broader opportunities for bond formation by having three phases in equilibrium, the transitions that can occur as one of the intermediate IMC phases disappears by continued diffusion, and Class II reactions that eliminate formation of a phase that limits the temperature of use. The systems described here are all Sn-based and include (1) Ag-Cu-Sn, (2) Cu-Ni-Sn, and (3) Ag-Bi-Sn, Bi-Cu-Sn, and Bi-Ni-Sn. The systems Ag-Bi-Sn, Ag-Cu-Sn, and Bi-Cu-Sn have been well-studied in terms of their thermodynamics and interdiffusion/reaction paths and the thermodynamics framework can be straightforwardly applied. The two systems involving Ni – Cu-Ni-Sn and Bi-Ni-Sn – offer interesting opportunities for TLPB even though there are gaps in knowledge of their thermodynamic transitions and interdiffusion/reaction path behavior.

7.2.1 Ag-Cu-Sn

Section 2.3.1 above gives an introduction into the lowest temperatures reactions relevant to TLPB for the Ag-Cu-Sn system [57]. To consider the whole range of possible processing and temperatures, there are nine ternary invariants in the Ag-Cu-Sn system that must be considered, as follows:

- L ↔ Cu_6Sn_5 + Ag_3Sn + (Sn) occurs at approximately 216 °C
- L + Cu_3Sn ↔ Cu_6Sn_5 + Ag_3Sn occurs at approximately 356 °C
- L + ζ-Ag ↔ Cu_3Sn + Ag_3Sn occurs at approximately 459 °C
- L + (Ag) ↔ ζ-Ag + Cu_3Sn occurs at approximately 574 °C
- L + (Ag) + $Cu_{10}Sn_3$ ↔ Cu_3Sn occurs at approximately 611 °C

- L + (Ag) + $\gamma \leftrightarrow Cu_{10}Sn_3$ occurs at approximately 613 °C
- L $\leftrightarrow \gamma + Cu_{10}Sn_3 + Cu_3Sn$ occurs at approximately 550 °C
- L + (Cu) $\leftrightarrow$ (Ag) + γ occurs at approximately 636 °C
- L + $\beta \leftrightarrow$ (Cu) + γ occurs at approximately 642 °C

These ternary invariants change the processing and use temperature limits relative to the binaries. For example, if Cu_6Sn_5 is present in the isothermally solidified bond, the maximum use temperature is below 356 °C rather than 415 °C where Cu_6Sn_5 becomes unstable in the Cu-Sn binary.

7.2.2 Cu-Ni-Sn

The Cu-Ni-Sn ternary systems has been studied at low temperatures for transient liquid phase bonding due to several unique characteristics of the binary phase diagrams. The Cu-Ni system has a well-known isomorphous phase diagram as illustrated in Fig. 4. Both Cu and Ni also exhibit solid solubility in binary intermetallic compounds, such as (Cu, Ni)$_6$Sn$_5$. Trace amounts of Ni have been shown to suppress the IMC's allotropic phase change upon cooling, i.e., stabilizing the high-temperature hexagonal (Cu, Ni)$_6$Sn$_5$ [61]. In addition to phase stability, Cu-Ni alloying may also affect reaction kinetics since it is known to affect the rate of solid Cu-Ni substrate dissolution in molten Sn and IMC growth rates [36]. The implications of Cu-Ni characteristics encountered in recent experiments are discussed in conjunction with several examples.

The ternary invariants for the Cu-Ni-Sn system have not yet been determined, but the phases in equilibrium at various temperatures have been determined by Schmetterer et al. [74]. At 400 °C, liquid Sn appears to be in equilibrium with (Cu, Ni)$_6$Sn$_5$ and (Ni,Cu)$_3$Sn$_4$. At 500 °C, Sn liquid is in equilibrium with (Cu, Ni)$_3$Sn and (Ni, Cu)$_3$Sn$_2$, while at 700 °C, Sn liquid is in equilibrium with γ and (Ni, Cu)$_3$Sn$_2$. There are ternary phases reported.

Greve et al. demonstrated Sn-Cu-Ni bonding of Ag-metallized, Si dies to Ni substrates using a paste [27]. The pastes included pure Cu, Ni, and mixtures of pure Cu/Ni powders HTP particles in combination with a Sn-3.5Ag LTP. Processing occurred in an inert atmosphere at 300 °C for approximately 30 min. Pressure was applied (0.2 MPa) to minimize void formation. The Ni-Sn and (Cu, Ni)-Sn microstructures exhibited few voids. Residual Sn in the structure indicated incomplete conversion of the liquid Sn. Elevated die shear testing indicated that the bonds withstood 10 MPa well beyond the melting temperature of Sn-3.5Ag. Bonds formed from the mixture of Cu and Ni withstood the 10 MPa die shear test up to approximately 435 °C which is consistent with the melting temperature of Cu_6Sn_5, 415 °C. The failure temperature implies that the bonding was supported by a matrix of IMC despite pockets of unreacted Sn. Similarly, bonds withstood up to 600 °C (test limit) in the case of Ni HTP with Sn, consistent with the melting temperature of Ni_3Sn_4, 798 °C. The microstructures of the mixed Cu and Ni exhibited competing formation of Cu_6Sn_5 and Ni_3Sn_4 at the Cu and Ni interfaces, respectively. The

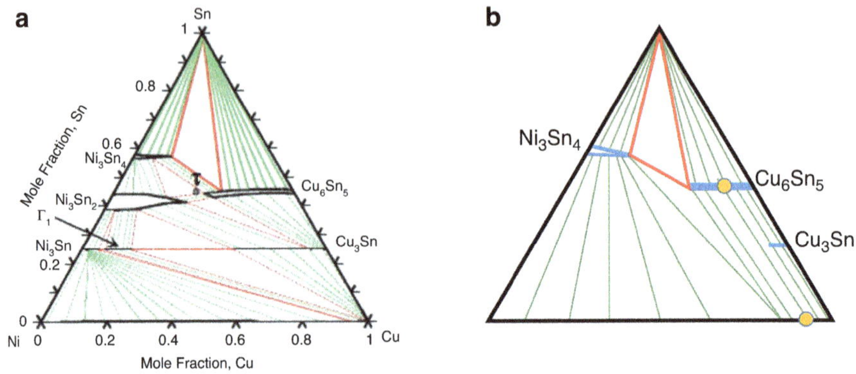

Fig. 19 (**a**) Calculated metastable equilibria in the Sn-Cu-Ni systems at 240 °C. The possible τ-phase is indicated and dotted lines in the Cu-Ni side of the diagram beyond the formation of (Cu, $Ni)_6Sn_5$ and (Ni, $Cu)_3Sn_4$ reflect the uncertainties in the phase equilibria. Plot for 240 °C based on 200 °C isotherm from [1] without the Cu-Ni miscibility gap. (**b**) Schematic reaction diagram represents the phases observed in diffusion couples composed of Cu-Ni HTP and Sn at around 240 °C. The gold circles indicate the final phases remaining after complete consumption of the Sn for a Cu-10 wt. % Ni HTP alloy [72]

phases and compositions of IMCs are known to shift with aging in reactions with BGAs where one substrate is Cu and the other is Ni.

As opposed to mixing Cu and Ni powders, Choquette and Anderson fabricated bonds using a Cu-Ni-Sn TLPB system incorporating a Cu-10Ni wt. % HTP alloy powder with a Sn-0.7Cu-0.5Ni LTP powder (SN100C, Nihon Superior). Bonds were processed using a reflow process with 30s above liquidus (227 °C) to a peak temperature of 250 °C. The microstructure in Fig. 13 illustrates the layer of (Cu, $Ni)_6Sn_5$ (mid-gray) between Sn (light phase) and the Cu-10Ni particles. Note the lack of (Cu, $Ni)_3Sn$ between the remaining Cu-10Ni particles and the (Cu, $Ni)_6Sn_5$ layer. Evidence presented by Baheti et al. suggested that at 200 °C (Cu, $Ni)_3Sn$ is not thermodynamically stable when Ni concentrations are equal to or greater than 7.5 wt. % Ni. Difficulty of nucleating the thermodynamically stable phase was also suggested by Baheti et al. [1]. Vuorinen et al. [80] instead attributed the lack of (Cu, $Ni)_3Sn$ to a kinetic origin based on experiments at 240 °C. Rapid growth of the (Cu, $Ni)_6Sn_5$ phase appears to prevent formation of the thermodynamically stable (Cu, $Ni)_3Sn$ [80]. Evidence of (Cu, $Ni)_3Sn$ in equilibrium with Cu-Ni alloys at 240 °C presented by Lin et al. supports the kinetic explanation [42].

The most important feature of this system is, therefore, that it can be treated as two pseudo-binaries. This can be understood by comparing the 240 °C isothermal section of the phase diagram (Fig. 19a) with what we refer to as a "reaction diagram" (Fig. 19b). This isothermal reaction diagram schematically shows the phases observed during the short annealing times required for LPDB systems. Many binary and ternary phases present in the equilibrium phase diagram are missing in the diffusion couples as discussed above. The Cu-Ni HTP alloys in contact with Sn form either (Cu, $Ni)_6Sn_5$ or (Ni, $Cu)_3Sn_4$ depending on the HTP alloy composition, and at low Ni concentrations, $(Cu, Ni)_3Sn$ is also observed. This diagram helps to visualize the resulting phases forming from diffusion couples in various HTP and LTP composition regimes. Further development of such diagrams may prove useful for TLPB design applications.

7.2.3 Sn-Bi LTP with Cu, Ni, or Ag

Alloys in the Sn-Bi system are attractive as the LTPs due to the low-temperature eutectic reaction in Sn-Bi occurring near 139 °C at 57 wt. % Bi–43 wt. % Sn. The Sn-Bi system is a simple eutectic system with no IMCs. The element Bi also forms simple binary eutectics with Ag ($T_e = 262.5$ °C) and Cu ($T_e = 270.6$ °C), without any IMCs. The Ni-Bi phase diagram contains two IMCs, NiBi and $NiBi_3$, with two peritectic reactions and the $NiBi_3$-Bi eutectic reaction ($T_e = 270.6$ °C). Given the deep Sn-Bi eutectic, the maximum effective addition of Bi is limited only by the invariant reactions and the reaction paths for isothermal solidification. In the Ag-Bi-Sn and Bi-Cu-Sn systems, there is very low solid solubility of Bi in the IMCs and, with no Ag-Bi or Cu-Bi IMCs, (Bi) is the only Bi-containing phase which melts at 271 °C. The Bi-Ni-Sn system contains both Ni-Sn and Bi-Ni IMCs which make the phase diagrams and reaction paths more complex.

7.2.4 Ag-Bi-Sn

The ternary invariants for Ag-Bi-Sn are all eutectics [60, 62]:

- Liquid ↔ Ag_3Sn + (Sn) + (Bi) at 137 °C
- Liquid ↔ Ag_3Sn + ζ-Ag + (Bi) at 252 °C
- Liquid ↔ (Ag) + ζ-Ag + (Bi) at 263 °C

The implication for Ag-Bi-Sn TLPB is that bonds containing Ag_3Sn will contain liquid above 252 °C, therefore limiting their processing and use to between 137 and 252 °C. If the bond contains only the ζ-Ag intermediate phase along with (Ag) and (Bi), the bond is stable up to 263 °C.

7.2.5 Bi-Cu-Sn

There are nine ternary invariants for the Bi-Cu-Sn, with the most important for isothermal solidification being [58]:

- Liquid ↔ Cu_6Sn_5 + (Sn) + (Bi) at 137 °C
- Liquid + Cu_3Sn ↔ Cu_6Sn_5 + (Bi) at 200 °C
- Liquid + $Cu_{44}Sn_{11}$ ↔ Cu_3Sn + (Bi) at 270.2 °C
- Liquid + (Cu) ↔ $Cu_{44}Sn_{11}$ + (Bi) at 270.3 °C

The implications for these invariants are that a bond containing (Sn) forms a liquid above 137 °C, Cu_6Sn_5 above 200 °C, Cu_3Sn above 270.2 °C, and $Cu_{44}Sn_{11}$

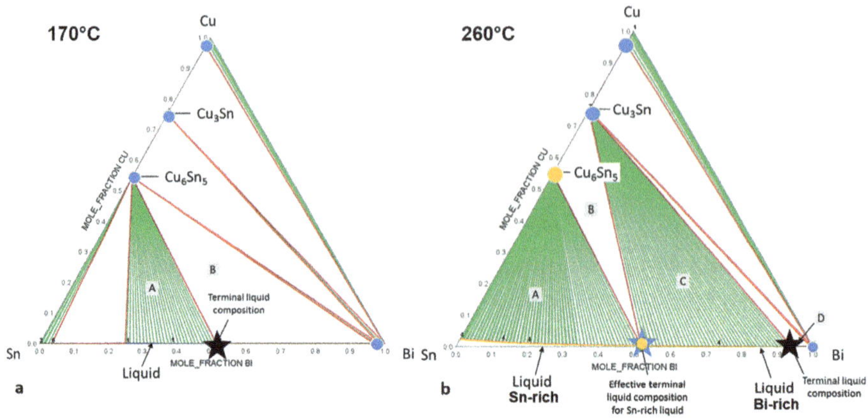

Fig. 20 Calculated isothermal sections of the Cu-Sn-Bi ternary phase diagram at (**a**) 170 °C and (**b**) 260 °C illustrate different phases in equilibrium with Cu saturated Sn-Bi liquid based on temperature and composition. At 170 °C, all compositions of liquid are in two-phase equilibrium with Cu_6Sn_5 as opposed to 260 °C where liquids between the stars are instead in equilibrium with Cu_3Sn (TCSLD3 for ThermoCalc)

above 270.3 °C, respectively. The other five invariants can be found in the NIST Webpage for Phase Diagrams and Computational Thermodynamics but were not included here because they involve reactions at temperatures above the Bi melting temperature, where all of the Bi is in the liquid, and bonds cannot be isothermally solidified.

Initial research on Cu-Sn-Bi for TLPB focused on eutectic Sn-Bi paired with Cu filler particles at low temperature to take advantage of the Sn-Bi eutectic temperature of 137 °C, allowing significantly lower processing temperatures than conventional Sn-Pb or SAC soldering. The Sn component of the LTP is intended to react with Cu to form IMC, while the rejected Bi solute forms a solid precipitate. D'Hondt and Corbin performed DSC and EDS analysis on mixtures of Sn and Bi with Cu filler particles as well as eutectic Sn-Bi with Cu filler particles at temperatures below 200 °C [19]. Analysis of the microstructures and heat flow profiles revealed a "persistent liquid-forming reaction" above 200 °C, the Class II reaction between the solid Bi phase and Cu_6Sn_5. This acts as limit for Criterion 4. However, analysis of computed isothermal sections of the ternary phase diagram shows that high-Bi non-eutectic compositions can be utilized to avoid the formation of Cu_6Sn_5, as shown in Fig. 20. Increasing Bi content and processing temperatures result in the formation of Cu_3Sn and avoids Cu_6Sn_5 formation entirely. The Criterion 4 limit is then elevated to the melting temperature of Bi, 271 °C which is in the lower range of melting temperatures for high-Pb, Sn-Pb solders [72].

7.2.6 Bi-Ni-Sn

The system Bi-Ni-Sn is different from Ag-Bi-Sn and Bi-Cu-Sn in that Bi-Ni IMCs form, in addition to the Ni-Sn IMCs. There are multiple metastable NiBi and metastable and stable ternary phases that have been reported in the literature. The binary invariants for the Bi-Ni system are [46]:

- Liquid ↔ Bi_3Ni + (Bi) at 271 °C
- Bi_3Ni ↔ liquid + BiNi at 465 °C
- BiNi ↔ liquid + (Ni) at 637 °C

For the ternary Bi-Ni-Sn, there are two low-temperature ternary invariants identified using ThermoCalc:

- Liquid ↔ Ni_3Sn_4 + (Sn) + (Bi) at 137 °C
- Liquid + Ni_3Sn ↔ Ni_3Bi_2 + Bi_3Ni at 300 °C

At temperatures below 271 °C, all Sn-Bi TLP alloys can isothermally solidify with the formation of Ni_3N_4 and (Bi). The Class II invariant reaction can be seen by comparing the crossing tie lines in Fig. 21a, b, limiting isothermally solidified bonds containing Ni_3Bi_2 and Bi_3Ni to use below 300 °C. Another important feature to note in both sections is that, above 271 °C, all Sn-Bi liquids containing less than approximately 94 mole% Bi in contact with Ni form Ni_3Sn_4 which is not in equilibrium with any Ni-Bi IMC at any temperature. This means that the bonds cannot isothermally solidify unless all Ni_3Sn_4 is replaced by Ni_3Sn_2 which can be in equilibrium with Bi_3Ni. Interdiffusion experiments by Wang et al. have shown that the situation is more complex than appear from the ternary phase diagram. At 300 °C

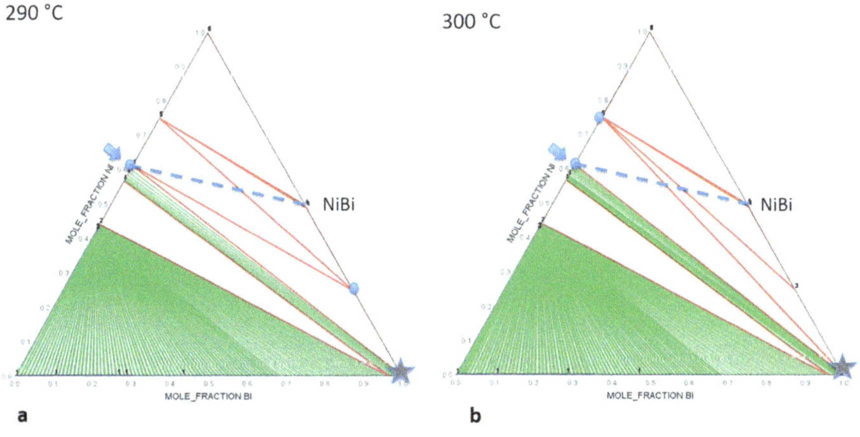

Fig. 21 (**a**) Calculated ternary isotherm showing the terminal liquid (star) in contact with Ni_2Sn_3 and Bi_3Sn (circles) while the dotted line shows the non-equilibrium phases observed during reaction at below the Class II reaction temperature. (**b**) Above the Class II reaction temperature, Ni_3Sn_2 and Ni_3Sn are in equilibrium with liquid. The reaction path was found to contain Ni_3Sn_2 in contact with Ni-Sn

and > 5 wt. % Sn in the Sn-Bi alloy in contact with Ni, Ni_3Sn_4 is indeed observed. With a Bi-2 wt. % Sn alloy at 300 °C, Ni_3Sn_2 was observed at interface with the TLP along with NiBi phase. As shown by the dotted lines in Fig. 21, Ni_3Sn_2 is not in equilibrium with NiBi. Additional interdiffusion experiments with a limited supply of Sn-Bi TLP are needed to determine whether there are processing regimes where Bi-Ni-Sn bonds can solidify above the melting temperature of Bi.

8 Comparison with Competing Technologies

A thorough discussion of TLPB design must compare it to competing bonding approaches for high-temperature interconnects such as high-temperature solders, sintered Ag, and Ag-filled epoxies. Several authors have reviewed and compared bonds using different technologies; however, the individual technologies may not have been optimized, making comparisons generally qualitative. Here we will briefly introduce several of these techniques, noting features and limitations of each, and present the few comparisons in which both technologies have been optimized.

8.1 Solders

As previously discussed, high-Pb solders have remained commercially important and exempt from RoHS regulation due to multiple characteristics. Pure Sn and Pb melt at 232 and 328 °C and form a simple binary eutectic as shown in binary phase diagram in Fig. 1. Alloys using predominately Pb are therefore used for high-temperature applications (>260 °C) and hierarchical soldering. The reliability characteristics are also well known due to their extensive use. High-melting temperatures, thermal fatigue resistance, electromigration resistance, long-term reliability, and resistance to the formation of IMC's are key characteristics of Pb-Sn alloys that have led to the continued exemptions for Sn-Pb [51].

8.1.1 Au-Sn

Au-Sn solder alloys are currently utilized in high-temperature applications such as microwave devices, laser diodes, RF power amplifiers, and flip-chip bonding [20, 32, 35, 49]. The thermal conductivity of Au-Sn, 57 W/mK, exceeds that of Pb-Sn, 35 W/mK for Pb-5Sn [13, 14]. However, Au-Sn is much stiffer compared to "soft" solders such as Sn-Pb which limits the use to small die area applications. Additionally, diffusion barriers such a Ni, Pd, or Pt may be needed to prevent the formation of intermetallics.

8.1.2 Au-Ge

Au-Ge solder may be used for relatively high-melting temperature applications since it is a simple binary system with a eutectic reaction near 360 °C with ~30 at. % Ge. The melting temperature may be further reduced using small alloying additions such as In, Sb, or Sn [13]. Au-Ge also exhibits greater thermal conductivity, 44 W/mK, than Pb-5Sn, 35 W/mK. Au-Ge is also relatively stiff, but small additions of Sb (5 at. Sb) have been shown to improve ductility [13]. However, Sb is toxic but much less toxic than Pb. Another limitation is that Ge cannot be easily electrodeposited. Corrosion is also a concern due to the large electromotive force between Au and Ge [13].

8.1.3 Au-In

Au-In has been demonstrated by Chin et al. for interdiffusion bonding of GaAs at processing temperatures as low as 200 °C while withstanding temperatures as high as 450 °C [39]. At low temperatures, Au-In forms seven IMC phases between Au and In, with AuIn2, AuIn, and Au7In3 observed in the reaction zone between Au and In even at the shortest reaction times [84].

8.1.4 Bi-Ag

Bi-Ag is another simple binary eutectic system, with the binary eutectic at 261. Elemental Bi melts at 270 °C, but it is relatively brittle and has a low thermal conductivity (9 W/mK) relative to other solders. The addition of Ag and other dopants have been shown to improve ductility [49]. Increasing Ag content has also been shown to improve the wetting performance of Bi on Cu substrates. Various dopants have been explored in literature [49]. Commercial ternary Bi-Ag-X solder alloys have been developed using Sn as a dopant to promote wetting by forming Cu_6Sn_5, limiting operation to approximately 200 °C [83].

8.1.5 Zn-Al

Zn-Al eutectic near 380 °C, 6 wt. % Al, exhibits a promising melting range without any IMCs, but the alloys are very stiff and brittle [77]. Various dopants have also been demonstrated [49]. However, poor wetting and oxidation of Zn and Al are significant bonding and reliability concerns [13].

Table 4 High-melting temperature solder alloys and relevant properties [51, 77]

Solder composition	Melting temperature (°C)	Thermal conductivity (W/mK)	CTE (ppm/°C)	Modulus (GPa)	0.2% yield strength at RT (MPa)
Ag-20In	695				
Au-12Ge	356	44	13	72.7	185
Au-20Sn	280	57	16	59	275
Au-3Si	363	27	12	83	220
Bi-11Ag	262–360	9 (for 100%Bi)		37.2	~33
Bi-11Ag-0.05Ge	262–360			37.2	~33
Pb-5Sn		23	30		14
Sn-5Sb		48	23		~40
Zn-(10–30)Sn	360	100–110	30		~30
Zn-(4–6)Al-(Ga, Ge, Mg, Cu)	~380				

8.1.6 Zn-Sn

Zn-Sn is a simple binary eutectic without any IMCs. Unlike other "hard" solders, it is extremely ductile and inexpensive compared to Au alloys. These alloys exhibit excellent wetting characteristics [77]. The thermal conductivity of Zn-Sn alloys, in the range 100–110 W/mK for 10–30 weight percent Sn, exceeds that of Pb-5Sn (35 W/mK) [77]. Zn-Sn is also extremely ductile. Reliability testing showed no significant reduction in shear strength after three reflow cycles up to 260 °C, demonstrating potential for hierarchical soldering applications. Exposure to 85 °C 85% RH for 1000 h only resulted in several microns of oxide at the free surface. Shear strength was maintained up to 2000 thermal cycles (−40 to 125 °C) using a TiN barrier layer. Due to formation of Zn-Cu IMCs, the TiN layer is needed to suppress IMC formation. Successful barrier layers have been demonstrated for soldering at 360–380 °C for 60s [77]. Solder alloy properties are summarized in Table 4 based on data published by Menon et al. and Suganuma et al. [51, 77].

8.2 Sintered Ag

Silver sintering for die attach applications has garnered significant interest due to the excellent electrical and thermal conductivity of pure Ag (419 W/mK). Sintering also allows for relatively low processing temperatures (100–300 °C) compared to the melting temperature of Ag (961.8 °C) [64, 76]. The sintering process is used with nano- or microscale Ag particles mixed into a paste using an organic binder. Joining of particles and densification occurs by diffusion such that the total surface area and corresponding surface energy are reduced. Pressure may also be applied during processing to promote densification; however, complete densification may not be desirable due to the stiffness of Ag. Instead some remaining porosity is often engineered into the bond to improve compliance. More recently, nanoscale Ag particles have been used to develop bonds that do not require the application of pressure.

Parameters such as sintering pressure, processing temperature profile, choice of organic paste constituents, particle size distribution, processing atmosphere, and the type of surface metallization are known to affect resulting bond characteristics. Paknejad and Mannan recently performed a review of the Ag sintering literature in which they extracted these parameters and compared performance based on shear strength as well as thermal cycling performance [64]. Average initial shear strength was shown to increase with increased pressure, but a decline is observed for pressures in excess of 7.5 MPa. A clear trend was not observed comparing initial shear strength to peak process temperature. The combination of parameters must be optimized for a given paste formulation as opposed to simply assuming that increased processing pressures or temperatures result in greater shear strength. Surface metallization was found to have some effect on average initial shear strength, particularly for pressureless bonding. Long-term annealing and thermal cycling performance of sintered Ag bonds are also sensitive to surface metallization composition and roughness [76]. Silver and other noble metals are known to function well as bonding surfaces for Ag sintering. Other metals such as Ni must be thick enough to prevent oxidation of the underlying Cu. Oxidation increases CTE mismatch between the sintered Ag bond and the substrate which causes early thermal fatigue failure. Comparison of shear strength after thermal cycling favored lower sintering pressures. These results suggest a trade-off between increased initial shear strength with greater pressure and increased thermal cycling lifetimes due to greater porosity with lower processing pressures. Practical optimization of sintered Ag bonds will likely require application specific testing.

8.3 *Electrically Conductive Adhesives*

Electrically conductive adhesives (ECA) utilize a polymer, typically epoxy, to provide the mechanical connection and filler particles, such as Ag flake, to provide electrical and thermal connectivity. Conventional ECAs utilizing epoxy benefit from epoxy's excellent adhesion, low shrinkage, and resistance to mechanical shocks at reasonable cost [52]. A recent review by Amoli et al. addresses various approaches involving particle scale and morphology for improving ECA performance. Silver flake has been used extensively in ECAs due to the excellent thermal and electrical conductivity of Ag; however, metals such as copper and nickel have used as well. The use of ECAs is limited by their electrical and thermal conductivities: ECAs offer much lower thermal conductivity (~10 W/mK) and electrical conductivity (10^-3 to 10^-4 ohm cm) compared to the other technologies discussed here. Since thermal and electrical conductivity in ECA's depends upon a network of touching particles, contact resistance between particles limits the overall performance. The intuitive solution would be to increase the fraction of filler relative to the epoxy content. However, increasing the filler content beyond the percolation threshold no longer affects conductivity significantly. The percolation threshold for a given ECA represents the concentration in which a continuous linkage of particles occurs. Increasing filler concentration beyond the percolation threshold provides little advantage while degrading mechanical performance due to the reduction of the epoxy network. Many parameters such as filler size, shape, and dispersion affect the percolation limit. Recent developments have focused on the inclusion of micro- or nanoscale particles. For instance, Wu et al. demonstrated a 10 wt. % reduction in the percolation limit using nanoscale particles as opposed to micron-sized Ag particles [81].

8.4 Comparisons of Mechanical Properties with TLPB

The first demonstration of the potential of TLPB was in 1966 by Leonard Bernstein of Fairchild Semiconductor who introduced "solid-liquid interdiffusion" bonding for semiconductor device assembly to solve limitations related to hierarchical soldering. Bernstein identified five distinct steps that take place during processing: wetting, alloying, liquid diffusion, gradual solidification, and solid diffusion, a subset of the processing steps identified here in this chapter. Mechanical properties of Ag-In, Au-In, and Cu-In lap shear bonds were measured using lap shear samples fabricated from 200 to 450 °C for 10–120 min. Successful bonding in some of the samples was demonstrated by heating lap shear bonds under load above the melting temperature of the initial low-melting temperature component. The paper concluded that a viable proof of concept utilizing these techniques had been demonstrated. Multicomponent alloys incorporating Sn were suggested for future work [5]. Bernstein and Bartholomew published an additional paper in 1966 in which bonds were fabricated using a greater variety of materials, including HTP bonding surfaces included gold, copper, nickel, Dumet, Kovar, nickel-plated molybdenum, Au-Pt paste on ceramic, and nickel-plated metallized ceramic. After processing at 300 °C for 60 min with pressure applied, high tensile strength was observed in the Ag-In bonds. Thermal cycling from liquid nitrogen to boiling water was performed on some of the bonds for 150 cycles without showing obvious damage [6].

8.4.1 Cu-Sn

Cu-Sn Greve et al. demonstrated that Cu-Sn bonds processed at 280 °C for 30 min using Cu particles – 40wt. %Sn LTP exhibited an average die shear strength of 14.6 MPa at 400 °C and 10.5 MPa at 600 °C [26]. The Cu-Sn, core-shell structures developed by Chen et al. exhibited an average die shear strength of 29.4 MPa at 400 °C [12]. However above that temperature liquid forms due to presence of Cu_6Sn_5.

8.4.2 Cu-Ni-Sn

Cu-Ni-Sn Greve et al. demonstrated Sn-Ni and Sn-Cu-Ni bonding of Ag-metallized, Si dies to Ni substrates using TLPS pastes. The pastes included pure Cu, Ni, or mixtures of pure Cu/Ni powders HTP particles in combination with a Sn-3.5Ag LTP. Processing occurred in an inert atmosphere at 300 °C for approximately 30 min. Pressure was applied (0.2 MPa) to minimize void formation. The Ni-Sn and (Cu, Ni)-Sn microstructures exhibited few voids. Residual Sn in the structure indicated incomplete conversion of the liquid Sn; however, elevated temperature die shear testing indicated that the bonds withstood 10 MPa well beyond the melting temperature of Sn-3.5Ag. Bonds formed from the mixture of Cu and Ni withstood the 10 MPa die shear test up to approximately 435 °C which is consistent with the melting temperature of Cu_6Sn_5, 415 °C. The failure temperature implies that the bonding was supported by a matrix of IMC despite pockets of unreacted Sn. Similarly, Ni-Sn bonds withstood up to 600 °C (test limit) consistent with the melting temperature of Ni_3Sn_4, 798 °C. The microstructures of the mixed Cu and Ni exhibited competing formation of Cu_6Sn_5 and Ni_3Sn_4 at the Cu and Ni interfaces, respectively. The phase and composition of IMCs are known to shift with aging in reactions with BGAs where one substrate is Cu and the other is Ni [26, 27]. As noted above in Sect. 7.2.2, additional experimental work is required to establish the temperature of stability.

8.4.3 Ag-Sn TLPB Compared with Sintered Ag

Ag-Sn TLPB compared with Sintered Ag Bajwa and Wilde compared the performance of Ag-Sn multilayer TLPB preforms to optimized sintered Ag before and after 1000 thermal cycles (−40 to 150 °C) [4]. Average shear strength for SiC Schottky diodes (1.66 mm × 1.52 mm) mounted on AlN was 45 and 39.8 MPa for sintered Ag and Ag-Sn TLPB, respectively. After thermal cycling, average die shear force was reduced to 40 MPa and 37.5 MPa, respectively. These results indicate similar bond strength in direct comparison between TLP and Ag sintering which is consistent with values reported elsewhere. Paknejad and Mannan's review of sintered Ag die attach materials indicated initial die shear strengths up to 60 MPa. Depending on processing conditions and particle size used, the typical initial die shear strengths ranged from 10 to 30 MPa after sintering. These results are similar to the die shear forces measured by Pan and Yeo for their hybrid epoxy-Cu-Sn-Bi TLPS materials varied from 46 MPa at 25 °C to 31 MPa at 140 °C. These experiments provide compelling evidence that the mechanical performance of TLP bonding is competitive with sintered Ag and the hybrid technology.

9 Novel Processing Techniques and Application of Design Criteria for TLPB

A brief review of these candidate systems is given here in order to provide further background on TLPB implementation and to highlight the improvements and best practices developed by other researchers. These experiments show that TLPB development, particularly Cu-Sn TLPB, has moved beyond proof of concept toward a processing optimization. The practical importance of design parameters and constraints discussed earlier in this chapter is demonstrated in this section.

9.1 Process Optimization

Luu et al. [47] assessed processing parameters for Cu-Sn bonds for wafer-level bonding. Die and cap wafers incorporating a thin membrane were bonded under vacuum to evaluate hermetic sealing capability. Emphasis was placed on optimizing the deposited Sn and Cu layer thickness in coordination with the heating and pressure profiles. During the heating process, Cu-Sn IMC forms before the melting temperature of pure Sn. Ensuring an adequate remaining Sn layer at the time of melting was identified as a key feature of successful bonding. The application of pressure was necessary to redistribute liquid Sn and to break up oxides at the interface. However, excess pressure resulted in squeeze out of Sn. A bonding profile utilizing 1.5 MPa pressure, bonding temperature of 270 °C, and a 1.5 μm Sn thickness resulted in a 100% yield after dicing and an 80% sealing yield [47].

Liu et al. [45] investigated intermetallic formation in Cu-Sn SLID for wafer-level bonding and 3D integration. Rather than forming a complete bond, Si was electroplated with Cu (5 μm) and Sn (3 μm) layers. The Sn layer remained a free surface to simplify characterization steps. IMC growth was consistent with literature, but the fine Cu grain size due to deposition method was shown to increase interdiffusion rates. At small scales, ramp rates must be fast enough to prevent complete conversion of the Sn layer to IMC before liquid formation. However, fast ramp rates were shown to cause droplet formation of the molten Sn, leading to uneven IMC formation [45].

Bosco and Zok examined the critical interlayer thickness for Cu-Sn-Cu bonds. Preliminary experiments using a 10 μm Sn interlayer heated to 550 °C at a rake of 5 °C per minute under slight pressure exhibited significant void formation. The cavities were attributed to the formation of IMC during the relatively slow heating process. A series of solid-state (<232 °C) IMC growth experiments were conducted to determine the minimum Sn interlayer thickness such that sufficient Sn remains as it melts to accommodate redistribution. Based on these experiments, the critical interlayer thickness was determined as a function of heating rate. Subsequent Cu-Sn-Cu bonds starting with 20–30 μm of Sn resolved the porosity issues apparent in the 10 μm layer thickness [8].

Garnier et al. also investigated the use of Cu-Sn TLP bonding for 3D integration [25]. Four different configurations of Cu-Sn bonds linking Al pads between dies were compared. The pads were 25 μm in diameter with a 50 μm pitch. Typical steps were used to prepare for electrodeposition of 1–2 μm of Cu on both top and bottom wafers. A 5 μm Sn layer was electrodeposited on the top wafers. Half of the bottom wafers received a 3 μm Sn layer. A 1-min thermal treatment at 200 °C was applied to half of each of the corresponding wafers to establish an existing layer of IMC at the Cu-Sn interface. Bonding occurred at 250 °C for 1 min under pressure. Cross sections of the as-fabricated bonds indicate complete conversion liquid Sn to IMC. Measurements of electrical resistance and mechanical shear force were collected before and after 500 thermal cycles from −40 to 125°C at a ramp rate of 14 °C/min. Configurations initiated with Sn on both sides outperformed those with only Sn on the top. However, differences in performance may be attributable to the variation in total bond line thickness, 5 μm Sn vs. 8 μm Sn, rather than defects arising from the initial lack of Sn on the bottom side. Most interesting, the configurations that underwent a thermal treatment to initiate IMC formation before bonding also outperformed their counterparts although the mechanism is unclear. Cross sections of the thermally treated bonds appear to maintain a great bond line thickness which may be attributable to reduction in liquid Sn at bonding. These results suggest that thermal treatment before bonding to initiate IMC formation may provide benefits in other bonding configurations.

These experiments demonstrate the importance of maintaining proper ratios of Cu and Sn. Consideration of IMC formation both before and after melting of the LTP must be considered. TLPB has been shown to be more sensitive to ramp rates compared to other processes. Inclusion of organic binders and solvents complicates this issue since slower ramp rates are needed to accommodate off-gassing. Careful design and process optimization on an application by application basis may be necessary.

9.2 Novel Processing and Geometry

Ehrhardt et al. presented two methods for forming Cu-Sn bonds. The more conventional approach utilized a mixture of 8–45 μm diameter Cu powder, SAC405 solder, and a solvent without reducing properties. A die was pressed into the printed paste using an unspecified pressure. An initial soldering step above the melting temperature of the majority Sn component was performed first. Without a fluxing agent, the Cu oxide prevents formation of IMC. This allows for redistribution of the Sn without the formation of an initial IMC network that prevents rearrangement and densification. During a second step, an activation gas was used to reduce the Cu oxides and initiate isothermal solidification. In a second process, a die is pressed into a printed mixture of Cu and a flux-free solvent. The same gas is used to reduce Cu oxides to avoid the use of flux. A source of solder is placed adjacent to the die on the substrate. Upon melting, capillary forces draws the Sn-rich liquid into the bond which infiltrates the Cu particles. Minimal void formation and complete conversion of the Sn to IMC was demonstrated at a processing temperature near 250 °C using these techniques. Passive thermal cycling from −55 to 125 °C resulted in vertical crack formation that propagated through the Si die. In other instances, crack propagation from the edge of the bond and failed adhesion of the die-side metallization was observed. This work demonstrates that novel configurations and processing techniques can be used to improve bond quality, specifically void formation [21]. Hongtao Chen et al. demonstrated a novel core-shell approach to TLPB using Cu particles electroplated with Sn. Cu-Sn preforms were fabricated by electroless plating of 30 μm Cu particles with 2–3 microns of Sn. The cleaned powder was then pressed into approximately 400 μm thick preforms using 30 MPa of pressure with a 1-min dwell. Complete conversion of the Sn layer to Cu_6Sn_5 and Cu_3Sn was demonstrated using a 12-min reflow process with a peak temperature of 250 °C. Bond fabrication using the preforms requires application of <0.5 MPa to ensure complete contact at the interfaces. The average shear strength at 400 °C was 29.35 MPa. Electrical resistivity was 6.5 μ*ohm*cm. Thermal conductivity was 128–154 W/mK. Thermal cycling results were also promising. The use of preforms made from core-shell particles was demonstrated to be a cost-effective method to fabricate durable, void-free TLP bonds under reasonable processing conditions [11].

Bajwa and Wilde demonstrated TLPB preforms fabricated by electroplating alternating layers of Ag and Sn. The overall composition of the preforms was approximately 80% Ag. Three layer and nine layer preforms were made by depositing the layers onto a steel plate. Bonds were produced from the preforms by heating to 240 °C at 35 °C/min and using a contact pressure of 5 MPa. The shear strength of resulting joints was 35 MPa even after thermal shocking. SiC Schottky diodes were mounted using both the Ag-Sn preforms and Ag sintering. Electrical and thermal performance was comparable to sintered Ag. Thermal cycling from −40 to 150 °C for 1000 cycles also showed promising results. The use of alternating layers was demonstrated to be a practical method for forming durable, void-free TLP bonds with performance comparable to sintered Ag [4].

10 Final Comments

In this chapter, we have described approaches for creating transient liquid phase technologies that meet specific thermodynamic and kinetic criteria for well-defined ranges of compositions and processing conditions appropriate for their use in specific applications. In addition, we have discussed our current understanding of (1) what it will take to achieve the manufacturability required for a Pb-free technology that could replace high-Pb solders in current applications, (2) how these can be extended to even higher temperature applications, and (3) how transient liquid phase bonding compares with other alternative technologies. The Sn-based LTP systems presented in this chapter continue to dominate research in this field; however, it is possible to use Sn-free LTP compositions based on In, Zn, and Bi as the basis for other technologies, as shown in the earliest reported TLPB research for electronics.

Acknowledgments The authors are grateful for the support from the NSF Cooling Technologies Research Center at Purdue University (NSF I/UCRC Grant IIP 0649702) and for important insights and advice on TLPB from Shailesh Joshi and Eric Dede of Toyota.

References

1. V.A. Baheti, S. Islam, P. Kumar, et al., Effect of Ni content on the diffusion-controlled growth of the product phases in the Cu (Ni)– Sn system. Philos. Mag. **6435**, 1–15 (2016). https://doi.org/10.1080/14786435.2015.1119905
2. V.A. Baheti, S. Kashyap, P. Kumar, et al., Effect of Ni on growth kinetics, microstructural evolution and crystal structure in the Cu(Ni)–Sn system. Philos. Mag. **97**, 1782–1802 (2017). https://doi.org/10.1080/14786435.2017.1313466
3. V.A. Baheti, S. Kashyap, P. Kumar, et al., Bifurcation of the Kirkendall marker plane and the role of Ni and other impurities on the growth of Kirkendall voids in the Cu–Sn system. Acta Mater. **131**, 260–270 (2017). https://doi.org/10.1016/j.actamat.2017.03.068
4. A.A. Bajwa, J. Wilde, Reliability modeling of Sn-Ag transient liquid phase die-bonds for high-power SiC devices. Microelectron. Reliab. **60**, 116–125 (2016). https://doi.org/10.1016/j.microrel.2016.02.016
5. L. Bernstein, Semiconductor joining by the solid-liquid-interdiffusion (SLID) process. J. Electrochem. Soc. **113**, 1282–1288 (1966). https://doi.org/10.1149/1.2423806
6. L. Bernstein, H. Bartholomew, Applications of solid-liquid interdiffusion (SLID) bonding in integrated-circuit fabrication. Trans. Metall. Soc. AIME **236**, 405–412 (1966)
7. N.S. Bosco, F.W. Zok, Strength of joints produced by transient liquid phase bonding in the Cu-Sn system. Acta Mater. **53**, 2019–2027 (2005)
8. N.S. Bosco, F.W. Zok, Critical interlayer thickness for transient liquid phase bonding in the Cu-Sn system. Acta Mater. **52**, 2965–2972 (2004)
9. E. Bradley, C.A. Handwerker, J. Bath, et al., *Lead-Free Electronics: iNEMI Projects Lead to Successful Manufacturing* (John Wiley and Sons, Inc., Hoboken, New Jersey 2007)
10. N. Budhiman, B. Jensen, S. Chemnitz, B. Wagner, High temperature investigation on a nickel–tin transient liquid-phase wafer bonding up to 600°C. Microsyst. Technol. **23**, 745–754 (2017). https://doi.org/10.1007/s00542-015-2738-6
11. H. Chen, T. Hu, M. Li, Z. Zhao, Cu-Sn core – shell structure powder preform for high-temperature applications based on transient liquid phase bonding. IEEE Trans. Power Electron. **32**, 1–1 (2016). https://doi.org/10.1109/TPEL.2016.2535365
12. S.-W. Chen, C.-H. Wang, S.-K. Lin, C.-N. Chiu, Phase diagrams of Pb-free solders and their related materials systems. J. Mater. Sci. Mater. Electron. **18**, 19–37 (2006). https://doi.org/10.1007/s10854-006-9010-x
13. V. Chidambaram, J. Hattel, J. Hald, High-temperature lead-free solder alternatives. Microelectron. Eng. **88**, 981–989 (2011). https://doi.org/10.1016/j.mee.2010.12.072
14. V. Chidambaram, H.B. Yeung, G. Shan, Reliability of Au-Ge and Au-Si eutectic solder alloys for high-temperature electronics. J. Electron. Mater. **41**, 2107–2117 (2012). https://doi.org/10.1007/s11664-012-2114-6
15. H.S. Chin, K.Y. Cheong, A.B. Ismail, A review on die attach materials for SiC-based high-temperature power devices. Metall. Mater. Trans. B Process Metall. Mater. Process. Sci. **41**, 824–832 (2010). https://doi.org/10.1007/s11663-010-9365-5

16. S.M. Choquette, I.E. Anderson, Liquid-phase diffusion bonding: Temperature effects and solute redistribution in high temperature lead-free composite solders. Int. J. Powder Met. **51**, 1–10 (2015)
17. K. Chu, Y. Sohn, C. Moon, A comparative study of Cn/Sn/Cu and Ni/Sn/Ni solder joints for low temperature stable transient liquid phase bonding. Scr. Mater. **109**, 113–117 (2015). https://doi.org/10.1016/j.scriptamat.2015.07.032
18. H.Y. Chuang, T.L. Yang, M.S. Kuo, et al., Critical concerns in soldering reactions arising from space confinement in 3-D IC packages. IEEE Trans. Device Mater. Reliab. **12**, 233–240 (2012). https://doi.org/10.1109/TDMR.2012.2185239
19. T. D'Hondt, S.F. Corbin, Thermal analysis of the compositional shift in a transient liquid phase during sintering of a ternary Cu-Sn-Bi powder mixture. Metall. Mater. Trans. A **37**, 217–224 (2006). https://doi.org/10.1007/s11661-006-0166-z
20. J. Doesburg, D.G. Ivey, Microstructure and preferred orientation of Au-Sn alloy plated deposits. Mater. Sci. Eng. B Solid-State Mater. Adv. Technol. **78**, 44–52 (2000). https://doi.org/10.1016/S0921-5107(00)00515-8
21. C. Ehrhardt, M. Hutter, H. Oppermann, K. Lang, A lead free joining technology for high temperature interconnects using transient liquid phase soldering (TLPS), in *Electronics Components & Technology Conference*, 2014, pp. 1321–1327
22. R.J. Fields, S.R. Low, G.K. Lucey, Physical and mechanical properties of intermetallic compounds commonly found in solder joints, in *The Metal Science of Joining*, (TMS, Cincinnati, 1991), pp. 165–174
23. J. Flanagan, E. Anderson, H. Bae et al., Low temperature lead-free assembly via transient liquid phase sintering, in *IPC APEX EXPO*, San Diego, 2012
24. D.R. Frear, Issues related to the implementation of Pb-free electronic solders in consumer electronics. Lead-Free Electron. Solder A Spec. Issue J. Mater. Sci. Mater. Electron. 319–330 (2006). https://doi.org/10.1007/978-0-387-48433-4_21
25. A. Garnier, C. Gremion, R. Franiatte et al., Investigation of copper-tin transient liquid phase bonding reliability for 3D integration, in *Proceedings – Electronic Components and Technology Conference*, 2013, pp. 2151–2156
26. H. Greve, L.Y. Chen, I. Fox, F.P. McCluskey, Transient liquid phase sintered attach for power electronics. Proc. Electron Compon. Technol. Conf., 435–440 (2013). https://doi.org/10.1109/ECTC.2013.6575608
27. H. Greve, S.A. Moeini, F.P. Mccluskey, Reliability of paste based transient liquid phase sintered interconnects, in *Proceedings – Electronic Components & Technology Conference*, 2014, pp. 1314–1320
28. J. Harris, M. Matthews, Selecting die attach technology for high- power applications. Power Electron Tech., (2009) https://www.powerelectronics.com/dc-dc-converters/selecting-die-attach-technology-high-power-applications. Accessed 1 June 2017
29. M. He, A. Kumar, P.T. Yeo, et al., Interfacial reaction between Sn-rich solders and Ni-based metallization. Thin Solid Films **462–463**, 387–394 (2004). https://doi.org/10.1016/j.tsf.2004.05.062
30. T.C. Illingworth, I.O. Golosnoy, T.W. Clyne, *Modelling of Transient Liquid Phase Bonding in Binary Systems-A New Parametric Study* (Technische Universiteit Eindhoven, Eindhoven, 2007)
31. IPC the ACEI, *IPC-4552 Amendment 1 Specification for Electroless Nickel/Immersion Gold (ENIG) Plating for Printed Circuit Boards* (IPC, Bannockburn, 2012)
32. D.G. Ivey, Microstructural characterization of Au/Sn solder for packaging in optoelectronic applications. Micron **29**, 281–287 (1998). https://doi.org/10.1016/S0968-4328(97)00057-7
33. J.W. Jang, D.R. Frear, T.Y. Lee, K.N. Tu, Morphology of interfacial reaction between lead-free solders and electroless Ni–P under bump metallization. J. Appl. Phys. **88**, 6359 (2000). https://doi.org/10.1063/1.1321787
34. M.J. Kammer, A. Muza, J. Snyder, et al., Optimization of Cu – Ag core – shell solderless interconnect paste technology. IEEE Trans. Compon. Packag. Manuf. Technol **5**, 910–920 (2015)
35. W.K.W. Kim, Q.W.Q. Wang, K.J.K. Jung, et al., Application of Au-Sn eutectic bonding in hermetic RF MEMS wafer level packaging. 9th Int. Symp. Adv. Packag. Mater. Process Prop.

Interfaces (IEEE Cat No04TH8742) 2004 Proc. **35**, 215–219 (2004). https://doi.org/10.1007/BF02690529
36. T.M. Korhonen, P. Su, S.J. Hong, et al., Reactions of lead-free solders with CuNi metallizations. J. Electron. Mater. **29**, 1194–1199 (2000). https://doi.org/10.1007/s11664-000-0012-9
37. S. Kumar, C.A. Handwerker, M.A. Dayananda, Intrinsic and interdiffusion in Cu-Sn system. J. Phase Equilibria Diffus **32**, 309–319 (2011). https://doi.org/10.1007/s11669-011-9907-9
38. T. Laurila, V. Vuorinen, J.K. Kivilahti, Interfacial reactions between lead-free solders and common base materials. Mater. Sci. Eng. R Rep. **49**, 1–60 (2005). https://doi.org/10.1016/j.mser.2005.03.001
39. C.C. Lee, C.Y. Wang, G. Matijasevic, Au-In bonding below the eutectic temperature. IEEE Trans. Compon. Hybrids Manuf. Technol. **16**, 311–316 (1993). https://doi.org/10.1109/33.232058
40. J.F. Li, P.A. Agyakwa, C.M. Johnson, Interfacial reaction in Cu/Sn/Cu system during the transient liquid phase soldering process. Acta Mater. **59**, 1198–1211 (2011). https://doi.org/10.1016/j.actamat.2010.10.053
41. J.F. Li, P.A. Agyakwa, C.M. Johnson, Kinetics of Ag3Sn growth in Ag-Sn-Ag system during transient liquid phase soldering process. Acta Mater. **58**, 3429–3443 (2010). https://doi.org/10.1016/j.actamat.2010.02.018
42. C.-H. Lin, S.-W. Chen, C.-H. Wang, Phase equilibria and solidification properties of Sn-Cu-Ni alloys. J. Electron. Mater. **31**, 907–915 (2002). https://doi.org/10.1007/s11664-002-0182-8
43. Y.C. Lin, T.Y. Shih, S.K. Tien, J.G. Duh, Suppressing Ni-Sn-P growth in SnAgCu/Ni-P solder joints. Scr. Mater. **56**, 49–52 (2007). https://doi.org/10.1016/j.scriptamat.2006.08.062
44. A. Lis, M.S. Park, R. Arroyave, C. Leinenbach, Early stage growth characteristics of Ag3Sn intermetallic compounds during solid–solid and solid–liquid reactions in the Ag–Sn interlayer system: Experiments and simulations. J. Alloys Compd. **617**, 763–773 (2014). https://doi.org/10.1016/j.jallcom.2014.08.082
45. H. Liu, K. Wang, K.E. Aasmundtveit, N. Hoivik, Intermetallic compound formation mechanisms for Cu-Sn solid–liquid interdiffusion bonding. J. Electron. Mater. **41**, 2453–2462 (2012). https://doi.org/10.1007/s11664-012-2060-3
46. Y. Liu, S. Liu, C. Zhang, et al., Thermodynamic assessment of the Bi–Ni and Bi–Ni–X (X = Ag, Cu) systems. J. Electron. Mater. **45**, 1041–1056 (2016). https://doi.org/10.1007/s11664-015-4272-9
47. T.-T. Luu, A. Duan, K.E. Aasmundtveit, N. Hoivik, Optimized Cu-Sn wafer-level bonding using intermetallic phase characterization. J. Electron. Mater. **42**, 3582–3592 (2013). https://doi.org/10.1007/s11664-013-2711-z
48. W. MacDonald, T. Eagar, Transient liquid phase bonding. Annu. Rev. Mater. **22**, 23–46 (1992). https://doi.org/10.1146/annurev.ms.22.080192.000323
49. V.R. Manikam, C. Kuan Yew, Die attach materials for high temperature applications: A review. Compon. Packag. Manuf. Technol. IEEE Trans. **1**, 457–478 (2011). https://doi.org/10.1109/tcpmt.2010.2100432
50. P. McCluskey, H. Greve, Transient liquid phase sintered joints for wide bandgap power electronics packaging, in *Pan Pacific Conference Proceedings*, 2014, pp. 1–10
51. S. Menon, E. George, M. Osterman, M. Pecht, High lead solder (over 85%) solder in the electronics industry: RoHS exemptions and alternatives. J. Mater. Sci. Mater. Electron. **26**, 4021–4030 (2015). https://doi.org/10.1007/s10854-015-2940-4
52. B. Meschi Amoli, A. Hu, N.Y. Zhou, B. Zhao, Recent progresses on hybrid micro-nano filler systems for electrically conductive adhesives (ECAs) applications. J. Mater. Sci. Mater. Electron. **26**, 4730–4745 (2015). https://doi.org/10.1007/s10854-015-3016-1
53. A.S. Moeini, H. Greve, P.F. McCluskey, Reliability and failure analysis of Cu-Sn transient liquid phase sintered (TLPS) joints under power cycling loads. WiPDA 2015 – 3rd IEEE Work Wide Bandgap Power Devices Appl., 383–389 (2015). https://doi.org/10.1109/WiPDA.2015.7369306

54. K.-W. Moon, W.J. Boettinger, U.R. Kattner, et al., Experimental and thermodynamic assessment of Sn-Ag-Cu solder alloys. J. Electron. Mater. **29**, 1122–1136 (2000). https://doi.org/10.1007/s11664-000-0003-x
55. NIST, NIST phase diagrams & computation thermodynamics: Pb-Sn system (2002), https://www.metallurgy.nist.gov/phase/solder/pbsn.html. Accessed 1 June 2017
56. NIST, NIST phase diagrams & computation thermodynamics: Cu-Sn system (2002), https://www.metallurgy.nist.gov/phase/solder/cusn.html. Accessed 1 June 2017
57. NIST, NIST phase diagrams & computation thermodynamics: Ag-Cu-Sn system (2002), https://www.metallurgy.nist.gov/phase/solder/agcusn.html. Accessed 1 June 2017
58. NIST, NIST phase diagrams & computation thermodynamics: Bi-Cu-Sn system (2002), https://www.metallurgy.nist.gov/phase/solder/bicusn.html. Accessed 1 June 2017
59. NIST, NIST phase diagrams & computation thermodynamics: Ag-Sn system (2002), https://www.metallurgy.nist.gov/phase/solder/agsn.html. Accessed 1 June 2017
60. NIST, NIST phase diagrams & computation thermodynamics: Ag-Bi-Sn system (2002), https://www.metallurgy.nist.gov/phase/solder/agbisn.html. Accessed 1 June 2017
61. K. Nogita, T. Nishimura, Nickel-stabilized hexagonal (Cu, Ni)6Sn5 in Sn-Cu-Ni lead-free solder alloys. Scr. Mater. **59**, 191–194 (2008). https://doi.org/10.1016/j.scriptamat.2008.03.002
62. H. Ohtani, I. Satoh, M. Miyashita, K. Ishida, Thermodynamic analysis of the Sn-Ag-Bi ternary phase diagram. Mater. Trans. **42**, 722–731 (2001)
63. Ormet Circuits I, Ormet family of sintering pastes (2012), https://www.ormetcircuits.com/d/parts/parts.php. Accessed 1 June 2017
64. S.A. Paknejad, S.H. Mannan, Review of silver nanoparticle based die attach materials for high power/temperature applications. Microelectron. Reliab. **70**, 1–11 (2017). https://doi.org/10.1016/j.microrel.2017.01.010
65. B. Pan, C.K. Yeo, Transient liquid phase sintering (TLPS) conductive adhesives for high temperature automotive applications. SAE Int. J. Mater. Manuf. **7**, 320–327 (2014). https://doi.org/10.4271/2014-01-0797
66. M.S. Park, S.L. Gibbons, R. Arroyave, Prediction of processing maps for transient liquid phase diffusion bonding of Cu/Sn/Cu joints in microelectronics packaging. Microelectron. Reliab. **54**, 1401–1411 (2014). https://doi.org/10.1016/j.microrel.2014.02.023
67. M.S. Park, S.L. Gibbons, R. Arróyave, Phase-field simulations of intermetallic compound growth in Cu/Sn/Cu sandwich structure under transient liquid phase bonding conditions. Acta Mater. **60**, 6278–6287 (2012). https://doi.org/10.1016/j.actamat.2012.07.063
68. Parliament E, Directive 2011/65/EU of the European Parliament and of the Council on the Restriction of the Use of Certain Hazardous Substances in Electrical and Electronic Equipment (RoHS) – Recast, 2011
69. Parliament E, Directive 2012/19/EU of the European Parliament and the Council on Waste Electrical and Electronic Equipment (WEEE) – Recast, 2012
70. A. Paul, *The Kirkendall Effect in Solid State Diffusion* (Technische Universiteit Eindhoven, Eindhoven, 2004)
71. K.J. Puttlitz, K.A. Stalter, *Handbook of Lead-Free Solder Technology for Microelectronic Assemblies* (Marcel Dekker, Inc., New York, 2004)
72. K.N. Reeve, J.R. Holaday, S.M. Choquette, et al., Advances in Pb-free solder microstructure control and interconnect design. J. Phase Equilibria Diffus **37**, 369–386 (2016). https://doi.org/10.1007/s11669-016-0476-9
73. J. Roman, T. Eagar, Low stress die attach by low temperature transient liquid phase bonding. Int. Soc. Hybrid Microelectron. Symp. Proc. (1992). https://doi.org/10.1007/978-3-642-05463-1
74. C. Schmetterer, J. Vizdal, A. Kroupa, et al., The ni-rich part of the ni-P-Sn system: Isothermal sections. J. Electron. Mater. **38**, 2275–2300 (2009). https://doi.org/10.1007/s11664-009-0854-8
75. H. Shao, A. Wu, Y. Bao, et al., Microstructure characterization and mechanical behavior for Ag3Sn joint produced by foil-based TLP bonding in air atmosphere. Mater. Sci. Eng. A **680**, 221–231 (2017). https://doi.org/10.1016/j.msea.2016.10.092

76. K.S. Siow, Are sintered silver joints ready for use as interconnect material in microelectronic packaging? J. Electron. Mater. **43**, 947–961 (2014). https://doi.org/10.1007/s11664-013-2967-3
77. K. Suganuma, S.J. Kim, K.S. Kim, High-temperature lead-free solders: Properties and possibilities. JOM **61**, 64–71 (2009). https://doi.org/10.1007/s11837-009-0013-y
78. W.J. Tomlinson, H.G. Rhodes, Kinetics of intermetallic compound growth between nickel, electroless, Ni-P, electroless Ni-B and tin at 453 to 493 K. J. Mater. Sci. **22**, 1769–1772 (1987). https://doi.org/10.1007/BF01132405
79. K.N. Tu, A.M. Gusak, M. Li, Physics and materials challenges for lead-free solders. J. Appl. Phys. **93**, 1335–1353 (2003). https://doi.org/10.1063/1.1517165
80. V. Vuorinen, H. Yu, T. Laurila, J.K. Kivilahti, Formation of intermetallic compounds between liquid Sn and various CuNi x metallizations. J. Electron. Mater. **37**, 792–805 (2008). https://doi.org/10.1007/s11664-008-0411-x
81. H.P. Wu, X.J. Wu, M.Y. Ge, et al., Effect analysis of filler sizes on percolation threshold of isotropical conductive adhesives. Compos. Sci. Technol. **67**, 1116–1120 (2007). https://doi.org/10.1016/j.compscitech.2006.05.017
82. G. Zeng, S. McDonald, K. Nogita, Development of high-temperature solders: Review. Microelectron. Reliab. **52**, 1306–1322 (2012). https://doi.org/10.1016/j.microrel.2012.02.018
83. H. Zhang, N. Lee, Reliability of BiAgX as a drop-in solution for high temperature lead-free die-attach applications. J. Surf. Mt. Technol. **26**, 28–32 (2013)
84. L. Deillon, T. Hessler, A. Hessler-Wyser, M. Rappaz, Growth of intermetallic compounds in the Au-In system: Experimental study and 1-D modelling. Acta Materialia **79**, 258–267 (2014)

Die-Attach Materials for Extreme Conditions and Harsh Environments

Z. Shen and O. Fanini

1 Introduction

The demand for increased energy supplies and more electric vehicles has prompted many industries to call for electronics that can work in harsh environments such as extreme high temperature. High-temperature electronics are currently playing a very important role, as evidenced by a continually growing demand for these components in various fields. High-temperature electronic product surveys, conducted by research and development organizations across multiple industries listed below, reveal the typical maximum operating environment temperature requirements for these products shown in parenthesis [1]. The industries included in the data collection survey showed the following results: oil and gas logging while drilling (175 °C), oil and gas directional drilling (175–200 °C), oil and gas surface wireline logging (200 °C), aviation-distributed engine control (225 °C), aviation power and actuator control electronics (350 °C), oil and gas subsurface reservoir monitoring (175 °C), geothermal drilling and subsurface monitoring (USA 300 °C; Europe 250 °C), automotive-distributed electronics (150 °C), automotive electrical power systems and engine compartment (300 °C), nuclear power plant-distributed sensors (300–500 °C), nuclear power plant controls (350 °C), nuclear power plant control room (125 °C), steel mill sensors (1800 °C), and cement plant kiln sensors (1550 °C).

An overview of high-temperature solders is provided in Table 1 to promote a broader understanding of lead-based and lead-free high-temperature attachments currently utilized in electronic packaging [2]. Table 1 shows the main high-temperature solder alloy systems used in the industry with their respective solidus and liquidus temperature characteristics determining their temperature utilization

Z. Shen · O. Fanini (✉)
Baker Hughes & GE Company, Houston, TX, USA
e-mail: Zhenzhen.Shen@bhge.com; otto.fanini@bhge.com

K. S. Siow (ed.), *Die-Attach Materials for High Temperature Applications in Microelectronics Packaging*, https://doi.org/10.1007/978-3-319-99256-3_10

Table 1 High-temperature (HT) solder alloy systems summary

HT solder alloys	Composition (wt%)	Solidus temperature (°C)	Liquidus temperature (°C)
Au alloy system			
Au-Sn	Au-20Sn	280 (eutectic)	
Au-Si	Au-3.15Si	363 (eutectic)	
Au-Ge	Au-12Ge	356 (eutectic)	
Bi alloy system			
Bi-Ag	Bi-2.5Ag	263 (eutectic)	
Bi-Ag	Bi-11Ag	262	360
Cd alloy system			
Cd-Zn	Cd-16.6Zn-5Ag	249	316
Cu alloy system			
Cu-Sn	Sn-(1–4)Cu	227	~400
Cu-Sn	Sn-Cu (particles compound)	~230	
High wt% Pb alloy systems			
Pb-Ag	Pb-1.5Ag-1Sn	309	309
Pb-Ag	Pb-2.5Ag	304	304
Pb-In	Pb-5In	292	314
Pb-In	Pb-25In	264	250
Pb-Sb	Pb-4Sb	252	299
Pb-Sb	Pb-15Sb-10Sn	239	270
Pb-Sb	Pb-24Sb-12Sn	239	330
Pb-Sn	Pb-2Sn	316	322
Pb-Sn	Pb-5Sn	300	314
Pb-Sn	Pb-10Sn	268	301
Pb-Sn	Pb-20Sn	183	279
Pb-Sn	Pb-30Sn	183	258
Pb-Sn	Pb-35Sn	183	248
Sn-Sb alloy system			
Sn-Sb	Sn-5Sb	237	240
Sn-Sb	Sn-8.5Sb	241	248
Sn-Sb	Sn-10Sb	243	257
Sn-Sb	Sn-25Ag-10Sb (J-alloy)	233 (eutectic)	
Zn alloy systems			
Zn-Al	Zn-(4–6)Al(-Ga, Ge, Mg, Cu)	300 ~ 340	
Zn-Sn	Zn-(10–30)Sn	199	360

ranges. Currently, solder development efforts target the broad commercial availability of lead (Pb)-free high-temperature solder solutions driven by international environmental regulatory rules. Electronic processing cost reductions also develop methods and materials that can reduce or simplify processing steps, reduce temperature and pressure process requirements, and provide structural and chemical stability to support higher stress and aging processing steps in more complex packaging solutions.

Traditional designs for electronic devices include an additional active or passive cooling system. However, this approach may not be practical in some applications. Consequently, it can be more appealing to operate the electronic devices hot, thereby improving system reliability and reducing the cost. This choice leads to challenges in many aspects, especially packaging techniques. The primary functions of electronic packaging are to protect the semiconductor devices and other circuit components and to provide a means of electrical connection and a path for heat removal. High-temperature electronic packaging solutions exhibit exponentially improved reliability levels when operated in lower temperatures. These solutions also enable accelerated higher-temperature aging testing to determine specific lifetime and faster failure modes in a shorter period of time, improving overall development schedules. New assembly and packaging technologies have been developed for high-power and high-temperature applications. Although multiple packaging approaches are available, this chapter discusses the multi-chip module (MCM). In contrast to single-chip packaging, an MCM integrates multi-bare dies and passives within one package, increasing the module reliability by reducing the number of interfaces. In addition, the high density of an MCM enables further miniaturization of microelectronic packaging.

Two primary attachments exist in electronic packaging. A die attachment joins the die to the package or substrate, and a substrate attachment joins the substrate to the package. Figure 1 illustrates the structure of the die attachment in multi-chip module. Attachment materials enable basic solutions for electronic packaging problems by providing a mechanical structural joint, an electrical connection, and a thermal path for removing heat through the substrate and underlying structure. High-temperature packaging often encounters its most challenging tasks in the die and substrate attachments. The die-package, die-substrate, or substrate-package interfaces can develop stresses due to differences in the coefficient of thermal expansion (CTE) as well as mechanical force loading. These stresses can compromise the packaging structural integrity, reliability, and durability. Wider temperature cycle ranges in high-temperature applications result in higher cyclic stresses and lead to fatigue failure, also compromising packaging assembly reliability and durability.

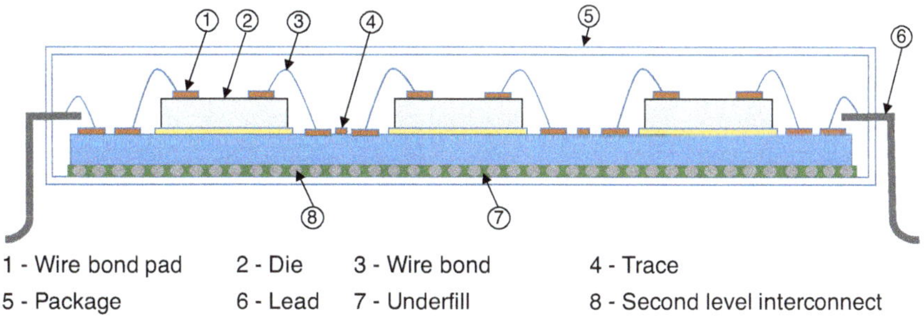

Fig. 1 Multi-chip module die-attachment assembly

Table 2 High-temperature die-attach material properties

Die-attach material	Melting temperature T_m (°C)	Maximum operating temperature T_{max} (°C)	Thermal conductivity K (W/mK)	Coefficient thermal expansion CTE (10^{-6}/°C)	Shear modulus G (Gpa)
Au88Ge12	356		52	12	
Au80Sn20	280		58	16	68
Epoxy compounds		300	3.5	26	
Silver filled glass	400	300	80	16	11.5
Thermally conductive tape		250	0.6	500	0.1
Silver (Ag) nano powder		500	240	19	9
SnPb37	183		51	25	12
Sn96.5 Ag3.5	221		78	22	
Cu-Sn TLP	>415		34–70.4		
Silver sintering (Nano scale)	961		100–240	18–23	

Table 3 Material properties of high-temperature interconnect materials

Material	Width/diameter (μm)	Melting point (°C)	Electrical conductivity 10^5 (Ω cm)$^{-1}$	Young's modulus (GPa)	CTE (ppm/K)
Al	25–300	660	2.3–2.8	68	24
Au	25–250	1064	4.5	73	14.2
Ag	25	961	6.3	76	19
Pt	250	1769	0.94	157	8.8

Figure 1 illustrates some of the important components of electronic packaging and assembly: die attach, interconnects (wire bond, bumps, etc.), substrates, and metallization layers (e.g., layer stacks for adhesion, contact, cap, diffusion barrier).

A summary of relevant high-temperature die-attach packaging materials' physical and mechanical properties (e.g., thermal conductivity, thermal expansion coefficient, relevant and critical to electronic packaging manufacturing process solution developments) is shown in Table 2. The material properties for some of the commonly used high-temperature interconnect materials are shown in Table 3. Substrates are an important element of microelectronic packaging, and the material properties for some of the high-temperature ceramic substrate materials used in the industry are shown in Table 4. The electronic package functionality, durability, and reliability often depend on contacting surfaces' adhesion, electrical and mechanical characteristics, electrical contact quality, and the material diffusion control during the manufacturing thermal profile and the operational lifetime. The structural and

Table 4 Material properties of high-temperature ceramic substrate materials

Material	Density (kg/m^3)	Elaastic modulus (Gpa)	Thermal conductivity (W/mK)	CTE (ppp/K)	Dielectric strength (kV/mm)
Al_2O_3 (96%)	3970	310	24	6	12
AlN	3260	345	150–180	4.6	15
BeO	3000	314	270	7	12
Si_3N_4	2400		70	3	10

Table 5 High-temperature metallization layer stacks (adhesion, contact, cap, diffusion barrier) and corresponding thicknesses

Metallization stacks	Thickness (μm)
Ti/Cu/Ni	0.2/0.8/10
Ti/Cu	0.5/0.3
Ti/W/W-AuSn	0.1/0.05–0.2/0.2
Ti/Mo/Pt	0.08/0.1/1
Ti/Pt	0.03/0.25
Ni/Au	0.3/1.0
Ni/Au	2.5/1.3–7.6
Ni/Au	5/1

material distribution stability often depends on intermediary metallization layers placed with this specific objective. Some of the high-temperature metallization layer stacks (adhesion, contact, cap, and diffusion barrier) utilized with the corresponding thicknesses are shown in Table 5.

In general, die and substrate attachment materials can be classified into five major groups: (i) solders and brazes, (ii) polymers, (iii) liquid-phase bonding, (iv) Ag-glass, and (v) sintered nanoparticles. This chapter discusses solders and polymer attachment materials and methods. Hard solders like AuSn (gold-tin), ZnAl (zinc-aluminum), AuGe (gold-germanium), AuSi (gold-silicon), and AgBi (silver-bismuth), as well as soft solders like tin-silver-copper/Sn-Ag-Cu (SAC) solder alloy compositions, high melting point (HMP), high-Pb content alloys, PbSn (lead-tin), and PbAg, are important for high-temperature microelectronic packaging and assembly processes to achieve high-temperature reliability and durability [3–5]. Polymer-based attachments include sintered Ag adhesives and cyanate ester. Sintered Ag die attachments are also included in the discussion on their high-temperature reliability.

2 Attachment Solders

Currently, two prevailing material types are used for die attachment in commercial packaging and multi-chip modules: solder alloys and conductive epoxies. However, both material types are seldom suitable for high-reliability and high-temperature applications. The Ag-filled epoxies used in commercial devices typically fail at temperatures near 200 °C. Metal-filled conductive epoxies have inferior electrical and thermal properties. Commonly available silver-tin (AgSn) solder alloys are also not recommended solutions for high-temperature applications due to their relatively low melting points. High-Pb (lead) content soft solders, like Pb95Sn5, could be used for high-temperature applications, but the implementation of the Restriction of Hazardous Substances (RoHS) directives has significantly increased the number of restrictions on products containing lead. Solder alloy systems like AuSn, AuGe, ZnAl, ZnSn (zinc-tin), BiAg, and SnSb (tin-antimony) are potential lead-free and RoHS-compliant replacements for PbSn (lead-tin) and high-Pb content solders and attachments for high-temperature electronic applications. Hard solders from the gold series alloys (Sn, Ge) have been used for ceramic board to heat sink junctions, vapor deposition on chip backsides, composite semiconductors die bonding, and ceramic packages hermetic seals. SnSb (Sn-10wt%Sb) lead-free solder pastes can be produced, exhibiting variable melting points (VMPs) with proper choice of flux composition and tin powder size. A VMP is achieved limiting re-melting during the reheating process steps, reducing the manufacturing temperatures and costs [6–11].

Five types of high-temperature die-attachment materials with high melting points (>250 °C) are also worth mentioning: high-lead (Pb95.5Sn2Ag2.5) solder paste, gold-tin (Au80Sn20) solder paste, pressure-less Ag sintered paste, pressure-assisted sintered Ag paste, and gold-germanium (Au88Ge12). Electroless nickel/immersion gold (ENIG) is commonly used for printed circuit boards (PCB) that require soldering and structural mechanical contacting. For high-temperature applications, electroless nickel/electroless palladium/immersion gold (ENEPIG) is used to mitigate the diffusion between nickel (Ni) and gold (Au) under prolonged temperature exposure. ENEPIG has excellent solderability for most solders and forms high-reliability wire bonds [12].

Parameters and requirements affecting high-temperature die-attachment development and performance come from three areas: process-related (bonding temperature, bonding temperature, conditioning environment), material-related (melting point, electrical conductivity, thermal conductivity, high-temperature chemical and mechanical structural cycling stability), and assembly-related (bending stress, die shear strength, active power cycling capability, passive power cycling stability, crack rates of stressed components, and die-attach layers). A small number of established bonding technologies can fulfill the functional and reliability requirements of high-power applications. One commonly used technology in high-reliability silicon die-attachment applications is eutectic silicon gold (AuSi) die bonding. The bond is achieved when the gold (Au) surface is scrubbed against silicon (Si) under temperature and pressure. A small amount of Si mixes with the Au, creating a eutectic AuSi composition (Au97.15/Si2.85) with a melting temperature of 370 °C. The AuSi eutectic, though, cannot be used for standard SiC die attachment because its lattice structure does not enable Si to mix with Au.

Silicon carbide (SiC) device assemblies can utilize hard solder gold alloys because they have a melting temperature above 250 °C. Good solder candidates for high-temperature applications are eutectic alloys tin-gold (Au80Sn20; 280 °C) and germanium-gold (Au88Ge12; 356 °C). These solders are characterized by a fair processing window, good wetting on Au pads metallization, high thermal conductivity, and high-strength solder joints. These eutectic compositions transform between solid and liquid phases at a specific temperature without going through two-phase equilibrium.

A reliable, high-temperature die attachment using eutectic gold-germanium (Au88Ge12) on a copper surface was evaluated [13]. A slight decrease in the shear strength values with aging (over 1000 h at 250 °C and 200 °C) was observed. Alternatively, aluminum zinc (ZnAl5) eutectic solder showed negligible microstructural evolution, resulting in a strong decrease of the shear strength with time. Because of rapid oxidation, a significant scientific effort seems to be required for the development of a reasonable application of this base alloy [13].

Some hard solders also demonstrate a good potential for use in harsh environments. Thick-film Ag and silver palladium (AgPd) conductors are selected as the economical options for substrate materials. One reliability concern with the use of Ag-based material is Ag migration under the presence of electrical potential and moisture. The use of proper encapsulation or coating can mitigate the exposed Ag and lower the risk of Ag migration. Another migration concern is the wire bonding reliability with aluminum (Al) wire [14]. Silver bismuth (AgBi) on an Ag-based thick-film substrate has been studied [15]. The coating showed good high-temperature storage and thermal cycling reliability with SiC die attachment. The test results showed stabilized shear strength from 100 to 2000 h at 200 °C in air. This study showed a cross section of the stabilized intermetallic (IMC) layer of AgSn through aging with scanning electron microscope (SEM) image sections marked with elemental analysis (at%) results of the Ag metallization after 2000 h of high-temperature storage.

Similar results were found in the thermal cycle samples. However, AgBi solder is not recommended for attachment of nickel-tin (NiSn)-based surface-mount (SMT) components, because the excess Sn from the terminals results in increased Sn intermetallics that decrease the reliability in high-temperature applications (>200 °C).

Increased research and demand exist for alternative solders to replace high-lead, high-temperature solders, although they are one of the most commonly used solders for high-temperature, high-reliability environments. Lead-free solders and other recommended joining materials have the following desirable properties as follows:

- Stable microstructure during high-temperature aging
- Less brittle IMC in the joint
- Cost-efficient
- Good wettability
- Compliant mechanical properties of high ductility and fatigue resistance
- Good thermal and electrical conductivity

Typical HMP solder used in harsh environments are Pb-5Sn, Pb-10Sn, and Pb-5Sn2.5Ag (wt%). The study of the effect of Sn concentration on the joint reliability has been reviewed in high-temperature solder development efforts. The concentration of Sn strongly influences the interface IMC growth. The metallization on the substrate side is Cu, and the microstructure evolution at 350 °C shows a spalling process at 20 min in low-Sn solder, while this process continues over 600 min in high-Sn solder. A rich Pb layer forms and continues to grow. The assumption is that Pb from the solder penetrated the Cu3Sn layer through the grain boundary, resulting in a slower spalling rate of Cu3Sn.

Tin-lead (Sn-Pb) solders, also called soft solders, are commercially available with tin concentrations varying between 5% and 70% by weight. Increasing the tin concentration increases the solder's tensile strength and shear strength, and dispensing flow processing capabilities are achieved under various temperature processing profiles. Historically, lead has been widely believed to mitigate the formation of tin whiskers. This element has been removed from commercial processing because of worldwide environmental regulations. Tin/lead (63/37) pastes have been printed at 220 °C for adhesion and interconnection of microelectronic assembly functions. Tin-lead (SnPb) solder with no flux produced good adhesion to gold and clean copper substrates. SnPb-doped solder formulation achieved good print adhesion to aluminum substrates. For example, solder printing has achieved thousands of bumps per substrate layer, 500 bumps per second, approximately 20–80 μm in diameter, and a standard deviation of approximately 8 μm [16].

The thermal conductivity of ZnSn is among the best for high-temperature solders (100 W/mK), and it has excellent ductility and tensile strength [5, 6]. This alloy has excellent mechanical and thermal properties as a die-attachment solder for high-temperature applications. Bulk ZnSn alloys exhibit ultimate tensile strength (UTS) of 65 MPa at room temperature and 30 MPa at 125 °C, which is higher than what was achieved with Pb-5Sn.

There are two major concerns of using ZnSn in high-temperature environments. One concern is the solder's stability during the multiple reflows. Because this alloy has the solidus temperature of 199 °C and liquidus temperature of 360 °C, the die may not hold strongly during the reflow with a peak temperature of 260 °C. Multiple reflows of SAC soldering usually do not result in shift or movement of the dies. X-ray examination performed before and after the reflow indicated the size of the voids and die location had not changed.

The second concern is the corrosion on the ZnSn alloy system under high humidity. When this alloy system was exposed under 85 °C and 85% relative humidity conditions (85C/85RH) for 1000 h, a few micrometers-thick ZnO layer was detected after this exposure, due to the high oxygen affinity. The oxygen affinity of Zn and Sn, combined with low alloy wettability, raise solder capability concerns for ZnAl's industrial use in high-temperature applications, although there is no published data showing the growth of the oxide layer with time and temperature.

Another key factor is the fatigue resistance in thermal cycling. A set of die attachments using An20Sn, Pb5Sn, and Zn30Sn were run side by side under thermal cycling (−40 °C to 125 °C) for 2000 cycles. The die attachment with AuSn showed the highest fatigue resistance; at retained 90% of the initial shear strength at the end of the test. The shear strength of the other two decreased after 500 cycles. The failure analysis indicated the crack formed in the intermetallic layer, reducing the shear strength of Zn30Sn. The Pb5Sn, had a severe crack inside the solder joint, resulting in low shear strength. The difference in the failure modes between the two solders implied that the ZnSn solder thermal cycle reliability can be improved by adding a barrier layer to reduce diffusion. A follow-up study utilizing titanium nitride (TiN) as a thin-film barrier layer on metal stacked on both sides showed dramatically improved thermal cycle reliability.

Research efforts have investigated zinc-aluminum (ZnAl) and tin-antimony (SnSb) alloys as high-Pb alternative for high-temperature applications [5, 6]. ZnAl and SnSb are tin alloy systems that require process steps with high temperature to form electrical interconnections, increasing the damage and degradation risk to other deposited structural and electronic circuit components and material elements used in other packaging process steps. Sn-5Sb (Sn-5wt%Sb) reaches complete melting above its liquidus temperature (245 °C) with limited IMC growth and results in attractive creep and mechanical properties, outperforming known Bi-alloy systems. Progressively increasing the alloy Sb percent weight like Sn-10Sb (Sn-10wt%Sb) and Sn-43Sb (Sn-43wt%Sb) gradually increases the alloy liquidus temperature to approximately 100 °C higher with limited IMC growth. For ZnAl alloy systems, the Al composition ranges from 4% to wt. 6%, and Zn-6Al eutectic, for example, melts at 381 °C. Trace alloying elements like Ga, Ge, Mg, or Cu can be added to the ZnAl alloy, reducing its liquidus temperature between 300 °C and 380 °C. This trace element addition introduces an important, well-known process trade-off between liquidus temperature adjustment and IMC formation. IMCs tend to be significantly brittle, which can compromise the mechanical integrity, durability, and reliability of the packaging attachments.

The AuGe alloy system is an attractive lead-free alternative to PbSn solder and attachment solutions because it does not exhibit intermetallic phases [5, 6]. High-temperature applications have broader operating temperature range requirements, generating higher thermal stresses than can be mitigated with solder attachments with higher ductility. This mitigation can be accomplished with the AuGe alloy system by the addition of Sb to the AuGe eutectic alloy, substantially increasing ductility and reducing its melting temperature. The resulting Au-0.24Ge-0.05Sb alloy example remains an attractive alternative after packaging-aging process, despite the formation of very hard IMCs (AuSb2). Indium (In) can be introduced to the AuGe alloy, enabling the system to form Au-0.18Ge-0.10In which is attractive for high-temperature optoelectronic packaging applications. The requirements list for this applications include high-strength (solid solution strengthening), low-elastic modulus, and microstructure stability at high temperatures, verified with bulk AuGe solder only. Other potential additional interfacial reactions with substrates need more research. Other challenges for the AuGe alloy system industrial adoption are the cost of gold and Ge processing difficulties like electrodeposition and elemental stability and handling during high-temperature processing.

3 Transient Liquid-Phase (TLP) Bonding

Transient liquid-phase (TLP) bonding technology, also known as the solid-liquid interdiffusion process (SLID), can be applied to different metallic systems such as gold-tin (AuSn), silver-tin (AgSn), nickel-tin (NiSn), and indium-gold (InAu), for example [13]. The method uses two metals with different melting points. At a processing temperature, the metal with a lower melting point diffuses into the other metal, forming a new alloy with a higher melting point that is suitable for high-temperature applications—typically above 250 °C. The final TLP-processed joint has a higher operation temperature than the processing temperature. For example, the process enables the use of a low-temperature solder like tin-bismuth (SnBi) in higher-temperature applications (200 °C), which are above their melting point (139 °C).

Three main methods exist for processing TLP. One method uses preforms that require pressure. The other two methods use mixed powder and paste, for which pressure is not necessary. The main advantages of using a TLP bonding process are as follows:

- Lower processing temperature than operation temperature. This is important for temperature-sensitive devices, where too much thermal input can cause material damage.
- Joint properties are similar to the base material properties. In some cases, no interface remains after the bonding process because of the significant diffusion at profile temperatures.
- Homogenized bond region structure.
- Less required surface preparation; flux agent is not needed.

However, TLP bonding is not suitable for all applications. The process can be time-consuming and expensive, depending on the materials. The bonding process involves isothermal solidification due to diffusion and bond homogenization at the bonding process temperature, both of which require high temperatures and long bonding times. In addition, pressure is needed in most cases to achieve a void-free joint. Transient liquid-phase diffusion bonding (TLPDB) is an attachment process applied for bonding ceramic and metallic systems that cannot be bonded by conventional fusion welding techniques that can compromise the attached material characteristics during the attachment process. The TLPDB process has found applications in the aerospace industry (e.g., turbine engines), nuclear plants, and packaging attachments in the microelectronic industry. During this process, diffusion of the melting point reduction element occurs in the interlayer into the two attached substrate lattices. If sufficient diffusion occurs, the ending joint will remain solid and strong—well above the original process melting and diffusion process temperatures suitable for high-temperature applications. Partial transient liquid phase (PTLP) is a TLP process variation mainly used for joining ceramic substrates.

3.1 Transient Liquid Phase (TLP) with Tin

A reliability comparison was made of a TLP joint on a different metal finish with an Ag sintered joint [13]. The Sn film was sandwiched between different metal plating materials (Au, Ni, Ag, Cu) to form TLP joints for reliability tests. The thermal cycling (0–350 °C) tests in this comparison study demonstrated that the TLP attachment methods are more reliable than silver sintering and well-suited for high-temperature power applications. The substrate type also plays a role in overall reliability performance, silicon nitride (Si_3N_4) with brazed Cu and aluminum nitride (AlN); direct-bond-aluminium (DBA) is better than direct-bond-copper (DBC).

3.2 Transient Liquid Phase (TLP) with Gold-Tin (AuSn)

AuSn has been widely used in the TLP bonding process. The AuSn binary phase diagram (Sn atomic percentage or at% versus temperature) is shown in Fig. 2 [17]. Phase diagram charts show conditions when thermodynamically distinct phases (i.e., solidus, liquidus) occur and coexist at equilibrium. These charts describe lines of equilibrium or phase boundaries marking conditions for multiple phases to remain at equilibrium or transition across these lines between one state and the other. Solidus (solid state) and liquidus (liquid state) phase transitions occur

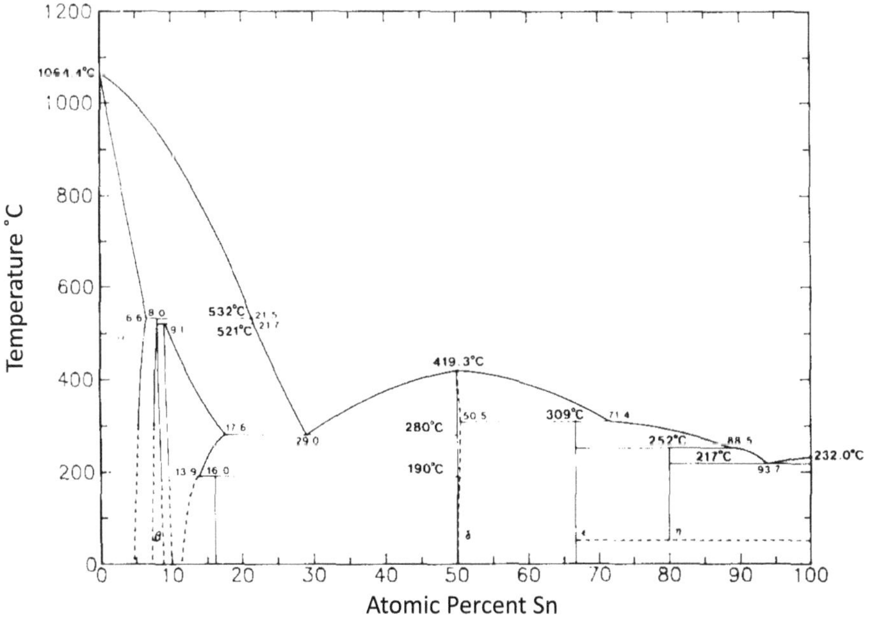

Fig. 2 AuSn binary phase diagram [17]

along these lines of equilibrium as materials' weight percentages, pressure, and temperature conditions vary. Manufacturing solder process development strongly relies on these phase diagrams to obtain desired assembly and packaging structures and characteristics after melting and solidification controlled by a sequence of process temperature profile steps. Key features of the AuSn binary phase diagram are the steep slope of the liquidus lines in the eutectic composition Au-rich side (e.g., 71 at% Au/29 at% Sn; 80Au-20Sn by weight) and the formation of numerous intermetallic compounds at room temperature. The phase diagram assists in determining and quantifying the presence of intermetallic growth (IMC) after the soldering process step, which has a very important effect on the solder joint mechanical characteristics, durability, and reliability in high-temperature applications. A criterion for high-temperature solder selection is the presence of no or little intermetallic formation for the softness and ductility of the alloy.

For example, as the AuSn and a gold-plated terminal come into contact during a soldering process step, the resulting soldering joint will have a higher Au concentration, and its melting temperature moves up along a steep slope in the phase diagram. This relatively small Au concentration change results in a melting point in the phase diagram much higher than 280 °C, due to the diffusion and leaching of the terminal's Au metallization into the solder. This solder joint will then be less disturbed by secondary reflows or heated assembly steps of other components. Experiments and industrial practice have shown AuSn eutectic alloy to be a reliable high-temperature solder despite the formation on intermetallic as indicated by its binary phase diagram.

A reduced-size eutectic AuSn attachment covering approximately 30% of the die area was used to limit the amount of AuSn [15]. The packaging attachment utilized a layered metalized thick-film substrate with a first layer of platinum gold (PtAu) thick-film and a second layer of thick-film Au to provide Au for diffusion with the AuSn. The final Sn concentration was <10 at%, and the die attachment was evaluated at 500 °C. A study evaluated the AuSn TLP bonding on platinum palladium gold (PtPdAu) thick-film substrate, with the top layer of pure thick-film Au [18, 19]. A similar study performed the high-temperature evaluation with single-printed and double-printed PtPdAu layers, with aging in a 500 °C oven. The shear strength did not show a significant difference after 2000 h of aging. The TLP AuSn bonding provides a method to achieve die attachment for applications up to 500 °C. The test results indicated that using a limited volume of eutectic AuSn with the excess Au provided by the die can achieve high bonding reliability.

3.3 *Transient Liquid-Phase Sintering (TLPS)*

A continued demand for chip size and electronic interconnection and assembly miniaturization improvements for higher-temperature applications supported industry development demand trends for new bonding materials to replacing

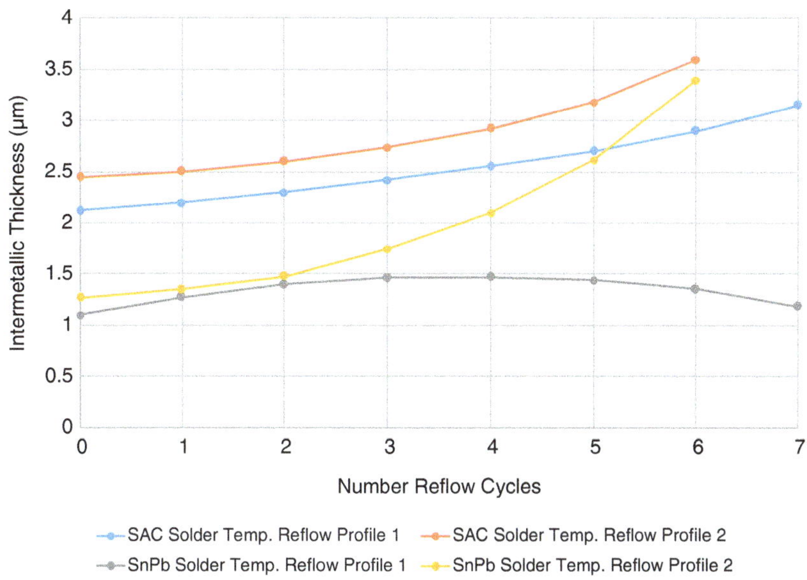

Fig. 3 Intermetallic thickness growth effects with increasing reflow cycles

lead-free solder, offering higher repeated bonding capability along with higher bonded substrate integration levels. Conductive bonding applications have been successfully addressed by solder. Solders have limitations when they are repeatedly subjected to bonding reflow processes that require precision control of bonding positions due to re-melting after previous bonding and connections steps. Tracking reflow effects on the final assembly quality due to the overall packaging process exposure are important. Figure 3 illustrates intermetallic growth effects developed with the increasing number of reflow cycles.

High-temperature TLPS copper and tin-based materials can be bonded at the same process temperatures as solders but do not re-melt after bonding. The temperature-induced processing reaction of the molten tin alloy with copper particles during the sintering forms an intermetallic compound. This compound has a process compatible and convenient high melting temperature (re-melting temperatures >400 °C), preventing it from melting during temperature processing reflows and repeated bonding. TLPS has significant shape and forms stability due to stress-relieving thermoplastic deformation resin. The resin results in low deformation and cracks, low bonding material elasticity, and high-temperature cycle reliability. TLPS offers relatively low-cost, simplified processing for high-temperature engines, and energy conversion machine applications [20].

3.4 Liquid-Phase Diffusion Bonding (LPDB)

Liquid-phase diffusion bonding (LPDB) is accomplished when isothermal solidification interaction of an adjacent high melting temperature (HTP) substrate and a low melting temperature phase (LTP) is used to attach substrates in a ternary copper-nickel-tin (CuNiSn) alloy system, for example [6]. Liquid forms when LTP is heated above its solidus and wets the HTP and adjacent substrate surfaces. This approach illustrates how a metastable reaction can be exploited to create a pseudo-binary LPDB using a CuNiSn LTP alloy and Cu-10Ni wt% binary solid solution HTP. Despite the presence of post-processing unreacted Sn, shear tests of Si dies metallized with Ag bonded to Ni substrates with LPDB showed test measurement levels of 10 MPa at 435 °C, with HTP formed by Cu and Ni particles and at least up to 600 °C with HTP formed by Ni particles only. IMCs form, but they are contained and do not compromised mechanical attachment integrity. Further developments are needed to optimize the CuNiSn PDB system by analyzing the LTP/HTP ratio, HTP particle size distribution and packing, processing temperature profile effects on liquid consumption rate, and volume location and distribution of residual liquid and understanding the resulting volumetric distribution and variation of the mechanical properties of the LPDB attachment structure.

4 Polymer-Based Attachment Materials

Key requirements for microelectronic packaging are die protection, direct electrical interconnections, thermally compliant interconnections, and easy assembly. This section discusses materials and processes for high-temperature die-attachment techniques that have shown technology developments in recent years. Some noteworthy advancements are anisotropic conductive film, high-temperature material jetting process, high-temperature eutectic gold alloys, inorganic Ag-filled glass conductive paste, micro-sized Ag particle-filled organic paste, and Ag nanoparticle sintering.

4.1 Anisotropic Conductive Film (ACF)

Anisotropic conductive film (ACF) enables electronic packaging to achieve desirable goals like simple device structures, light weight, very fine pitch capability, simple manufacturing process, alternate lead-free process, small product size, lower bonding temperatures, and high-temperature application packaging rating [21]. ACF consists of a composite of adhesive matrix and conductive particles that establish thermal and electric conduction between the chip and the substrate. These particles are solid metal or metal-coated polymers, homogeneously dispersed in a thermosetting epoxy-based adhesive matrix. The particles protect the metallic contacts and strengthen joints. The particle volume fraction varies between 0.5% and 5%. A typical ACF has 3.5 μm diameter and a volume concentration of up to $3.5 \times 10^6/mm^3$. During the ACF bonding process, the temperature and pressure cause multiple changes:

- Resin expansion and shrinkage
- Conductive particle changes from spherical to ovoid shapes
- Adhesive melting that forms a viscous fluid

The adhesive fluid distributes the conductive particles, forming a conduction path between the chip bump and the substrate pad. This path reduces the contact resistance. The three-dimensional cross-linkage of the polymer molecules then consolidates the rigidity of the conduction paths. Thermal cycling and ACF resin thermal setting control the extent of the ACF epoxy curing. This degree of curing (85–90%) controls the reliability; adhesion strength; microstructural integrity; physical, electrical and mechanical properties of ACF connections; and attachments. Beyond 90% degree of curing, the ACF has shown an increase in contact resistance. Unlike solder joints, ACF joints form a mechanical contact of conductive particles without a firm metallurgical connection. The ACF matrix components' dissimilar coefficient of thermal expansion (CTE) can lead to conduction gap formation in the ACF joints, especially if improper thermal curing is applied. Another aspect is the formation of a conduction gap formed in the ACF-pad interface rather than in the ACF-bump interface. The Au-coated bump is soft, and it can deform easily. This deformation can prevent formation of the ACF contact gap. In the other extreme is the Cu pad with Ni electroplating. This pad is stiffer and does not yield to temperature-induced structural stresses that can prevent development of an ACF gap contact.

4.2 *Electrically Conductive Adhesive*

Electrically conductive adhesive (ECA) is suitable for controlled temperature assembly requirements when the product contains heat-sensitive devices. ECA avoids the use of soldering processes to establish electrical connections free of flux residues [22]. ECA is generally more expensive than solders, and ECA thermosetting behavior during curing complicates microelectronic assembly placement alignments (pick and place, use of stencils, etc.), application viscosity control, and re-work besides shorter shelf-life and storage requirements. ECA interconnection resistivity changes due to aging were significant for curing temperatures below 200 °C, but were not significant for curing temperatures above 250 °C. ECA curing temperatures ranging between 50 °C and 265 °C resulted in volume resistivities ranging between 50×10^{-5} and 2×10^{-5} ohm-cm.

Electronic packaging and assembly applications have a variety of ECA formulation choices. The ECA interconnections' quality and reliability application-specific requirements drive the optimization of the electronic system using ECA formulation variations, processing condition's controls (e.g., lamination temperature and pressure), and bonding surfaces pre-treatment. Objective application constraints have driven the selection of materials. Processing conditions and treatments can produce packaging ECA interconnections able to resist the rigors of various harsh environmental stresses, e.g., temperature and humidity.

A variety of ECA formulations exist with micro-Ag-filled epoxies, modified with nanoparticles (e.g., nanoparticles of Ag); micro-sized Ag particles, conducting polymers; and low melting particles (LMP) for interconnection applications [22]. The sintering temperature is reduced with the incorporation of nanoparticles into the ECAs. LMP particles melt at low temperature and reduce the inter-particle resistance. The volume resistivity of ECAs decreases with increasing curing temperature. ECAs have been utilized even when processed at temperatures above their commonly accepted maximum rated temperature. Electrical stability improvement of ECA-to-metal joints has been demonstrated by the use of several treatments, prior to bonding of the metal surface to which the ECA is mated. This improvement may be attributed to a greater degree of sintering at elevated temperatures. ECA formulations with conducting polymers have shown good mechanical strength and electrical conductivity.

4.3 Non-conductive Epoxy

High-temperature, non-conductive (NC) epoxy has been developed to withstand continuous exposure to temperatures between −40 °C and 230 °C. Non-conductive epoxy typically exhibits dielectric temperature-dependent characteristics with volume resistivity values of approximately 1.0×10^{12} ohm-cm at 25 °C and typical dielectric constant values between 3.5 and 6 at 1 kHz and 25 °C. Non-conductive epoxy formulation can be customized to produce specialized epoxies with low viscosity, low CTE range (10–125 ppm/°C), room temperature curing, and low-outgassing specs compliance, for example. Potential beneficial characteristics are excellent chemical resistance, physical structural and adhesive strength, thermal shock resistance, impact resistance, and wide operating ranges. Their dielectric strength, also known as breakdown voltage, is measured in volts/mil and can reach typical values of 500 V/mil at room temperature. It is a material widely used in electronic assembly and packaging. When used as a structural matrix material, NC epoxy can be reinforced by fibers like Kevlar, glass, carbon, and boron. Some assembly aspects deserve attention in a fabrication process application consisting of a die with gold bumps bonded with a non-conductive adhesive (NCA) epoxy, for example. Mechanical compression brings all the bumps into contact, thermal curing of the epoxy adhesive undergoes chemical shrinkage, and mechanical compression removal leaves formed bump contacts that cool down to room temperature [23]. The uneven bump contact stress distribution due to the metal contact bumps' height variability, combined with a nonuniform distribution of cured epoxy, can generate initial residual stresses that are further increased by CTE mismatches that can compromise assembly lifetime contact reliability during operating temperature cycles.

Some of the most important applications are in the electronic industry such as cabling, medical electronics, hybrid avionics, solar cells, PCB, hybrid RF microwave, fiber-optic components, electronic components (resistors, capacitors, coil, transformers, optical device packaging, connectors, etc.), medical ultrasound, surface-mount device (SMD) staking, chip underfill, wafer passivation, glob-top encapsulation, etc. This broad range of industrial applications makes non-conductive epoxies a very important assembly and packaging material.

4.4 Cyanate Ester

Cyanate ester (CE)-based resin is a type of polymer that contains a chemical organic compound of the OCN (cyanate) group. Because of its special chemistry structure, it has very high thermal stability and inherent hydrophobicity. CE-based resin has been used widely for high-temperature electronics, as die attachments, circuit boards, and underfill.

There are many advantages of cyanate ester-based resins over epoxy. First, the moisture absorption is relatively low, two to three times less than that of epoxy. However, it also has the unique ability to trap moisture and act like a moisture absorber in the electronic packaging. Cyanate ester has a fully cross-linked polymer structure containing a triazine structure after curing, providing high glass transition temperature up to 290 °C that is suitable for high-temperature applications. Thirdly, the weight loss of CE is significantly less than epoxy, according to study of chemical performance. After curing, the weight loss can be as low as 0.4%, while most of the epoxy typically demonstrates 1.9–4.4% weight loss [24, 25]. Fourthly, it has a very low dielectric constant, making it potentially a good candidate for RF application. Lastly, cyanate ester has a low modulus of approximately 400 MPa, which is in favor for die attachment in electronic packaging, especially for reliability in thermal cycling for large dies. All these benefits that cyanate ester provides make it a valuable material in automotive, aerospace, and downhole applications. There are two major disadvantages of CE. One disadvantage is the outgassing in moisture environments where CO and CO_2 are produced; for hermetic package, the outgassing can produce pressure. Consequently, a bake-out process is necessary before sealing. Another disadvantage is the CTE of CE. Although the CTE of most epoxy can be tuned, the CTE of CE is very difficult to customize. For Si or SiC die attachments, the CTE mismatch can be a major concern for large area attachment, especially for harsh environment applications.

Reliability studies have been conducted to evaluate tensile bond strength and shear strength of CE, Ag-filled epoxy, and AuSi. AuSi was tested and exhibited the highest tensile bond strength, cyanate ester is the second highest, and the test results for Ag-filled epoxy had the lowest. High-temperature reliability tests of die attachment using epoxy and cyanate ester have also been conducted. Die-attachment samples were subjected to 500 thermal cycles (−65 °C to 150 °C) and showed no degradation in die shear performance [26]. In this study, QFP devices were constructed and subsequently exposed to moisture conditioning at 85 °C/85% RH for 72 h and soldering conditioning, two passes at 240 °C for 20 s. Cross-sectional analysis indicated that only packages with cyanate ester die attachments gave the least number of cracks and no die delamination [27]. The reliability of CE is influenced by the curing profile, such as curing temperature, time, and catalyst. These process attributes must be carefully selected to obtain a reliable joint.

Besides die attachment, CE also been used in PCB boards and underfill material. In the most recent research, it also can be used in 3D printing technology as conductive and dielectric.

4.5 *Silver-Glass-Based Material*

Silver-glass (Ag-glass) adhesives have been used for years in the semiconductor industry as an alternative to gold-silicon (AuSi). Typical manufacturing process temperatures range from 400 °C to 450 °C, and the recommended maximum continuous operating temperature is approximately 300 °C. Comparative experimental study conducted at 500 °C has shown Ag-glass micro-contacts have shorter lifetime than nano-Ag attachment contacts. Higher manufacturing process temperatures developed thermal stresses in the package assembly during cool down. This higher processing temperature resulted in more cracks and delamination than nano-Ag in this comparison study. Ag-glass CTE ranging from 16 to 26 ppm/°C did not match well CTE for Si (2.6 ppm/°C) and SiC (3.2 ppm/°C) die CTEs. The stresses on these dies are minimized due to the lower elastic modulus in the Ag-glass material. Experiments have shown an optimal mixture ratio of Ag and glass to be 80/20, respectively, with regard to stability and electrical performance of the contact material. This optimal glass-silver mix material is supplied with 80% flake and 20% glass blended in an organic medium. The mechanical strength of Ag-glass die bond was shown to be stable at 400 °C for approximately 300 h. Ag-glass die attachment is a potential solution for SiC die bonding for application temperatures up to 400 °C and deserves further research investigations [26]. Silver-filled glass (Ag-glass), due to its high thermal conductivity (60 W/mK) and high-temperature tolerance, is a promising candidate for high-temperature and high-power applications [28, 29].

5 Leadframes

To lower microelectronic packaging costs for high-volume and high-temperature applications, leadframes with high-temperature ratings and die-attachment epoxy-dispensing advancements are required. The epoxy-dispensing process became a very common die-attachment process in high-volume, standardized packaging solutions for the lower-temperature semiconductor assembly involving the consumer's electronic market. For example, flat no-leads packages such as quad-flat no-leads (QFN) and dual-flat no-leads (DFN) are limited to use epoxy die-attachment methods because of the temperature processing constraints of the leadframe tape (the tape withstands less than 300 °C). However, epoxy-based dispensing die-attachment methods have significant quality issues, despite a long production history. Packaging process-induced field failures like intermittent epoxy shorting and dispensing volume variability like epoxy overflow or insufficient epoxy have not been corrected with permanent and sustainable corrective actions and solutions. This has resulted in efforts for alternative epoxy-based methods and material developments for higher-temperature applications. Some of the common root causes identified were clogged or deformed nozzles, pressure changes inside the syringe at different content levels, variations of epoxy material mechanical properties (e.g., viscosity, modulus), etc.

A promising alternative to die-attachment methods is wafer back coating (WBC). This methodology involves wafer backside epoxy paste printing, to be at a later step and sawn with the wafer [30]. At the die-attachment machine, the separated die (with epoxy-coated or other die-attachment materials underneath) is then bonded directly to the carrier leadframe, combining and reducing packaging process steps. The WBC method can potentially reduce processing steps and the human and machine setup dependency level that is critical to epoxy handling in the individual die-attachment process. Unfortunately, the WBC material is mostly available only for non-conductive and medium-conductive applications. The high-conductive WBC has limited availability for the semiconductor industry production mode environment, mainly due to material requirements and development challenges.

The main challenge for developing a high-conductive WBC material is the large percent weight content that must be filled with Ag or another conductive filler to achieve the desired high-conductivity level. Such a high-Ag filler content can cause uneven surfaces or interfere with the bonding resin agent, resulting in poor adhesion. Crack lines have been detected propagating from the leadframe surface, through the WBC interface, and then cracking the wafer backside surface. Some WBC breakthrough actions have included optimization of the Ag filler size mixture, catalyst types (accelerate cross-linking), overall process stability in the epoxy manufacturing process and assembly printing, masterbatch robustness, hardener types (promote cross-linking reactions), Ag filler types, molecule chain length, process controls, etc. The use of nanomaterials such as nano-Ag for the wafer backside attachment can potentially improve conductive WBC process performance.

The WBC development and adoption process are an example illustrating the epoxy-based process improvement with attractive solutions for dispensing quality issues, production bottleneck reduction and throughput improvement, competitive cost, and use of maximum die size possible (i.e., zero clearance required at die edge) for better functionality packing capacity. These changes enable important competitive advantages to operate a robust assembly processes, a streamlined process flow with lower costs, and more repeatable production scale processes and reliability performance.

6 Encapsulant and Adhesive Selection Process

The use of multi-chip modules (MCM) is of great importance for electronic products manufactured in medium-to-small volumes. MCM packaging for high-temperature applications is well developed. It has used high-temperature electrical circuit elements that have been proven in other packaging technologies like printed circuit boards, often including naked dies, conductive attachments via interconnections, surface-mount passive components, multiple-layer interconnections, wire bonding, etc. MCM offers advantages like size reduction, lower exposure to external contaminants with hermetic packages, improved signal integrity and better reliability, overall top-level product size reduction, top-level assembly simplification, etc. Thick- and thin-film substrate choices are available for MCM construction based on the complexity, application temperature, and required reliability.

One of the MCM packaging challenges is the encapsulant selection, which typically is a polymer material. MCM encapsulants can be applied primarily by two methods: the glob-top method and the dam (high-viscosity barrier) and fill (lower-viscosity flow) two-step method [31].

Regardless of the encapsulant choice, high-temperature and harsh environment MCM requirements must focus on adhesion between the glob-top and the MCM internally mounted die surface interface and mismatches in the coefficient of thermal expansion (CTE) of contacting dissimilar materials. Compatibility selection of contacting dissimilar materials must also consider the modulus of elasticity (E-modulus) and glass transition temperature (Tg) that can also contribute to delamination complication problems.

Understanding the material properties (e.g., glob-top) and the process parameters that influence the end-product reliability in an application-based harsh environment can be assisted by systematic research and development methods such as design of experiments (DOE) and laboratory metric scoring methods like simple additive weighting (SAW) and the technique for order preference by similarity to ideal solution (TOPSIS).

Laboratory metrics of thermal and mechanical properties must be used for high-temperature and harsh environment material selection and packaging process development [32]. These testing methods have been studied by employing thermo-mechanical analysis (TMA), dynamic mechanical analysis (DMA), differential scanning calorimetry (DSC), energy dispersive X-ray (EDX) spectroscopy, infrared (IR) thermography, digital imaging correlation (DIC), thermogravimetric analysis (TGA), moisture absorption, and shear stress (e.g., die shear) tests. Typically, selection evaluation focuses on the following parameters: lowest CTE mismatch possible of packaging interacting construction materials, high elastic modulus (e.g., 1034 MPa at 260 °C) for various reliability stress tests, good interfacial adhesion at room temperature and elevated temperatures, lowest moisture absorption rate (e.g., 0.11%), and lowest outgassing of volatile materials (e.g., 0.8%).

7 3D Integration Challenge

Electronic products die and interconnection stacking, combined with underfilling methods by capillary-type underfill, are the main conventional microelectronic assembly processes for 3D integration. Dealing with increased 3D electronic complexity requires further integration, assembly process simplification, improved processing throughput, and increased 3D stacking yields to address the growing electronic thin die stack packaging requirements [33, 34]. For example, high-throughput, 3D die stacking adhesive bonding solutions with wafer-level underfill material have been demonstrated with 30 μm-pitch micro-interconnections. Die-to-interposer modules assembled with 3D stacking adhesive bonding schemes have demonstrated their reliability performance at high temperatures.

Micro-jet printing has evolved into a highly cost-effective, flexible, customizable, and precise process. The process provides capabilities that include small geometric construction, high-speed printing, environmentally friendly and versatile microelectronic manufacturing, and packaging of high-temperature-rated application devices. Micro-jet printing has been developed toward additive manufacturing and 3D micro-dispensing technologies delivering attachments (thermoset, thermoplastic, or ultra-violet curing), adhesives, carbon-filled epoxy resistors, isolation and protection layers, dielectrics, optical epoxies, particle-filled systems, solders, traces and vias, fluxes, substrates, 3D structures, and structural layers.

The micro-jet printing or dispensing process can be continuous flow or drop-driven on demand. The droplet dispensing could be driven by static charge, acoustic energy, piezoelectric, and pressure pulses. Materials for high-temperature applications can be dispensed by this method which requires temperature control of the dispensing head, chamber and substrates, substrate cleanliness and wettability, an inert gas chamber to prevent oxidation, droplet fluid stability control, and heat and solvent viscosity control [16].

Manufacturing quality, lifetime durability, and reliability assessment are temperature tested and accelerated primarily with recommended pre-conditioning test steps, thermal and humidity storage test, die shear test, and temperature cycling test. For example, development and testing activities of 3D integration solutions could focus on the following areas: (i) the material evaluation of wafer-level underfill material evaluation, (ii) underfill film die-to-interposer bonding, (iii) die-to-interposer stack durability and reliability characterization, and (iv) micro-interconnection issues failure analysis.

The 3D packaging process development, due to its increased structural complexity, must rely more heavily on additional material and processing modeling of material dispensed and structure formed, lifetime reliability and durability modeling, characterization and qualification test methodology, flow and material curing behavior, surface attachments, and interacting interfaces of dissimilar materials possibly involving different packaging technologies.

Material development and selection, process development with multiple sequential attachment, and dispensing process steps with cumulative and multiple exposures to temperature and dissimilar material interfaces all require characterization and thorough understanding of process and material temperature stability during the processing and operational use of the packaged electronics. The 3D integration of electronic modules and equipment for high-temperature applications and harsh environments is at the frontier of packaging development challenges.

8 Conclusion

Material and packaging processing developments work hand-in-hand toward the advancement of microelectronic packaging advancement for high-temperature and harsh environment applications. Interdependent packaging process and material developments mutually interact by leading, enabling, and justifying each other at each progress step. Historically, the continuous process and materials' improvements have enabled advancements impacting system level costs, manufacturing throughput, miniaturization levels, devices' functionality volume density and integration levels, quality and reliability modeling and characterization, progressively smaller assembly geometry registration and feature sizes, process steps simplification and temperature and pressure process requirements, faster dispensing and mounting rates, higher yields, and longer life with reduction of operational hourly costs. These advancements are constrained by processing costs, optimized process, and materials' capabilities and variability. Materials supply chain aspects, handling and storage, and materials and packaging processing, and optimized operating widows are also important areas of concern.

High-temperature manufacturing processing has been simplified and developed for less-costly processing steps like at lower processing temperature profiles while operating reliably at higher-temperature environments from 150 °C to 300 °C under high acceleration and impact levels in the oil and gas industry with increased lifetime reliability. Reliability and quality aspects of product's lifetime durability performance are important components for the business model success and corresponding commercialization delivery of the ultimate customer experience and realized value in the competitive marketplace environment.

References

1. R. Norman, *First High Temperature Electronics Products Survey 2005*. Sandia National Laboratories, SAND2006-1580, April 2006
2. K. Suganuma, S.-J. Kim, K.-S. Kim, High-temperature lead-free solders: properties and possibilities. J. Miner. Met. Mater. Soc. **61**(1), 64–71 (2009)
3. V.R. Manikam, K.Y. Cheong, Die attach materials for high temperature applications: a review. IEEE Trans. Compon. Packag. Manuf. Technol. **1**(4), 457–478 (2011)

4. P. Zheng, A. Wiggins, R. Johnson, R. Frampton, S. Adam, L. Peltz, Die attach for high temperature electronics packaging, in *International Conference and Exhibition on High Temperature Electronics 2008*, HiTEC, 2008
5. G. Zeng, S. McDonald, K. Nogita, Development of high temperature solders: review. Microelectron. Reliab. **52**, 1306–1322 (2012)
6. I. Anderson, S. Choquette, Pb-free solders and other joining materials for potential replacement of high-Pb hierarchical solders, in *Pan Pacific Microelectronics Symposium (Pan Pacific)*, 2018, Waimea, HI, 5–8 Feb. 2018
7. W. Sabbah, S. Azzopardi, C. Buttay, R. Meuret, E. Woirgard, Study of die attach technologies for high temperature power electronics: silver sintering and gold–germanium alloy. Microelectron. Reliab. **53**, 1617–1621 (2013)
8. V. Chidambaram, J. Hattel, J. Hald, High-temperature lead-free solder alternatives. Microelectron. Eng. **88**(6), 981–989 (2011)
9. J.G. Bai, Z. Zhang, J.N. Calata, T. Lei, G. Lu, Low-temperature sintering of nanoscale silver paste: a lead-free die-attach solution for high-performance and high-temperature electronic packaging, in *International Conference on High Temperature Electronics (HiTEC 2006)*, Santa Fe, NM, 15–18 May 2006
10. C. Hunt, L. Crocker, O. Thomas, M. Wickham, K. Clayton, L. Zou, R. Ashayer-Soltani, A. Longford, *High Temperatures Solder Replacement to Meet RoHS* (National Physical Laboratory, Middlesex, UK, NPL REPORT MAT 64, 2014)
11. Z. Shen, R. Wayne Johnson, E. Snipes, Lead free solder attach for 200°C applications, in *International conference on High Temperature Electronics (HiTEN 2013)*, Oxford, UK, 8–10 July 2013
12. L.C. Wai, S.W. Wei, H.H. Yuan, D.R. MinWoo, High temperature die attach material on ENEPIG surface for high temperature (250DegC/500hour) and temperature cycle (−65 to +150DegC) applications, in *IEEE 16th Electronics Packaging Technology Conference (EPTC)*, 2014, pp. 229–234
13. S. Egelkraut, L. Frey, M. Knoerr, A. Schletz, Evolution of shear strength and microstructure of die bonding technologies for high temperature applications during thermal aging, in *12th Electronics Packaging Technology Conference*, 2010, pp. 660–667
14. Z. Shen, A. Reiderman, High-temperature reliability of wire bonds on thick film, in *International Microelectronics assembly and Package Symposium 2017*, Raleigh, NC, Oct. 2017
15. Z. Shen, K. Fang, R. Wayne Johnson, M.C. Hamilton, Characterization of Bi–Ag–X solder for high temperature SiC die attach. IEEE Trans. Compon. Packag. Manuf. Technol. **4**(11), 1778–1784 (2014)
16. D.J. Hayes, W. Royall Cox, M.E. Grove, Micro-jet printing of polymers and solder for electronics manufacturing. J. Electron. Manuf. **8**(3 & 4), 209–216 (1998)
17. J. Ciulik, M.R. Notis, The Au Sn phase diagram. J. Alloys Compd. **191**(1), 71–78 (1993). (figure 8)
18. D. Hamilton, L. Mills, S. Riches, P. Mawby, Performance and reliability of SiC dies, die attach and substrates for high temperature power applications up to 300°C, in *International Conference and Exhibition on High Temperature Electronics Network (HiTEN)*, Cambridge, UK, 6–8 July 2015
19. R. Ping Hagler, W. Johnson, L.-Y. Chen, SiC die attach metallurgy and processes for applications up to 500°C. IEEE Trans. Compon. Packag. Manuf. Technol. **1**(4), 630–639 (2011)
20. K. Gurth et al., New assembly and interconnects beyond sintering methods, in *International Exhibition & Conference for Power Electronics, Intelligent Motion, Renewable Energy & Energy Market*, 2010, PCIM, pp. 232–237
21. M.J. Rizvi, Y.C. Chan, C. Bailey, H. Lu, A. Sharif, The effect of curing on the performance of ACF bonded chip-on-flex assemblies after thermal ageing. Soldering Surf. Mount Technol. **17** (2), 40–48 (2005)
22. L.J. Matienzo, R.N. Das, F.D. Egitto, Electrically conductive adhesives for electronic packaging and assembly applications. J. Adhes. Sci. Technol. **22**, 853–869 (2008)

23. D. Farley, A. Dasgupta, J.F.J.M. Caers, Characterization of non-conductive adhesives, in *Proceedings of ASME Inter PACK 2005*, San Francisco, CA, 17–22 July 2005
24. J.N. Hay, Chapter 6, in *Chemistry and Technology of Cyanate Ester Resins*, ed. by I. Hamerton (Blackie, New York, 1994), pp. 151–192, (CE1–12) and Appendices A.3 – A.8 (pp. 332–341)
25. I. Hamerton, High-performance thermoset-thermoset polymer blends: a review of the chemistry of cyanate ester bismaleimide blends. High Perform. Polym. **8**(1), 83–95 (1996)
26. D. Herr, N.A. Nikolic, R.A. Schultz, Chemistries for high reliability in electronics assemblies. High Perform. Polym. **13**, 79–100 (2001)
27. R. Kisiel, Z. Szczepański, Die-attachment solutions for SiC power devices. Microelectron. Reliab. **49**, 627–629 (2009)
28. T. Hongsmatip, B. Twombly, Dynamic mechanical analysis of silver/glass die attach material, in *1995 Proceedings of the 1995 45th IEEE Electronic Components & Technology Conference*, 21–24 May 1995, pp. 692–700
29. R. Zeiser, L. Lehman, V. Fiedler, J. Wilde, Reliability of flip-chip technologies for SiC-MEMS operating at 500°C, in *2013 Electronic Components & Technology Conference, ECTC 2013*, 28–31 May 2013
30. C.T. Tan, K. Naoya, Y.S. Lee, K. Tan, Breakthrough development of new die attach method with high conductive wafer back coating, in *2015 17th Electronics Packaging Technology Conference*, 2015, pp. 1–5
31. M. Lindgren, I. Belov, P. Leisner, Experimental evaluation of glob-top materials for use in harsh environments. Int. Microelectron. Packag. Soc. **2**(4, 4th Qtr), 253 (2005)
32. H.T. Wang, Y.C. Poh, An analysis on the properties of epoxy based die attach material and the effect to delamination and wire bondability, in *33rd International Electronics Manufacturing Technology Conference*, 2008, pp. 1–6
33. Y.-W. Huang, C.-W. Fan, Y.-M. Lin, S.-Y. Fun, S.-C. Chung, J.-Y. Juang, R.-S. Cheng, S.-Y. Huang, Development of high throughput adhesive bonding scheme by wafer-level underfill for 3D die-to-interposer stacking with 30μm-pitch micro interconnections, in *2015 Electronic Components & Technology Conference*, 2015, pp. 490–495
34. D. D'Angelo, S. Manian Ramkumar, P. Borgesen, Evaluation of a novel anisotropic conductive adhesive shear under multiple tin-lead and lead free reflow cycles for package-on package (POP) assembly, in *Proceedings of the ASME 2012 International Mechanical Engineering Congress & Exposition IMECE 2012*, Houston, TX, 9–15 Nov. 2012, pp. 1–9

Correction to: Transient Liquid Phase Bonding

J. R. Holaday and C. A. Handwerker

Correction to:
Chapter 9 in: K. S. Siow (ed.), *Die-Attach Materials for High Temperature Applications in Microelectronics Packaging*, https://doi.org/10.1007/978-3-319-99256-3_9

The original version of the book was inadvertently published with an incorrect phrase stating "Need more here. Combine with above?" in Page 237 under Section 8.1.3 Au-In. This phrase has been corrected to show "At low temperatures, Au-In forms seven IMC phases between Au and In, with AuIn2, AuIn, and Au7In3 observed in the reaction zone between Au and In even at the shortest reaction times [84]."

The updated online version of this chapter can be found at
https://doi.org/10.1007/978-3-319-99256-3_9

K. S. Siow (ed.), *Die-Attach Materials for High Temperature Applications in Microelectronics Packaging*, https://doi.org/10.1007/978-3-319-99256-3_11

Index

K. S. Siow (ed.), *Die-Attach Materials for High Temperature Applications in Microelectronics Packaging*, https://doi.org/10.1007/978-3-319-99256-3

S

www.ingramcontent.com/pod-product-compliance
Ingram Content Group UK Ltd.
Pitfield, Milton Keynes, MK11 3LW, UK
UKHW022213051026
15624UKWH00001B/144
9783319992556